PROBLEMS OF SEMANTICS

BOSTON STUDIES IN THE PHILOSOPHY OF SCIENCE

EDITED BY ROBERT S. COHEN AND MARX W. WARTOFSKY

VOLUME 66

LADISLAV TONDL

PROBLEMS OF SEMANTICS

A Contribution to the Analysis of the Language of Science

Translated from the Czech by David Short

D. REIDEL PUBLISHING COMPANY

DORDRECHT : HOLLAND / BOSTON : U. S. A.

LONDON : ENGLAND

Library of Congress Cataloging in Publication Data

CIP

Tondl, Ladislav.
Problems of semantics.

(Boston studies in the philosophy of science; v. 66).
Translation of Problémy sémantiky.
Bibliography: p.
Includes index.
1. Semantics (Philosophy) 2. Languages—Philosophy. I. Title. II. Series.
Q174. B67 vol. 66 [B840] 501s 81–10594
ISBN 90–277–0148–2 [149'.946] AACR2
ISBN 90–277–0316–7 (pbk.)

Published by D. Reidel Publishing Company,
P.O. Box 17, 3300 AA Dordrecht, Holland.

Sold and distributed in the U.S.A. and Canada
by Kluwer Boston Inc., Lincoln Building,
160 Old Derby Street, Hingham, MA 02043, U.S.A.

In all other countries, sold and distributed
by Kluwer Academic Publishers Group,
P.O. Box 322, 3300 AH Dordrecht, Holland.

D. Reidel Publishing Company is a member of the Kluwer Group.

Original title *Problémy Sémantiky*

Printed in The Netherlands

EDITORIAL PREFACE

Ladislav Tondl's insightful investigations into the language of the sciences bear directly upon some decisive points of confrontation in modern philosophy of science and of language itself. In the decade since his *Scientific Procedures* was published in English (*Boston Studies* 11), Dr Tondl has enlarged his original monograph of 1966 on the promise, problems and achievements of modern semantics: the main topic of his later work has been semantic information theory. A Russian translation, considerably expanded as a second edition, was published in 1975 (Moscow, Progress Publishers) with an appreciative critical commentary, in the form of a conclusion, by Professor Avenir I. Uemov of Odessa. Indeed many Soviet studies in the problems of the semantics of science show the same sort of philosophical curiosity about the relationship of meanings in scientific language to procedures in scientific epistemology that characterizes Tondl's work, as in the work of Mirislav Popovich (Kiev) and Vadim Sadovsky (Moscow) and their colleagues. But we know that interest in these matters is world-wide, ranging from such classical topics as sense and denotation, empiricist reduction, vagueness and denotational opacity, to the new and equally exciting topics of the semantics of non-unique preference choices, the nuances of informational synonymity, and the semantics of a picture shape (so briefly but beautifully sketched in Tondl's dense and promising last chapter). We are pleased to have had Tondl's kind cooperation in producing this English edition, actually a third edition, of his research about semantics.

* *

Once again we note with thanks the intelligent and graceful work of the translator of our two Tondl books, Mr David Short of the School of Slavonic and East European Studies at the University of London.

Center for the Philosophy and History of Science, Boston University

ROBERT S. COHEN
MARX W. WARTOFSKY

June 1981

TABLE OF CONTENTS

EDITORIAL PREFACE v

FOREWORD xiii

I. THE SEMANTIC PROBLEM – SOURCES AND THEMES 1

II. THE CONCEPT OF SEMANTICS AND PREREQUISITES FOR THE INVESTIGATION OF SEMANTIC PROBLEMS 10
1. The Concepts of Object Language and Metalanguage 11
2. The Semantic Level of Analysis and its Relations to the Syntactic and Pragmatic Levels 15

III. SEMANTIC CONCEPTS 31
1. Semantic Concepts and their Relations in Common Parlance 31
A. Express, state that 31
B. Signify 35
C. Denote 41
D. Represent 42
2. Semantic Concepts in Formalised Languages 43
A. The Concept of a Formalised Language 43
B. The Interpretation of a Formalised Language 47
C. The Introduction of Semantic Concepts by Definition 50
D. The Axiomatic Introduction of Semantic Concepts 57

IV. THE SEMANTICS OF LOGICAL CONCEPTS 61
1. Problems of L-Semantics 62
A. The Introduction of L-Concepts as Non-Defined Terms 64
B. The Concepts of 'Logical Range' and 'State Description' 66
C. The Concepts of 'Logical Content' and 'Semantic Information' 74
D. Meaning Postulates 83

2. The Semantics of Logical Concepts on the Basis of the Concept of Interpretation 91
A. The Concepts of Model and Interpretation 93
B. The Most Important Definitions of Logical Semantic Concepts on the Basis of 'Interpretation' 97
C. The Concept of 'Translation' 100

V. SENSE AND DENOTATION 103
1. Frege's Conception of Sense and Denotation 103
A. Sense, Denotation and the Identity of Names 103
B. The Concept of 'Name' 108
C. Frege's Extension of the Semantic Characterisation of Names 111
2. The Theory of Descriptions 114
A. The Concept of Denoting Phrase 115
B. The Differentiation of Descriptions 117
C. Individual Descriptions and Statement Functions 118
3. The Method of Extension and Intension 122
A. The Concepts of 'Class' and 'Property' 123
B. The Concepts of 'Extension' and 'Intension' 128
C. Extensional and Intensional Context 134
D A Possible Amplification of the Method of Extension and Intension 138
4. The Problem of Naming 143
A. The Name Relation 143
B. Antinomies of the Name Relation 149
5. Synonymity 151
A. The Concept of 'The Same Meaning' 151
B. Synonymity, Intension and Intensional Isomorphism 155
C. Synonymity, Pragmatic Intension and Pragmatic Criteria 158

VI. THE CRITERION OF SENSE 166
1. The Formulation of the Problem 166
A. Conceptions of a Criterion of Sense 166
B. The Criterion of Sense and the Problem of Linguistic and Sense Intension 167
2. The Operationist Criterion of Sense 170
A. Einstein's Definition of Simultaneity and the Operationist Criterion of Sense 171

B. Critical Comments on the Operationist Criterion of Sense 172
3. The Verifiability Criterion of Sense 177
A. Motives for the Original Version of the Verifiability Criterion 177
B. Different Degrees of the Verifiability Criterion 183
C. The Logical Nature of the Verifiability Criterion 186
4. The Translatability Criterion of Sense 187
A. The Concept of 'Having a Sense' as a Primitive Concept of the Semantic Metalanguage 187
B. Pragmatic Limitations of the Translatability Criterion 189
C. The Concept of 'Having a Sense' as a Many-Place Predicate 192
5. Sense and the Empirical 194
A. The Concept of 'Empirical Predicates' 194
B. Dispositional Predicates 198
6. 'Theoretical Concepts' and the Relativity of the Empirical Starting Point 202
A. The Problem of Scientific Empiricism 202
B. The Admissibility of Theoretical Concepts 208
C. Ways of Interpreting Theoretical Concepts 210
7. Problems of Sense and Reduction Procedures 217
A. A Critique of Empiricist Reductionism 217
B. The Ontological Aspect of Reduction; the Theory of Levels 220
C. The Semantic and Pragmatic Aspects of Reduction 224

VII. VAGUENESS 229
1. Vagueness and the Un-Sharpness of Boundaries 229
A. Vagueness and the Empirical 230
B. Vagueness and Theoretical Concepts and Constructions 232
C. Vagueness and the So-Called Fringe 233
2. Sources of Vagueness and Ways of Analysing Vagueness 235
A. The Pragmatic Aspects of Vagueness 235
B. The Semantic Aspects of Vagueness 239
3. Vagueness, Ambiguity and Denotational Opacity 242
A. Ambiguity 242
B. Denotational Opacity 243
C. Extremes and Graduated Differences 244

VIII. SEMANTICS AND SOME PROBLEMS OF ONTOLOGY 247
1. Semantics and Ontic Decision 247
A. The Use of Terms and Ontic Commitments 248
B. Linguistic Framework and So-Called External and Internal Questions 251
2. Nominalism, Platonism and Semantics 255
A. The Reification of Abstract Entities and the Problem of Nominalism 255
B. 'Praeter Necessitatem' 257
C. The Problem of Similarity and Identity 260
3. Analytical and Synthetic Aspects in the Language of Science 265
A. The Traditional Problems and Tasks of Semantics 265
B. Degrees of Analyticity 267
C. Some Methodological Problems 270

IX. AN OUTLINE OF THE EVALUATION OF THE RESULTS OF SCIENTIFIC ACTIVITY IN TERMS OF SEMANTIC INFORMATION 274
1. The Scope for Evaluating Scientific Results 274
2. Brillouin's Attempt at an Informational Evaluation of Scientific Laws 277
3. Linguistic Devices in Tasks of the Systematising Type 280
4. The Concept of 'Decision Base' and the Evaluation of a Decision Base 286
5. The Relevance of A Posteriori Data 293
6. Evaluation of the Goal Complex and the Concept of 'Epistemic Gain' 298

X. THE SEMANTICS OF PREFERENCE ATTITUDES 304
1. The Role of Preference and Preference Ordering 304
2. The Comparability Principle as a Presupposition for the Construction of a Preference System 306
3. Preferences of Things and Preferences of States of Affairs 309
4. Preference 'Ceteris Paribus' 315
5. The Concept of 'Preferable States of Affairs' as a Qualitative Concept 318
6. Preference as a Propositional Attitude 321
Conclusions 323

XI. THE PROBLEM OF INFORMATIONAL SYNONYMITY 325
1. The Traditional (Leibnizian) Criterion of Identity and the Problem of Semantic Identification 325
2. The 'Salva Veritate' Criterion 328
3. The Criterion of 'Salva Relatione' and the Concept of 'Informational Synonymity' 329
4. Informational Relevance and the Concept of 'Strict Informational Synonymity' 335

XII. AN OUTLINE OF THE SEMANTIC EVALUATION OF GRAPHIC COMMUNICATION 338
1. Introductory Remarks 338
2. Graphic Communication 340
3. The Semantics of a Picture Shape 345
4. Informational Synonymity and the Informational Evaluation of a Picture Shape 347
5. Informational Synonymity and the Time Factor 350

NOTES 354

REFERENCES 379

INDEX OF NAMES 386

INDEX OF SUBJECTS 388

FOREWORD
TO THE CZECH EDITION

The original version of this work was a course of lectures that I gave at the Philosophical Faculty of Charles University in 1958–59, along with some lectures delivered at the instigation of the Department of Logic and its then head, Professor Otakar Zich. I am beholden to Professor Zich for the invaluable advice and assistance that he offered me in my earlier work as well as for the care and attention which he devoted to the original form of the present work. At the same time I wish to extend my thanks to colleagues at the Czechoslovak Academy of Sciences Institute for the Theory of Information and Automation and to the heads of the Institute for their profound understanding in enabling me to continue in this work and in affording me to that end not only excellent conditions, but also the many stimuli that came out of our discussions and private conversations. My gratitude is also due to Academician Igor Hrušovský and Professor Vladimír Skalička, corresponding member of the Academy of Sciences, for their valuable advice and remarks on the manuscript of this work.

This is a work about language, in particular the language of science, yet the author is no philologist. It should not therefore be expected to furnish an analysis of every facet of the semantics of the language of science. Attention is centered here on those problems of the language of science that are of interest methodologically, together with certain philosophical and methodological side-tracks.

Although the problems of logical semantics have not long been a topic for scientific investigation, they have already fathered a considerable corpus of literature which cannot be overlooked. I have therefore been unable to avoid references to some of the most important findings and viewpoints.

Some of the assumed results of this study are based on my earlier publications or are a direct continuation from them. Other sections look back to works written jointly with A. Perez, namely the study *On the Role of Information Theory in Certain Scientific Procedures* (in 'Information and Prediction', Academic Press, 1965) and the study *Modely některých vědeckých procedur z hlediska logiky a teorie informace* (in 'Problémy kybernetiky', ČSAV publ. house, 1965). Yet another part of the work was presented at the International Congress of Logic and the Methodology of Science, held at Jerusalem in August 1964.

It is my duty to acknowledge with gratitude the great care devoted to reading the first version of the manuscript by Karel Berka and Pavel Materna, who also gave me the benefit of their critical comments and much valuable advice on how the work might be improved.

I. THE SEMANTIC PROBLEM – SOURCES AND THEMES

The terms 'semantics' and 'semantic problems' have recently featured with increasing frequency on the pages of technical and scientific texts. They do not, however, always mean the same thing, nor is the attribute 'semantic' understood in the same sense when variously associated with problems of logic, linguistics, psychology, sociology, psychiatry, philosophy, information theory and other disciplines. In other words, while these and many other fields of scientific endeavour doubtless do have their own semantic problems, the link between them is often hazy. There are several reasons why this is so.

Any more or less homogeneous group of problems that are being processed and resolved by a particular discipline or theory usually arises out of certain needs, be they of a practical or theoretical nature. There are certain sources, sets of stimuli or motives that lead to theorising. In the case of semantics and the problems it entails the situation is particularly complicated in that the sources and motives are conspicuously non-homogeneous. There are a number of quite different domains within which the semantic problem area has taken shape, but each domain has been treated independently of the others. While these domains are most frequently described as being associated with linguistics and logic, or with certain trends in philosophy (see for example [138]), they do also affect a number of other disciplines. In particular this includes psychology and psychiatry, esthetics and art theory, information theory and many other theoretical and technical disciplines whose subject has to do with processes of communication.

The link between semantic problems and linguistics would appear to be the most obvious. Beginning with the classical *Essai de sémantique* [15] by M. Bréal, who is generally credited with introducing the term 'semantics', semantics or semasiology (other terms with various interpretations also occur) has been considered a legitimate component of the science of language which studies language, and especially its elements – words – in terms of what is usually called their meaning. Naturally enough, there are a number of different meanings of 'meaning'. Despite the considerable differences among the various conceptions the attention of linguistically based works focuses on two aspects in particular: the word and its meaning, i.e. chiefly the relation of a word as an element of language to extralinguistic elements, and changes of meaning as conditioned historically in part by the evolution of language itself, and in part by the development of the domain of objects to which particular expressions of language relate, usually known as the '*univers de discours*' or '*universe*', but also changes in the psychology of the language's users. In these terms it is then possible to study various kinds of meaning changes, for example the narrowing, broadening or shifting of a meaning in relation to a given universe. Clearly, focussing attention on the word, its meaning, changes of meaning and the causes of those changes by no means exhausts the problems of linguistic semantics, although it is true, as Schaff states, that the specific nature of linguistic semantics does lie in investigating the history of meanings (of words).

In addition to the classical problems of linguistic semantics there are several other problems. The important ones include:

The differentiation of types of meaning in terms of the nature of the communicative act (for example, in terms of distinguishing between emotional and intellectual communication there is a difference between emotional and cognitive or referential meaning);

the differentiation of types of meaning by word-classes;

the problems of the transfer and preservation of meaning in translation, problems of meaning identity, the semantic problems of homonymy, synonymy, vagueness and ambiguity of words;

the semantic problems of other language expressions, sentences in particular;

the semantic problems of imperatives, questions, normative sentences and the like.

This is clearly not intended to be an exhaustive review of the semantic problems that occur even in linguistics alone. New stimuli for the study of semantic problems made themselves felt in the context of machine translation, new logical and mathematical methods for the study of natural languages, the analysis of the relations between natural and formal languages, and so forth.

Another major area within which a mass of essentially semantic problems have had to be solved is logic. It is worth noting in passing that even Aristotelian logic has had its semantic problems, and these were well known even in ancient times. Medieval logic was likewise aware of numerous problems that were fundamentally semantic in character, i.e. they concerned the relations of words or expressions of language to a particular universe. This is the case, for instance, with various conceptions to do with the use of words, suppositions etc. In the literature of the first half of the 19th century it was J.S. Mill in particular whose conception of connotation and denotation went some way towards perfecting traditional teaching on the scope and content of concepts.

A major upsurge in interest in semantic problems in logic came at the end of the 19th century in connection with a number of problems to which we shall be returning. What is interesting is that many of these problems, such as those to do with paradoxes, date right back to ancient times.

The kind of interest in questions of semantics that stems primarily from the needs and problems of information theory is rather different in character. The successes of the Shannon type of mathematical theory of information, which uses mathematical methods to study the patterns of the communication process at the syntactic level, led to the legitimate demand for a semantic theory of information to be constituted and for the means of measuring semantic information to be found. Up to the 1960s most attempts to work out such a theory, of which the most significant was that by R. Carnap and Y. Bar-Hillel [6], were still based on the means supplied by mathematical logic, especially the theory of logical probability.

On the subject of the philosophical sources for problems of semantics, two lines are pointed to[1]: the line of 'theoretical semantics', which grew out of Central European positivism and was represented in particular by the Vienna Circle (M. Schlick, R. Carnap and others), and the line of 'general semantics', which originated in the English-speaking world and is mostly represented by American writers. This division of the philosophical sources for semantic problems is, to say the least, inexact and a simplification. In fact at least six trends should be distinguished, each of which brought its own individual contribution to the solution of the problems of semantics. They are:

(a) The semantics of the Lwow-Warsaw school (T. Kotarbiński, K. Ajdukiewicz and others), associated with attempts to analyse and interpret some of the basic semantic categories, such as 'indicates', 'denotes', and with an account of the concept of 'sense' and others.

(b) Attempts at a so-called empirical theory of sense, influenced by L. Wittgenstein and developed especially by the Vienna Circle (M. Schlick, R. Carnap). The conception of a purely empirical criterion of sense on the basis of verification and verifiability did not lead to any satisfactory solution and has largely been abandoned.

(c) A somewhat different trend is represented by the later conceptions of some members of the Vienna Circle, Carnap in particular, already showing strong signs of the influence of A. Tarski. In the preceding period the logical analysis of the language of science had been elevated to being the main, and practically the sole, programme of philosophical searching, the analysis itself being understood as that of logical syntax, whereas now it was logical semantic analysis that moved to the fore.

(d) A special place in the catalogue of the philosophical sources of problems in semantics is held by the conceptions of Morris's semiotics [95] [96], which follow on from the earlier works of the American logician C.S. Peirce and from the traditions of American pragmatism. At the same time Morris's conceptions betray strongly the influence of the work of Pavlov, Bekhterev and the behaviourist trends in psychology. Morris is also to be credited with systematically drawing the distinction between the syntactic, semantic and pragmatic levels of language analysis.

(e) Separate traditions that developed independently of the above trends led up to the different versions of 'linguistic philosophy', which largely share with these other trends the idea that language is the main subject of philosophical investigation. In later periods the various conceptions of linguistic philosophy began understandably to merge in greater or lesser degree with the earlier currents. The most influential groups within linguistic philosophy have included the Oxford group (G. Ryle, A.J. Ayer and others). Quite close to linguistic philosophy come the conceptions of 'analytical philosophy' (e.g. the Cambridge school, certain conceptions to be found in America, and others), where the method of logical analysis is understood first and foremost as a method of linguistic analysis. The common features of all these currents – disregarding the many differences between them – may be stated in the following terms: Adherents of methods that rest on linguistic analysis in philosophy are not interested in the question 'What is it?', but in 'What does it mean?'. It is largely owing to this that the attention devoted to problems of semantics may sometimes have been exaggerated even to the point of making them the be all and end all of any thinking about nature, man and human thought.

(f) Morris's early semiotics, although it was born out of certain philosophical traditions, has been presented as an independent discipline, with just the same claims to independence and relative self-sufficiency as physics, biology, economics and other sciences. Most adherents of 'general semantics' have made similar claims. General semantics is not an independent current in philosophy, as its critics have sometimes suggested;

in philosophical terms, the views of the general semanticists are an eclectic mixture which contains nothing new or thought-provoking, being more an infusion of views taken over from a variety of different schools of thought. What was more important here were the stimuli that came out of psychology, in particular social psychology, psychiatry, neurology, linguistics, anthropology and in part also esthetics. The pioneering works in this field include C.K. Ogden and I.A. Richards's well-known study *The Meaning of Meaning*, first published as early as 1923 [103]. This work focussed attention on the emotional and intellectual effects of words, differentiated between the emotive and referential, or intellectual, use of language, and outlined a basic scheme of reference (Richards's referential triangle), amongst other things.

General semantics proper are however the work, first and foremost, of the group that includes A. Korzybski [75], S.I. Hayakawa [53], H. Walpole [159], S. Chase and others centred round the journal ETC. For these authors general semantics is above all a means of 'psychic and social therapy', an instrument of 'mental or social hygiene', a means of reconciling individual and social conflicts, and so forth. In many instances they have produced popular pseudoscientific writings drawing eclectically on Freudian psychoanalysis, Pavlovian reflex theory, psychiatry, linguistics, sociology, etc. Adherents of general semantics concern themselves with the 'magic' effects of words and with criticising the so-called Aristotelian system of language (which they identify by and large with two-value systems of logic); they wrestle with the 'tyranny' of words and put forward a series of devices that are connected with using natural languages in one way or another and are capable of resolving extralinguistic problems and conflicts.

There is no overlooking the fact that in scientific circles and among a majority of logicians, linguists as well as philosophers, the works of the general semanticists have been given a critical reception. A. Tarski for one, in the introduction to [149], deemed it essential to stand quite clear of the conception of semantics proposed by the general semanticists. Likewise M. Black [10] and other authors. Thus although criticism of 'semantic philosophy' by certain philosophers in the forties and fifties was considerably one-sided and affected by generalisation, not unlike the criticisms of cybernetics, genetics, mathematical logic and certain other modern scientific disciplines, it has to be said that the severe criticism of general semantics, based of course on sound analyses, was fully warranted.

A different set of sources gave rise to the findings of work undertaken in cultural anthropology, which in varying degree touches on problems of semantics. Out of the domain of cultural anthropology, comparative and historical ethnography and historical comparative linguistics a number of trends have been born, some of which have provided countless ideas for the study of semantic problems. For instance, this is the case of researches into the so-called primitive mentality and the relation between thought and language in primitive peoples. Interestingly enough, the resultant conceptions, which were formulated against the background of a huge wealth of material, are remarkably disparate. While L. Lévy-Bruhl, for example, arrives in [80] at a conception of the logical inferiority of the thought and language of primitives (introducing here the idea of 'pre-logical thinking', characterised by quite different patterns, such as the so-called law of participation), B. Lee-Whorf tries to show that primitive languages are in no way logically inferior to the languages of developed nations, while acknowledging the profound differences between the separate language groups [78]. From the point of view of semantic problems Whorf's (and also Sapir's [137]) hypothesis on 'linguistic relativism' is not without interest. This hypothesis suggests that language is a kind of picture of social reality and a guide to that reality at the same time. Language participates in forming our image of the world, our behaviour and reactions to various stimuli. This also leads Lee-Whorf to draw an analogy between this conception of language and Einstein's theory of relativity. The idea of interdependence and feed-back between the entities and processes of the objective world, the cultural level of thought, and the structure and elements of language is certainly a fruitful one, although it still remains to be seen just how far it can be taken.

A similar line was taken by other research projects in anthropology, sociology and ethnography, exemplified by the works of F. Boas [13], B. Malinowski [86] and others. While the aims of their research were clearly fairly broad and concerned more than just the semantic aspects of human communication in different cultures, many valuable stimuli came out of it and can be applied to the solution of questions of the semantics of natural languages.

Within psychology, semantic problems were for a long time somewhat in the shade. The main problems here have been to do with methodology, coupled with the vast, often quite fundamental differences between the various conceptions. Such differences include the opposition between the mentalist and behaviourist approaches to semantic problems. If we visualise the semanticisation of the means of communication as the making of decisions about the possibilities of assigning certain semantic functions to those means (or their elements), two ways, at first sight contradictory, offer themselves. (What is meant here by 'assigning' is assigning certain extralinguistic entities to individual elements or devices of human communication.) It is possible so to assign both entities of an objective nature (whether from the sphere of the objective world or from that of the objectively apprehensible behaviour of the users of those devices) and subjective entities.

The dispute between the mentalist and behaviourist approach to problems of semantics is connected with methods of research into meanings and their changes. It is interesting that the controversy concerned not only the psychological, but also the linguistic approach to the question of meaning. For example, L. Bloomfield [11] [12] protested strongly against the use of mentalist terminology in the analysis of language and its semantics (terms like 'concept' and 'idea') and described his own standpoint as physicalist or mechanist.

The linguistic side of the problem also comes in under those psychological researches, in part also the experiments, that concern the measurement of semantic functions and what C.E. Osgood described as the 'semantic differential' [104], on the basis of context in the basic types of language usage. These methods of 'measuring meaning' (in which the results are also processed in large part mathematically), although having more of the behaviourist than the mentalist approach in them, clearly take in what was in Morris's semiotics and logical semantics generally described as the 'pragmatic level' of human communication.

Since then certain linguists have begun to discriminate between 'linguistic meaning' and the so-called 'social cultural meaning'. The latter may of course feature in other cultural areas beyond just natural languages, but this is not the only reason why semantic problems have begun to be an object of attention in art theory and esthetics as well. Here too the issues are still fraught with controversy, many of the arguments being quite reminiscent of those about semantics in linguistics, psychology and certain other disciplines. Certain supporters of structuralist and descriptivist trends in linguistics have treated questions of semantics as something that goes beyond the capabilities of purely linguistic analysis, and hence as problems that are extralinguistic in nature. The assessment of semantic questions may be similarly regarded in art theory. Roughly speaking, the problem is as follows: In its traditional form linguistics has amounted to a complex of metalanguage statements about language as object. But the study of semantic problems inevitably assumes that respect will be paid to entralinguistic entities and to the relations of the language-object to those entities, whether those relations are subjective or objective in nature. There may therefore indeed be grounds for doubt whether the study of these relations falls within the competence of linguistics. And like doubts may arise in the analysis of semantic problems in art.

Certain other controversial questions in the semantics of art may also be analogous to those in linguistics, at least in some respects. It is a familiar fact that the semantic analysis of language begins with the word and ends with continuous speech; thus it is far from being concerned with only certain expressions of language, i.e. word and sentence in the main. Likewise in the context of art we may consider an individual work of art as a whole consisting of particular components, parts, elements etc. We are then faced with the important question whether art semantics concern only a work of art as a whole, or only certain kinds of works of art, and then in various degrees, or whether it concerns individual elements, means of expression etc. This is of course only a small part of the total area over which semantics may come into contact with the vast and varied field of problems in art theory and esthetics.

Our look at the variegated spectrum of places where certain semantic problems raise their heads – and we cannot have exhausted the possibilities, demonstrates that it is clearly not possible to treat problems of semantics with sufficient generality to take in all the areas mentioned. Semantic problems occur in all those areas naturally, but because of their variety it is equally natural that it is going to be exceedingly difficult to provide for the basic categories of semantics an interpretation sufficiently uniform for it to be equally powerful and valid in each area.

Furthermore it will be important to distinguish – in other fields besides linguistics – a number of levels at which the analysis of semantic problems may be carried out. For instance, if we take into account human communication on the basis of natural language, our decisions on the assignment of semantic functions may be made, as we have already said, with respect to various different elements of language, yet generally speaking there are certain limitations to the fineness or crudity with which we can discriminate those elements. The question presents itself whether one or other of the available levels may not be decisive from the point of view of semantic analysis.

Although in different forms, a similar situation over the problem of different levels of semantic analysis seems likely to occur in other domains where such analysis may be applied. Thus semantic problems are always relativised, one of the chief gauges of relativisation being the discrimination of levels.

It is obviously unthinkable to try to encompass within a single work all the manifold areas where semantic problems are of current significance. Therefore the account that follows will be primarily concerned with those ranges of semantic problems that have to do with the progress of modern logic and the evolution of methodological issues in science.

II. THE CONCEPT OF SEMANTICS AND PREREQUISITES FOR THE INVESTIGATION OF SEMANTIC PROBLEMS

In the evolution of logic in modern times, since the end of the last century, very many problems have been investigated that have a direct bearing on semantics or are intimately concerned with the relation between expressions of language and certain elements of the universe.

Within the corpus of problems that led directly to semantics the greatest importance accrues on the one hand to those of identity, substitution and hence the interpretation of Leibniz's law of identity ("Eadem sunt, quae sibi mutuo substitui possunt salva veritate".), and on the other hand to those problems connected with the resolution of the so-called semantic antinomies.

In the case of problems of identity, fundamental significance attaches to the work of G. Frege [45], who put forward one of the first solutions of the naming relation in logical terms. This work suggested the differentiation of two aspects or modes of meaning: sense and denotation.[1] Frege's conception offered a solution to what was described in later writings as the antinomy of synonymous names (see e.g. [23]) by drawing a differentiation within the concept of meaning. Frege's conception was partly taken up by B. Russell, although the latter treated the naming relation somewhat differently, especially in connection with the theory of descriptions. In more recent writings in logic Frege's conception has been most consistently followed by A. Church [37] [39]. One development of Frege's standpoint is the semantic conception of the Münster school, represented by the works of H. Scholz [58] [141] and G. Hasenjaeger, who like Frege evolved a platonizing conception of logical semantics.

While the foregoing works attempt to solve certain logico-semantic problems by carrying out a kind of refinement of the distinctions made in the sphere of what we assign to expressions of language, there is another body of work that, while overlapping on a number of fronts with the previous ones, attempts to solve other problems of logic by putting forward a refinement of distinctions within the domain of language itself, both in the form of a hierarchisation of expressions within language and in a hierarchisation of languages.

The need for a kind of refinement of distinctions within language itself arose primarily in connection with the broad circle of questions pertaining to paradoxes, some of which, as it has transpired, were of the nature of semantic paradoxes. Many paradoxes have been connected with set theory (the Burali-Forti paradox, Cantor's paradox, Russell's paradox, Richards's paradox and others), and of these some have been shown to be related to the classical paradoxes of antiquity (especially Epimenides's paradox of the Liar), while elsewhere it became necessary to draw a line between paradoxes of a logical and mathematical nature and those of a semantic nature. The endeavour to eliminate paradoxes gave rise to the theory of types (B. Russell) and later the so-called simple theory of types (L. Chwistek) and the theory of semantic categories (S. Leśniewski).

The distinction of object language and metalanguage, and the simultaneous tighter definition of 'semantics' from the point of view of logical semantics, are likewise connected with the problem of paradoxes. In this respect the works of A. Tarski, dating from the thirties, are of major significance [146] [147].

1. THE CONCEPTS OF OBJECT LANGUAGE AND METALANGUAGE

Tarski described semantics as "the totality of considerations concerning those concepts which, roughly speaking, express certain connexions between the expressions of a language and the objects and states of affairs referred to by those expressions" [146, p. 401]. As typically semantic concepts Tarski gives denotation, satisfaction and definition. It is to Tarski's credit that he demonstrated that the characteristic features of these concepts are shared by the concept of truth, at least in the sense in which truth is used in the logical analysis of the communication process.[2]

Semantic concepts are always relative, i.e. they must always be related to a particular language. There are very serious reasons for this relativity. If it is not duly respected, which is all too easy in the ordinary use of natural languages, the result may be paradoxes or antinomies. By antinomies are of course meant those connected with the use of semantic concepts in the sense given above – concepts that concern the relations of expressions of language to the objects to which the expressions are assigned. In this respect we are speaking of semantic antinomies as distinct from antinomies of other kinds.

Some of the semantic antinomies are already familiar from classical philosophy, while others were only formulated at a later date. What are mentioned most frequently are various formulations of the paradox of the Liar (Epimenides's paradox). In the literature we most commonly find the formulation of the antinomy in its most exact form as originated by J. Lukasiewicz.[3] In order to grasp the essence of the antinomy, let us look at the sentence

This sentence, printed on page 12 in a rectangle, is not true.

If – for brevity's sake – we replace the sentence by the symbol '*S*', we obtain:

(1) '*S*' is true if and only if the sentence, printed in a rectangle on this page, is not true.

In addition we may, on the basis of the meaning of '*S*', establish empirically that

(2) '*S*' is identical with the sentence printed in the rectangle on page 12.

Since identical expressions can be substituted for one another (by Leibniz's 'salva veritate' conception of identity), we may replace the part of the sentence in (1) that runs 'the sentence printed in a rectangle on this page' by the symbol '*S*'. This then gives us:

(3) '*S*' is true if and only if '*S*' is untrue.

Although sentence (3) has been derived in a way that is logically unimpeachable, it obviously represents a contradiction. It is for this reason that Tarski stresses the need to analyse the suppositions on which the antinomy is based and draw the appropriate conclusions therefrom. If we analyse the suppositions we find them as follows:

(I) We are assuming that the language in which the antinomy arises contains not only the expressions that occur in it, but also the names of those expressions *and* the predicate 'true'. We further assume that all the sentences in which that predicate may be adequately used can be expressed in that same language. Languages endowed with these properties are called by Tarski 'semantically closed languages' [149].

(II) We are assuming that the ordinary laws of logic are valid in language.

(III) Finally we are assuming that empirical premises of the kind in sentence (2) can be formulated in the given language.

Leaving aside assumption (III), which Tarski considers immaterial,[4] we still have assumptions (I) and (II) by means of which to reconstruct the antinomy of the Liar. Therefore any language that satisfies assumptions

(I) and (II) cannot be consistent. It hardly seems tenable to surrender assumption (II); this leaves us with only one way to eliminate the antinomy, namely by surrendering assumption (I), i.e. deciding not to use a semantically closed language. Tarski emphasised that this way is not really possible in the case of natural languages, i.e. ordinary non-formalised languages, which anyway do not have any accurately specified structure. Therefore the problem of consistency in respect of this kind of language is pointless.

Although Tarski agrees unreservedly that abandoning assumption (I) is the only way of overcoming the difficulties as stated, it is nevertheless possible to consider whether other ways might not also be possible. First and foremost we might consider the possibilities for modifying assumption (II), since it has not been made fully clear what is meant by 'the ordinary laws of logic'. If these are the laws of classical logic, including the principle of 'tertium non datur', it is questionable how they can be applied in any language. Or we might try to modify or generalise the principles in some way as to preserve assumption (I).

There are some serious problems connected with the empirical premiss, the role of which in the construction of the antinomy Tarski omitted altogether. There are two possible interpretations:

(a) If '*S*' is an abbreviation for the sentence 'This sentence, printed on page 12 in a rectangle, is not true', then the expression 'is an abbreviation' may also be interpreted as 'is a name'. In this case we have a naming relation and '*S*' and the sentence in the rectangle are expressions in different languages, i.e. metalanguage and object language. Then of course the two expressions cease to be mutually substitutable.

(b) If '*S*' is identical with the sentence in the rectangle, as is also claimed by Tarski – the sentence 'This sentence, printed on page 12 in a rectangle, is not true', we have to ask what kind of identity this is. It is clearly not true that whatever may be said of '*S*' may also be said of the sentence in the rectangle. We are accordingly offered the solution that '*S*' and the sentence 'This sentence, printed on page 12 in a rectangle, is not true' are identical in terms of meaning – they have the same meaning. In other words, it can be assumed that they are synonymous. It can of course be proved[5] that 'salva veritate' substitution is not an adequate guarantee of synonymity. Moreover, this guarantee can obviously not rest on empirical evidence.

Thus although objections may be raised against Tarski's approach, which is restricted to the surrender of the assumption of a semantically closed language, it must be taken as unequivocal that the expressions that we use to make statements about elements of a given universe cannot be confused with those used to make statements about those expressions. The need for this distinction, which carries with it the need to reject the idea of a semantically closed language, arises out of the problematical aspects of certain other semantic concepts as well. For there are other concepts to which it is possible to construct antinomies analogous to that of the Liar. One example is the Grelling-Nelson antinomy, which attaches to the concept 'denotes'.[6]

An example of an antinomy tied to the concept of definition is Richards's antinomy (see, e.g. [169], p. 10ff., [123], p. 213, [72], p. 38). The same group of antinomies may also include Berry's antinomy of the smallest whole number, which cannot be expressed by less than a particular number of syllables (that number depending on the particular language), and certain other paradoxes.

Having established that semantically closed language is inconsistent, we find that it is not enough to be satisfied with the differentiation of expressions within the framework of a single language. This further means that semantic concepts cannot be included in the same language as the other expressions to which the semantic expressions relate.

Thus we are faced with the necessity of splitting the hitherto unitary language into two languages.

The first of these is the language about which we speak. This language naturally enables us to express ourselves about the phenomena of a given universe, but it does not enable us to make statements of a semantic nature. Tarski called this language the 'object language' (język przedmiotowy), but the term found most widespread in logical writings is 'language-object' ($\mathscr{L}$), after the German version of Tarski's original work [147].

The second language is that in which we can formulate statements about sentences and other expressions of the object language. This is the 'metalanguage' ($\mathscr{M}\mathscr{L}$). Sentences in the metalanguage may contain semantic concepts.

This splitting of the originally unitary language is not absolute. This means that (a) the concept of metalanguage can only be used meaningfully with respect to a given object language, and (b) in introducing semantic concepts for the metalanguage we have to pass up to a metalanguage of a higher order. In this way we arrive at a whole hierarchically ordered system of languages. Each sentence containing any of the semantic concepts is then at least a component of the metalanguage, but it may well be a component of one or other of the higher order metalanguages.

It follows from the non-consistency of a semantically closed language and the differentiation between object language and metalanguage that sentences containing semantic concepts (relating to expressions of the object language) can be created in the metalanguage, and at the same time all sentences from the object language can be translated into the metalanguage. Thus the metalanguage must meet the following requirements:

(1) It must allow for all expressions of the object language to be expressible in it.

(2) It must contain semantic concepts in relation to the object language that cannot be formulated in the object language.

Herein also lie the grounds for the differentiation of two distinct components within the metalanguage: its semantic and non-semantic parts. The non-semantic part must furnish an expressive potential that is at least as rich as that of the object language. Thus there is some correspondence between the non-semantic part of the metalanguage and the object language, while no such correspondence can be sought in the case of the semantic part of the metalanguage.[7]

The fission of a unitary language into object language and metalanguage as a means of overcoming the difficulties of a semantically closed language brings us simultaneously to another necessary step – in the analysis of language we have to discriminate between the semantic level and the syntactic, and possibly also the pragmatic, level.

2. THE SEMANTIC LEVEL OF ANALYSIS AND ITS RELATIONS TO THE SYNTACTIC AND PRAGMATIC LEVELS

The differentiation of the semantic level of language analysis from the syntactic and pragmatic levels is ultimately the consequence of conceiving language as a means or instrument of communication. Any process of communication presupposes at least:

(1) a system of means of communication (in the case of communicating through language this is language in the general sense of the word; we shall leave aside for the time being the difference between language communication and communication in general). Externally, a system of means of communication is a system of individual discriminable expressions that can be transmitted along a particular channel of communication. Without this assumption, i.e. of the possibility of transmission, which information theory was the first to give an exact sense to, it is not possible to speak of a means of communication.

(2) Then there is always the sphere of the phenomena to which the

individual expressions relate, to which the expressions can be assigned. Thus we have a kind of system of assignments, a system which allows us to ascertain the relations between elements of the system of means of communication, or between individual expressions of language and elements that lie outside that system and therefore constitute a class of phenomena that we do not include in the system of the given language. It is usually pointed out that individual expressions can be applied in a process of communication if it is possible to decide that this kind of assignment is possible, in other words if the expressions used can be semanticised. What we have here is something analogous to the possibility of transmission.

(3) Finally account has to be taken of the systems between which communication is made. These systems have sometimes been denoted by the terms 'interpreter' or 'communicant' [33], while in the case of communication through language the convention is to refer to 'language-users'.

The individual levels of language analysis may now be constituted as follows: a) we may consider just system (1) and the relations within its framework, in which case we are talking about the syntactic level; or b) we may consider system (1) expanded by system (2), in which case we are talking about the semantic level; or c) we may consider system (1) and (2) expanded by system (3), in which case we are talking about the pragmatic level.

The distinction of these three levels goes back to the work of C.S. Peirce, while they were later given a more systematic justification by C.W. Morris in particular, who coined the term semiotics to cover the whole approach to language and communication generally that involves the three levels [95]. Morris's distinction of the three levels was in turn taken over by Carnap, who tried to refine the distinction even further, especially in the early sections of his *Introduction to Semantics* [22]. Carnap drew a distinction between empirical semiotics and 'pure' semiotics. Empirical semiotics aims at the empirical investigation of language systems that have emerged historically, while 'pure' semiotics analyses artificial language systems created on the basis of certain rules. These are the systems described by Tarski as formalised languages. Carnap himself was not greatly interested in the empirical branches of the different components of semiotics, i.e. empirical syntax, empirical semantics and empirical pragmatics. He thought they came within the purview of linguistics, and possibly also anthropology, psychology, sociology and other empirical disciplines. His attentions were focussed on the analysis of 'pure' syntax, the so-called logical syntax, and 'pure', or logical, semantics. In the case of the pragmatic level he initially doubted whether any other than an empirical approach was possible at all. This means that Carnap did not assume that it was possible to proceed at that level in a manner to match logical syntax and logical semantics. Yet it has to be added that, despite his scepticism,[8] there were attempts by others to set up a 'pure' or logical pragmatics, of which the most serious was R.M. Martin's *Systematic Pragmatics* [90].

The distinction of the three levels of language analysis outlined here in connection with the process of communication was originally drawn more from the point of view of the static appraisal of language and its expressions (or to put it in general terms: the elements of a system of communication). Attention was usually focussed on the word, considered as the basic element of language (more exactly of natural language), and its relations to certain entities considered as the basic elements of extra-linguistic systems. Two other systems, of some significance in the given context, were also assumed in addition to the language system: the system of things, their properties and/or relations, i.e. an objective system, and the system of ideas, thoughts or concepts, i.e. a system not immediately accessible to an objective approach.

Within this framework there were also a number of thoughts turning on semantic problems, for instance the early ideas of C.K. Ogden and I.A. Richards in the first chapter of their work on the meaning of 'meaning' [103]. The very title of that chapter – 'Thoughts, Words and Things' – points to the general connections within which this method of approach to the basic semantic relations and the levels of these relations moved. The basic assumption was of three distinguishable systems: that of things, that of ideas or thoughts, and finally that of words. One other assumption was even more fundamental: it was taken for granted that in all three systems it is possible to differentiate their basic elements in such a way that relations between elements of different systems can be reconstructed. In the broadest terms these are the chief features of the background thinking that led to the formulation of Richards's familiar 'triangle of reference'.[9] This triangle of reference indicates the relations between elements that differ by each of them belonging to a different system. These elements are:

(1) the symbol (in the terminology of Ogden and Richards; it means a sign in general, and in the case of natural language it is above all the 'word');

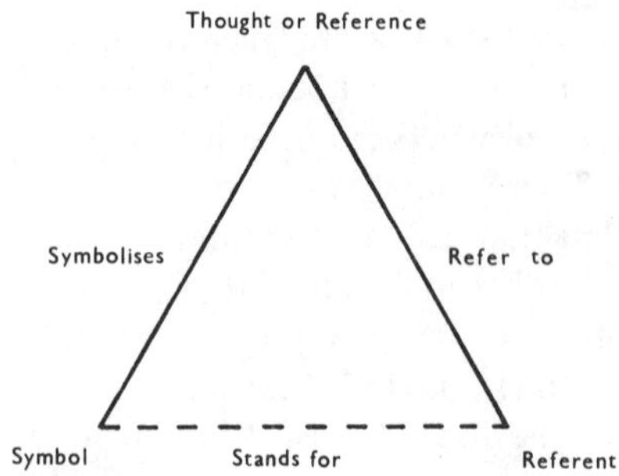

Fig. 1.

(2) the referent (in the given terminology it is an object or thing to which the particular symbol refers; thus it is something that corresponds to the term 'denotate', although the latter term will be used, as will be seen in due course, in a much broader sense than the original conception of 'referent' in Richards's diagram);

(3) the reference or thought, i.e. an assumed ideal entity that mediates between symbol and referent, between word and object. The relations between the elements are expressed schematically in Figure 1.

The solid lines in the triangular schema are held by the authors to express real relations. This means that the relation between symbol and referent (i.e. word and object, sign and denotate) is only an 'imputed relation' mediated by means of the elements of the thought process.

This schema is clearly not to do with communication as such. It merely assumes a user of the symbol or, to be more precise, 'states of awareness' of that user that mediate the relations between symbol and referent. Thus it is a very elementary interpretation of semantic relations, i.e. relations between language expressions and the elements of some extra-linguistic world. These semantic relations are then *relations of assignment* which contain as their part pragmatic relations, i.e. relations to the language user. In this schema the syntactic problem is entirely to one side so that it becomes doubtful whether the triangle of reference actually captures those relations that are characteristic of language. For it is an undisputed fact that the concept of language always assumes:

(1) a certain categorisation of elements and the possibility of combinations of those elements in such a way that elements of one category give rise to elements of another category;[10]

(2) certain rules for the interrelations of the various elements, e.g.

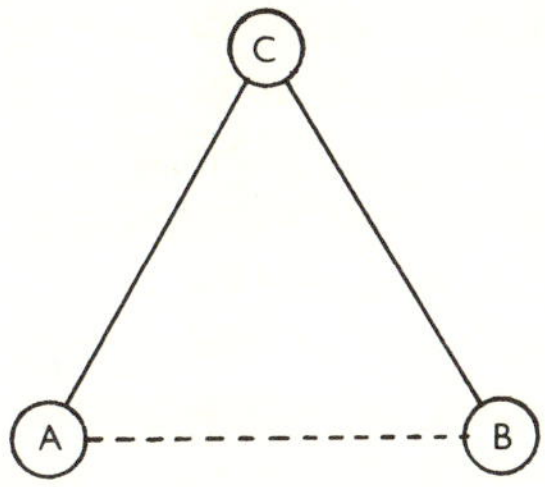

Fig. 2.

rules for their mutual substitution, for the derivation of one set of elements from others, and so forth.

Richards's original schema of reference may therefore be modified on the lines of Figure 2.

In this modified schema **A** represents the system of elements of a language; **A** has certain characteristic syntactic rules, it can be interpreted with respect to **B** and it is used in relation to the activity of individuals of system **C**. **B** represents a given universe, a certain extralinguistic domain of phenomena that are distinguishable by individuals of system **C**. System **C** is the system of users of system **A**.

It must of course be stressed that the 'reverse' procedure, which starts from the syntactic level and gradually enriches it, cannot be considered unjustified. A reverse procedure of that kind is inevitable in the construction of formalised languages.

From this point of view the procedure from the syntactic level to the semantic level represents a kind of specification of system **A** and in that respect the semantic approach is 'narrower' than the syntactic. In this way semantics appears as the study of modes of interpretation of system **A** in which the interpretation is given explicitly. In that light it becomes necessary to amend the original description of system **A**: System **A** is not of itself the system of the elements of a language. It is characterised merely by a specification of the relevant vocabulary and syntactic rules, and only in relation to **B**, i.e. in setting up a given interpretation, does it become a language in the proper sense of the word. It remains to say that a language always assumes a system of users of it. This means that in characterising the concept of language we have to go beyond the relations of **A** and **B**, however these are organised, and look as well at their organisation with respect to system **C**.

It follows further that in the analysis of language the syntactic, semantic or pragmatic level, if taken in isolation, only represent a kind of abstract viewpoint. But that is not to say that any one of these abstract viewpoints cannot be applied fairly independently. Their independence is naturally justified in the analysis of formalised languages.

Earlier views on the distinction of the three levels of language analysis basically failed to take account of those properties of language that are connected with communicating, with the process of transmitting, processing and storage of information. If we were to remain with just the differentiations of systems **A**, **B** and **C** as described so far, we would be omitting from the essential properties of language the fact that its components facilitate the *communicability* of certain items of information. In the foregoing description of system **A** we were effectively assuming only tacitly that system **A** could be adequately expressed by certain devices that are objectively apprehensible and distinguishable to a given circle of users, e.g. the sounds of spoken language, written or printed letters or other signs, numbers, gestures, cues, pictures etc.

If we pay due respect to the fundamental circumstance that language is a vehicle of communication, we can outline the distinction of the basic levels of language analysis by an expansion of the communication model. This communication model is commonly illustrated by diagrams like schema (1) (Figure 3).

Schema (1), which is also at the basis of Shannon's mathematical theory of information, actually represents the most abstract approach. In the mathematical analysis of this communication schema as put forward by Shannon's original version of information theory, no regard is paid to the interpretation of the message sent and received, which is understood in an abstract sense, i.e. as a given realisation of any random process. Accordingly therefore, no regard is paid either to any 'meaning' the message sent or received may have for the user of the given channel of communication. This also means that analysis of the communication process under these assumptions does not pay any respect to those aspects that are proper to the semantic and pragmatic levels.

It is possible to draw up a confrontation between this abstract schema

(1)
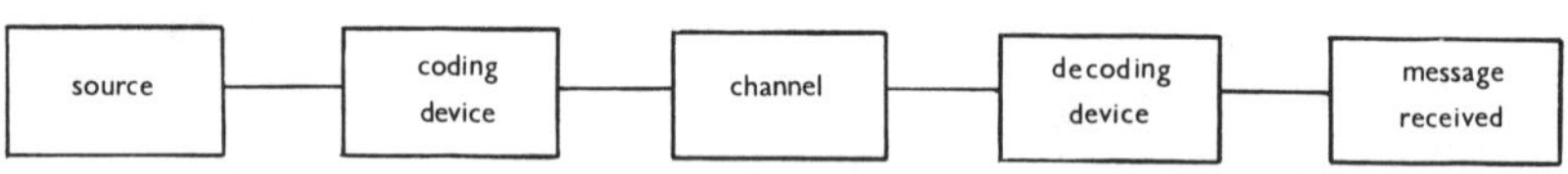

Fig. 3.

(2)

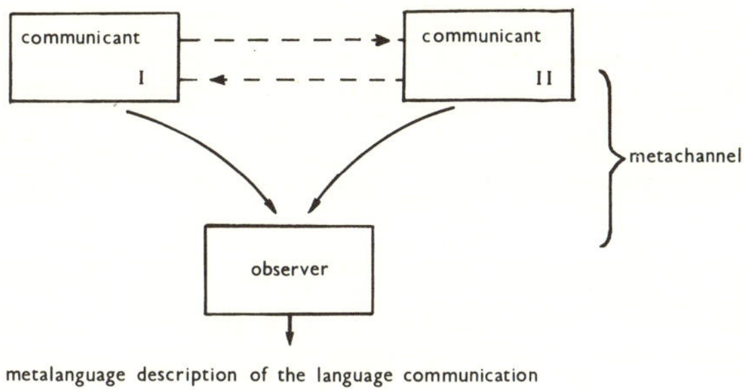

Fig. 4.

of the process of communication and a schema of communication by language. One diagram of this might be schema (2) (Figure 4).

Schema (2)[11] is actually a diagram of the conversation of two language users – the diagram of a dialogue. The language used by I and II in their conversation is an object language for the outside observer who is studying the use of the language. Thus two channels have to be assumed: the conversation channel, which with respect to the observer is the object channel, and the metachannel, i.e. the channel of the observer. Naturally, both channels may be assumed to be affected by noise, which may concern both the devices used in communication and observation, and whatever is connected with interpretation, the assignment of meanings. In this connection we speak of semantic noise. Imagine the conversation of two persons, person I, who speaks Czech, and person II, who speaks Slovak. Considering the great areas of similarity between Czech and Slovak, and accepting that the two persons are of similar intellectual capacity, the semantic noise in the object channel may be quite slight. Let us assume that the conversation is being observed by someone who cannot understand either Czech or Slovak. That being so there will obviously be very great semantic noise in the metachannel – the observer channel. Yet the observer will still be able to make out individual sounds or gestures – in the case of a spoken conversation, or the written letters, words and sentences – in the case of a written conversation.

For a more accurate differentiation of the separate levels of language analysis within the framework of this schema we shall have to confront Cherry's conversation diagram with the previously described schema of the communication process. This confrontation might seem a simple matter inasmuch as it might appear possible simply to write the communi-

(3) 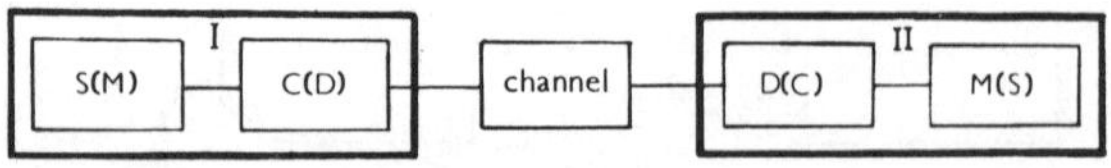

S-source
M-message received
C-coding device
D-decoding device

Fig. 5.

cation schema into the conversation schema in such a way that the decoding device is treated as a basic component of the equipment of the two communicants. For instance, in the case of a written conversation, both parties must be able to read and write, while in the case of a dialogue they must use a language which they both know. Ignoring the observer channel, the incorporation of the original communication schema into the conversation schema could be illustrated as in schema (3) (Figure 5): the thick frame is the conversation schema and the thin frame the original schema of the communication process. Because we are dealing with a conversation (here involving two persons, hence a dialogue), the symbols in brackets in the relevant blocks are for communication in the reverse direction. It may be assumed analogously that coding also applies in the observer channel, and then communicant I and communicant II are both supplied with the relevant coding equipment while a part of the observer's equipment is what corresponds to the decoding device.

However, this simple incorporation of one schema into another does not enable us to separate syntactic relations and the syntactic level from semantic and pragmatic relations. This calls for some alteration and expansion of the schema in its present state. The expansion will be clear from schema (4) (Figure 6).

In schema (4), which abstracts from the observer channel and shows only the object channel, **C** is the method of coding and **D** is the appropriate method of decoding. $\mathbf{S}^{\mathrm{I}}$ and $\mathbf{S}^{\mathrm{II}}$ are the subjective worlds of I and II respectively.[12] It should of course be stressed that $\mathbf{S}^{\mathrm{I}}$ and $\mathbf{S}^{\mathrm{II}}$ are objectively inaccessible to the outside observer. In other words $\mathbf{S}^{\mathrm{I}}$ and $\mathbf{S}^{\mathrm{II}}$ are not open to objective investigation and may only be reconstructed hypothetically. In a conversation (by which we mean an intellectual exchange, associated with what is usually referred to as referential meaning, as opposed to emotional meaning) it is further assumed that there is a certain universe, or domain of objects, to which the conversation relates.[13]

It is generally assumed that the universe is identical for I and II, whereas in fact that is only the extreme case. We must therefore posit not one

(4)

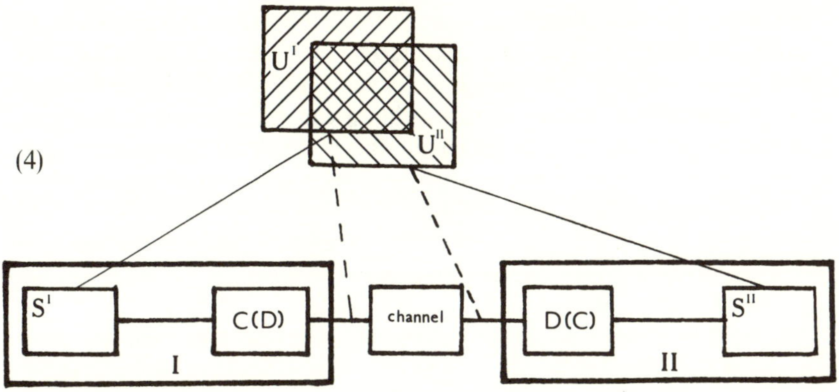

Fig. 6.

but two domains of objects, $\mathbf{U}^{\mathrm{I}}$ and $\mathbf{U}^{\mathrm{II}}$, that coincide in greater or lesser degree. This is expressed schematically in the intersection of $\mathbf{U}^{\mathrm{I}}$ and $\mathbf{U}^{\mathrm{II}}$. A postulate for a possible exchange between communicants I and II is therefore that this intersection be not equal to zero. If it is equal to zero, no effective exchange can take place.

On the basis of the coincidence of elements of $\mathbf{U}^{\mathrm{I}}$ and $\mathbf{U}^{\mathrm{II}}$ it becomes possible to clarify the terms 'understand' (in the case of just one of the communicants) and 'reach an understanding' (in the case of both of them). Postulates can be formulated which, if met, are a basis for mutual understanding. The first is narrower and in a sense stricter, the second is broader:

(1) If the conversation relates to an element that is simultaneously an element of $\mathbf{U}^{\mathrm{I}}$ and $\mathbf{U}^{\mathrm{II}}$, understanding may be said to have been reached if and only if I and II are capable of assigning a given expression of the language used in the conversation to the same element. In other words, for understanding to be reached there has to be an assumption of identity between $\mathbf{u}_i^{\mathrm{I}}(\mathbf{u}_i^{\mathrm{I}} \in \mathbf{U}^{\mathrm{I}})$ and $\mathbf{u}_j^{\mathrm{II}}(\mathbf{u}_j^{\mathrm{II}} \in \mathbf{U}^{\mathrm{II}})$.

(2) If the conversation relates to an element that is simultaneously an element of $\mathbf{U}^{\mathrm{I}}$ and $\mathbf{U}^{\mathrm{II}}$, understanding may be said to have been reached if and only if I and II are capable of assigning a given expression of the language used in the conversation to elements the difference between which does not exceed a certain degree. In other words, for understanding to be reached it has to be assumed that the risk associated with the difference between $\mathbf{u}_i^{\mathrm{I}}(\mathbf{u}_i^{\mathrm{I}} \in \mathbf{U}^{\mathrm{I}})$ and $\mathbf{u}_j^{\mathrm{II}}(\mathbf{u}_j^{\mathrm{II}} \in \mathbf{U}^{\mathrm{II}})$ does not (on average) exceed a certain level compatible with the aim of the conversation.[14]

Between the two postulates (the first will be called the postulate of identity of the referent and the second the postulate of admissible risk) there is clearly a kind of relationship, for the first is effectively the extreme case of the second: the identity postulate can be interpreted to mean that the risk associated with the (slight and hence, from the point of view of aims of the conversation in question, negligible) difference between $\mathbf{u}_i^{\mathrm{I}}$ and $\mathbf{u}_j^{\mathrm{II}}$ equals zero.

There are certain difficulties attached to the formulation of these postulates, and to overcome them requires the introduction of certain further aspects. We shall consider at least some of them:

(a) Both postulates are conceived in such a way that they capture only that aspect of meaning which is expressed by the concept of denotation. Yet it may be objected that effective mutual understanding is based in most cases on the identification of that aspect of meaning usually expressed by the concept of sense.[15]

(b) Postulate (1) assumes identification of two elements, and we might start out from Liebniz's formulation of the principle of identity. That formulation has as a criterion of identity of expressions their interchangeability within a sentence context such that the semantic characteristic, i.e. truth value, of the sentence is preserved. Substitution (and it should be emphasised that what is meant here is substitutability as a dispositional property, not an actual exchange) initially concerns elements of language and not elements of domains of objects. More important, however, is the fact that substitutability is not and cannot be absolute in any general case: it can only be substitutability within a given system of tasks or within a given type of decision. problem. In other words the identification of $\mathbf{u}_i^{\mathrm{I}}(\mathbf{u}_i^{\mathrm{I}} \in \mathbf{U}^{\mathrm{I}})$ with $\mathbf{u}_j^{\mathrm{II}}(\mathbf{u}_j^{\mathrm{II}} \in \mathbf{U}^{\mathrm{II}})$ is feasible only with respect to a given system of tasks, aims, decision problems etc. (In practice this means that, in a given problem, identity is not a bilateral relation between elements of $\mathbf{U}^{\mathrm{I}}$ and $\mathbf{U}^{\mathrm{II}}$, but a relation that is at least trilateral.) But in conversation – and here is the core of the problem – it cannot always be assumed that what we have described as a system of tasks, aims or decision problems is going to be the same for both communicants I and II. So the problem of identity re-emerges on a new, higher plane.

(c) Like problems may occur in connection with postulate (2). In this case the sphere of possible substitutability is not governed by the necessity of preserving the truth value, but by the fact if $\mathbf{u}_i^{\mathrm{I}}$ and $\mathbf{u}_j^{\mathrm{II}}$ are not the same their mutual substitution is accompanied by a risk the (average) level of which will not exceed a certain level that is compatible with the aim of the given conversation. In this connection the problems may be compounded further if that aim (and with it the compatible (average) level of risk) happens not to be the same for the two communicants. Again in this case the problem of possible substitution reappears at a higher level.

With these last remarks in mind we can now give a more exact definition of the individual levels of language analysis in connection with schema (4).

1. If we take into account only elements of language and the relations between them, including those relations that are connected with the transmission of certain texts along a channel that facilitates the exchange of information between I and II, and if at the same time we disregard any relations between those texts and the domain of objects $\mathbf{U}^{\mathrm{I}}$, or $\mathbf{U}^{\mathrm{II}}$, respectively, and further disregard any system of aims (whether the same or different as between I and II), then we are at the *syntactic level* of analysis of the language in which the conversation is conducted.

2. If we take into additional consideration the relation of those elements, or texts formed on the basis of the elements, to the domain of objects $\mathbf{U}^{\mathrm{I}}$, or $\mathbf{U}^{\mathrm{II}}$ respectively, and at the same time disregard the system of aims, then we are at the *semantic level* of analysis of the language in which the conversation is conducted.

3. If we then take into account the relations of the elements or texts to the system or systems of aims, then we are at the *pragmatic level* of analysis of the language in which the conversation is conducted.

Thinking back now to the earlier remarks on 'reaching an understanding' we can see more clearly the mutual relations of the different levels, as well as the potentials that each level offers. It is obvious that interpretational questions in conversation are insoluble at the syntactic level, and a move must be made to the semantic level. Questions connected with the goal of a particular conversation are soluble only on the pragmatic level. All this further means that effective mutual understanding requires that all three levels and their mutual relations be respected.

In an analysis of the conversation schema (schema (4)) it is also useful to consider some of the aspects that are connected with the observer channel. Two questions here are fundamental: one concerns the elements in the schema that are directly accessible to the observer and those that he can only reconstruct with varying degrees of accuracy; and one concerns the degree to which he is capable of distinguishing the three levels of language analysis.

The solution of these questions may be simplified by accepting the possibility put forward by C. Cherry [33]. Cherry points out that the obser-

(5) 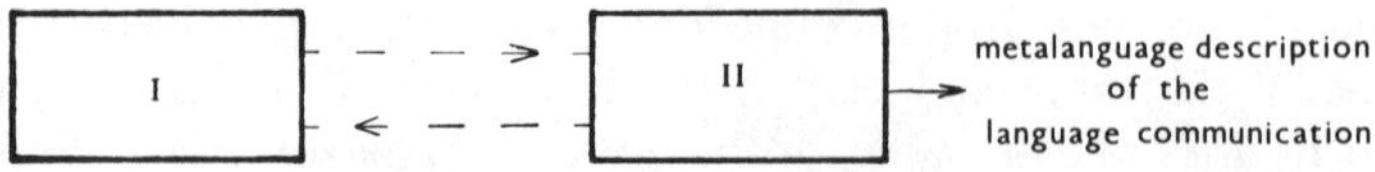

Fig. 7.

ver of a conversation may at the same time be one of the participants. He accordingly simplified his original schema (i.e. schema (2)) in such a way that the observer is not a third party, but one of the communicants. Schematically expressed this gives schema (5) (Figure 7).

In this case, which anyway matches common practice, the exchange of information between communicant I and communicant II takes place in the object language. Statements about this exchange, made by one of the communicants who is also the observer, are then in the metalanguage. In these circumstances, the observer has different prospects in relation to himself and in relation to the other person. While, for instance, $\mathbf{S}^{\mathrm{I}}$ is not immediately accessible to him, this is not so with $\mathbf{S}^{\mathrm{II}}$. If he is unable to determine $\mathbf{U}^{\mathrm{I}}$ exactly and if no scale of discrimination is accessible to him in $\mathbf{U}^{\mathrm{I}}$, his situation is quite different vis-à-vis $\mathbf{U}^{\mathrm{II}}$. If the aims of the conversation are not accessible to him from the point of view of communicant I, he can be more or less precise about his own aims. This is evidently the situation that commonly leads him to make judgements about $\mathbf{S}^{\mathrm{I}}$ on the basis of $\mathbf{S}^{\mathrm{II}}$ and so to reconstruct $\mathbf{S}^{\mathrm{I}}$ on the basis of $\mathbf{S}^{\mathrm{II}}$. (It goes without saying that communicant I is in an analogous situation.) This also means that in conversation and language communication generally we cannot avoid making the entities and systems that are at our disposal into a kind of yardstick for evaluating the entities and systems available to our partners.[16]

If we consider the situation where the observer of a conversation is different from either of its participants, i.e. he is not I or II, matters are more complicated. We have already emphasised that $\mathbf{S}^{\mathrm{I}}$ and $\mathbf{S}^{\mathrm{II}}$ are inaccessible to the outside observer. In the case of the domains of objects, there are, roughly speaking the following possibilities:

(a) $\mathbf{U}^{\mathrm{I}}$ and $\mathbf{U}^{\mathrm{II}}$ are inaccessible to the outside observer. In other words the observer cannot discern 'what' the parties to the conversation are talking about, what they are writing about, or what the meaning of the messages passing between them is.

(b) $\mathbf{U}^{\mathrm{I}}$ and $\mathbf{U}^{\mathrm{II}}$ are only partially inaccessible to the outside observer. In practical terms this means that the observer has at his disposal a certain $\mathbf{U}^{\mathrm{III}}$ that coincides in greater or lesser degree with $\mathbf{U}^{\mathrm{I}}$ and $\mathbf{U}^{\mathrm{II}}$.

(c) U^{I} and U^{II} are accessible to the observer. This means that the observer at his disposal a certain U^{III} that coincides with U^{I} or U^{II}, or, where U^{I} and U^{II} are identical, is as one with the universe of both communicants.

When it comes to the system of aims, tasks and decision problems connected with the given conversation, the following cases are possible:

(a′) The system or systems is unknown and inaccessible to the outside observer.

(b′) The system or systems is only known to the observer in part, or he is only partially capable of reconstructing it.

(c′) The observer knows the system, or in the case that the systems of aims, tasks and decision problems of the two communicants differ, he knows both systems.

If these cases are kept distinct, it is possible to differentiate the observer's prospects for investigating the syntactic, semantic and pragmatic relations that figure in the given conversation.

In case (a) the observer may investigate only the syntactic relations, but only provided the observer channel has the properties essential to that end. In the case, for example, of a spoken dialogue in a language unknown to the observer, we can at least assume that he is capable of distinguishing separate sounds, words etc. In the case of the written form of a conversation, we assume that the observer is capable of differentiating between the individual letters, parts of the text and so forth.

In case (b) the observer may also investigate in part some of the semantic relations. In other words, he can carry out the semantisation of certain elements of the language used in the conversation, certain texts etc. Semantisation goes proportionately further the more case (b) approaches case (c).

In connection with the task of semantisation from the point of view of the outside observer, what is of exceptional importance is what is usually described as experience. If the external observer does not know (or knows only in part) what I and II are talking about, i.e. if U^{I} and U^{II} are inaccessible or only partly accessible to him, he may nevertheless try to decide about the assignment of certain elements of the language used in the conversation, i.e. makes certain guesses about the universe of the conversation. These guesses may progressively improve and so improve his decision-making, i.e. he may get nearer and nearer to the real U^{I} or U^{II}. (If, for example, the observer hears the two parties discussing 'bodies' and 'spaces', he will not normally be able to provide an exact semantisation of the terms, i.e. ascertain their assignment to a fairly accurately delimited universe. Only as the conversation progresses will he discover that the speakers are mathematicians and hence that the 'bodies' and 'spaces' they are talking about have an abstract mathematical meaning, nothing at all to do with heavenly bodies or outer space as might have followed from a conversation between astronomers.) This is then a process that is generally speaking stochastic, and it could be mapped by adequate stochastic devices. In this connection it is worth mentioning some of the interesting problems and tasks it raises, such as the problem of the speed of improvement of decision-making, i.e. the rate at which the observer's guesswork brings him closer to the real universe, the problem of the optimum route for bringing him closer, and the requirements for that optimum route to be found, and so forth.

If we summarise these remarks on progressive semantisation, i.e. the reconstruction, ascertainment and verification of semantic relations, we may state that progress in semantisation depends on the observer's prospects for improving the quality of his decisions and hence also on the existence of certain real criteria for evaluating those decisions. What is therefore fundamental is the fact that the possibilities that we have described as (a), (b) and (c) are not strictly separated and that it is possible to cross from case (a) to case (b), and in the extreme instance to case (c).

The same basic situation as characterises the possibilities for analysing conversation at the semantic level, i.e. the prospects for semantisation from an outside observer's point of view, also applies to the pragmatic level. The case here is of course more complicated and less well defined, since what we have described as S^{I} and S^{II} ('subjective world') – the subjective aspects of a given system of aims, tasks or decision problems associated with a given conversation, are always, and in principle, inaccessible to the outside observer. This is not to say that some more or less reliable idea of the aims of the conversation cannot be gradually formed on the basis of outward signs. In other words, as in the case of the semantic level, it is possible to increase the accuracy of one's assumptions about the system of aims or tasks or decision problems connected with a given conversation. This kind of procedure may also contribute to the move from case (a′) to case (b′) and in the extreme instance to case (c′).

It might be possible to question whether an outside observer may ever be fully conversant with a set (or two different sets) of aims. It may be objected that such an eventuality is at variance with the assertion that what is described as S^{I} and S^{II} are inaccessible to the outside observer in principle. Such doubts and objections relate to certain problems of psychology and its methodology. Two kinds of circumstance may be distinguished:

(i) The communicant is fully aware of the aims of his own participation in a given conversation and is therefore capable of adequately expressing those aims by means of language. This means that for the highest degree of rationality we may postulate complete communicability of the states of the 'subjective world' to another person.

(ii) The communicant is not aware of the aims of his own participation in a given conversation and is therefore not able to express those aims adequately by means of language. The possibility remains, however, whereby the outside observer may be able, on the basis of objectively detectable behaviour, to ascertain those aims in a more exact form than is open to the person observed.

These two instances do not represent two possibilities that are disparate in principle, but rather two extremes between which a whole range of transitional cases are possible. This leads us therefore to raise a hypothesis on the existence of different levels of rationality of the human psyche. From this it follows that what we call rational awareness is in essence the highest level of the human psyche, fundamental to which is the possibility of adequate expression by means of language. This is certainly not to say that the lowest level, as associated, for example, with emotionality, cannot also be expressed by means of language. But the greater the accuracy with which certain elements of the awareness can be expressed by means of language and so reconstructed on the basis of these direct statements, the 'higher' is the level of the psyche concerned. In this connection the attributes 'higher' and 'lower' are clearly not to be understood as value judgements, nor is there any implication of a hierarchy in which the 'lower' elements of the psyche are subordinate to the 'higher'. Accordingly the hypothesis of different levels of the human psyche not only does not exclude a measure of autonomy of the lower levels, but it also accepts that in certain circumstances one or other of the lower levels may even become decisive in connection with the presence of certain aims.

This brief outline of some problems in psychology, and perhaps also social psychology, is not intended to have exhausted all aspects of the problem, and in any case it goes beyond the framework of the subject under investigation. It has merely served to show the kind of problems that may be encountered in any analysis of conversation at the pragmatic level.

It seems appropriate to conclude this chapter with a few words in justification of the apparently disproportionate attention paid to the conversation model in our argument for differentiating the different levels of language analysis. It might be asked how far the conversation model enables us to capture the characteristic properties of language.

It might seem that in clarifying the concept of language there is no need to consider the language users. But such an abstraction is admissible only under certain circumstances, or to be more exact, only for a certain limited approach to language analysis. Yet one fundamental property of languages in the form in which man uses them for social intercourse is that they serve to transmit messages, as an aid to human intercommunication, which is invariably connected with certain tasks, a system of aims or decision problems. This 'aim-orientedness' of language is determined moreover by factors that are primarily extralinguistic – social, psychological, biological and other factors. It was to include these other aspects in our clarification of the concept of language that we selected the model of a conversation between two persons, the *dialogue* model. This does not preclude a conception of language as a *monologue*, but the factual use of language is more a dialogue than a monologue and factual use always assumes an exchange of information, the extreme minimal number of language users involved being two. This was the chief reason for pausing for a more detailed look at the model of conversation. It is fully recognised that the exchange of information by means of language usually requires a group of more than two language users (or communicants in Cherry's terminology), and that there will be moreover a complete system of language users in certain social and historical conditions, with differentiated needs, aims, interests, and with different scales of values and so forth.

III. SEMANTIC CONCEPTS

1. SEMANTIC CONCEPTS AND THEIR RELATIONS IN COMMON PARLANCE

In evaluating the role of language as a means for the transmission of messages, a means of communication, we encounter a number of terms that from various points of view concern the relations of language and expressions of language to certain entities that are of an extralinguistic character with respect to the language used in any given communication (i.e. with respect to the object language). Whole strings of such terms are customarily quoted. But the first thing to be recalled here is that common parlance does not generally have clear-cut differences between many of the terms, which are often interchangeable, at least in some contexts.

T. Kotarbiński [73, p. 13ff.] lists the following as words that concern relations between language expressions and extralinguistic entities: express, state, denote, signify, stand for, represent.[1] Usually these terms take the form of two-place predicates, but occasionally more then two-place. It cannot be asserted generally that one of the arguments of these predicates is necessarily an expression of language. Everyday speech even tolerates cases where neither of the arguments is an expression of language, such as in the sentence: 'The smile on his face expressed his joy'. Then there are instances where both or all arguments have to be considered expressions of language, i.e. components of certain language systems, as in the sentence "'Das Haus' in German means the house".

In thc account that follows we shall adhere to those uses of these terms that do concern language expressions. In other words we assume that at least one of the arguments of a given two- or more-placed predicate is an expression from a certain language.

[*A*] *Express, State that*

Assuming that *express* is at least a two-place predicate, one argument of which is a language expression, then this term has to do with relations between the language expression and the language user, or some experience, idea or state of him or his psyche. Kotarbiński states that the relation may be either immediate or mediated. A child who has burnt his hand will yell 'it hurts', the words actually expressing a particular state or experience of the child. If I say 'it is raining today', I am expressing what I perceive, see or hear. If someone utters the sentence 'Napoleon died on St. Helena', that sentence expresses certain knowledge, or some idea, held by the person who utters it.

The mediated kind of relation applies, in Kotarbiński's view, if the language expression conveys a possible state of the language user, his possible experience, idea, state of the psyche and so forth, irrespective of whether the particular state features or not in a given specific situation. In other words a particular language expression expresses, with respect to the accepted interpretation or convention of the language, a mediated possible state of the language user, who, if the relevant state were to come about, is capable of expressing it immediately. The word 'yes' uttered by person X.Y. expresses his agreement or his positive answer to a question, but the word 'yes' can express like states of third parties.

Kotarbiński's differentiation of expressing as an immediate or mediated relation appears to concern two somewhat different but interlinked functions of language: the functions described by B. Russell as the *function of expression* and *the function of communication* [136, p. 72ff.]. The expressing function differs only slightly, in its most primitive form, from other forms of behaviour. For example, interjections and other such expressions that express emotional states or experiences have much in common with those kinds of behaviour that are clearly not connected with means of language. However, Russell explicitly states that this is not a property of just the 'elementary forms of language', associated solely with emotional states, but also of higher forms where a certain intellectual level may be assumed. The communicating function basically concerns the transmission of messages. As Russell demonstrates, language is able to fulfil the communicating function on account of its two fundamental properties: firstly it is always a social phenomenon, and secondly it is capable of expressing things that are not directly accessible to another language user and become accessible precisely because of being expressed in language (the objectivisation here is obviously not exclusively tied to language). The communicating function can be brought to bear because it is assumed, as Russell says, that it can interest another person or influence his behaviour. This none too happy turn of phrase appears to have something to do with the fact that the transmission of messages by means of language can only be considered if the two or more language users involved are capable of performing the semantisation of the means used and if that semantisation can be evaluated or assessed in terms of whatever aims or decision problems are connected with the particular act of communication.

Thus although the two functions of language cannot be separated unequivocally from each other, and even less isolated completely, it remains obvious that the term 'express', if conceived as a two- or more than two-place predicate, one of whose arguments is an expression of language, does capture rather different aspects according to whether it is the expressing function or the communicating function that is considered. The word 'yes' expresses agreement, or a positive stance or viewpoint adopted by a particular person if we consider the expressing function of language – the possibilities that person has for objectivising his states, experiences, attitudes or the states of his psyche that are not directly accessible to an outside observer, another person. But if we consider the communicating function, the word 'yes' expresses agreement, i.e. essentially a state of the psyche, that is communicable because the word 'yes' has a certain sense. In this respect, as we shall be seeing again, the term 'express' is connected with the term 'signify' and the concept of 'sense'.

Essentially similar relations to those surrounding 'express' are the basis of the term 'state about'. Here too, however, the above-mentioned constraint should be borne in mind: it is a two- or more than two-place predicate, at least one argument of which is an expression of language. An expression states something about an experience of a certain person, or about a given state of that person's psyche and so forth. We may speak here of direct statement-making. A language expression makes a mediated statement about an experience or state of a language user if, given the accepted interpretation of the language or a certain language convention, the relevant experience, state of the psyche etc., can be assumed on the expression's being used.

It is obvious that in differentiating the terms 'express' and 'state about' in the immediate and mediated senses there are different approaches to one and the same relation: in the case of immediate expressing or stating what we have in mind is one specific experience or psychic state of a person who is capable of using language. Such a state is therefore a (relative) starting point for certain concrete language behaviour – the use of certain specific language expressions. This (relative) starting point can of course be conceived in various ways. Russell, for example, uses to characterise this starting point the not very happy term 'belief', which may lead to misunderstanding, especially if 'belief' attracts any ideological connotations. He then points out that language from the very outset, or from the outset of thinking about language, has served the expression of 'beliefs' in certain more or less stable things or persons [136, p. 100]. Despite the unfortunate terminology there is certainly a rational core here: an expression of language expresses a particular state of a language user, a state which includes within itself a more or less active relation of the language user to certain entities or actions, including those that are proper to the language user himself.

If whatever corresponds to Russell's term 'belief' is the relative starting point for expressing or stating in the immediate sense, then the (relative) starting point in the case of mediated expressing or stating is a given language expression. This means that language is not only a means for expressing the states, experiences, ideas, attitudes etc. of language users, but also the creator, or part creator of these states and experiences – language enables their formation and differentiation.[2] Man can, for example, discriminate between feelings of pleasure and displeasure. This differentiation of these and other similar feelings is obviously very crude, but it can be refined in parallel to the ever increasingly fine differentiations created in language, until, for instance, we become able to discriminate between the senses of expressions like 'I am thrilled', 'I like', 'I do not like', 'I am appalled' and so on. The increasing fineness of discrimination in language can surely not be assessed in abstracto or merely in connection with the level of the psyche. It is associated with factors that are primarily pragmatic in character, being connected with the historically and culturally conditioned tasks where the relevant language expressions are applicable.

We are now in the area of the much debated problem of the relation between language and thought, or the problem of which came first. Russell, for example, does not share the view that there can be no thought without language while maintaining that 'beliefs' can exist even without language. This kind of assumption can scarcely be proved, yet there are based on it quite a number of conceptions, including the psychoanalytical conception of the subconscious. This subconscious is also a state of a being that may be a language user. If that being can express these states by means of language, the states become other states; the subconscious ceases to be the subconscious. So there are states of the psyche without language, without either actual or possible expression in language. But these and other similar viewpoints assume a particular hypothetical conception of the psyche, and one that is still scarcely verifiable. Hence Russell's standpoint and those like it can best be treated as the product of one conception of the psyche, though rather a different one from that put forward by psychoanalysis.

In opposition to this conception it is possible to erect the previously mentioned conception of hierarchically ordered levels of rationality. By that conception the question of the relation between thought and language cannot be resolved globally, but only with respect to one level of rationality, a different solution (the inseparability of language and thought) being clearly called for at the higher levels as opposed to the lower.

This means that the connectedness of language and thought is not the same at all levels of rationality. At the higher levels the connection may be assumed to be closer and more immediate than at the lower. This conception of the relation between language and thought may also be formulated from another angle: any use of language, whether expressive or communicational in character, objectivises certain thought processes. But language is proportionately more capable of adequately objectivising the processes the higher, i.e. the more rational, the level at which the processes occur. Looked at in this way, the possibility of adequate expression in language becomes a measure of the rationality of the psyche.

⌊B⌋ *Signify*

If expressions of language participate in the creation of certain states, allowing them to arise, take shape, become differentiated etc., this is only possible because the expressions have the important property commonly called 'sense'. In other words, this property rests in the fact that the expressions *signify* something.

Before proceeding to any detailed clarification of the term 'signify' we must emphasise that it does not mean the same as 'have a relation to a certain experience or a certain state of the language user'. The property whereby an expression of language signifies something, or has a sense, is not the same as the property of leading to certain experiences, or creating certain states of the psyche – 'beliefs' to use Russell's terminology. Thus if we take a language expression as a relative starting point, then the relation to what we have described as a language user's state, experience or 'belief' is not immediate, but mediated by what we usually call 'sense'. As early as 1892 Frege [45] was pointing out that the sense of an expression must not be identified with an experience, idea or what he called an

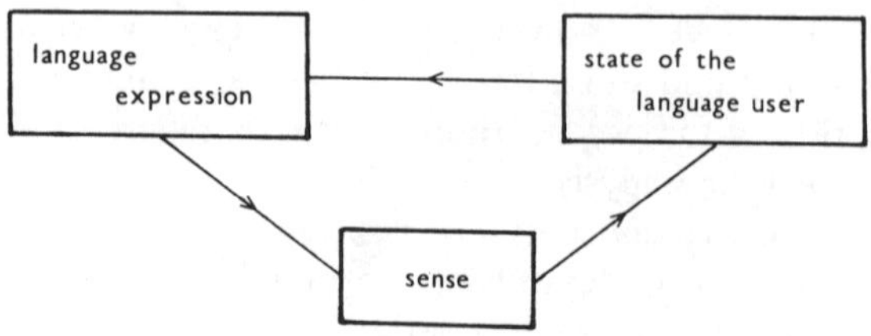

Fig. 8.

'internal image' that arises on the basis of sensory perceptions, memory and the inward or outward activity of a person. An internal image associated with an expression that has some sense may differ from person to person.

The mutual relations between a language expression and a given state of a language user is illustrated by the diagram in Figure 8. Here we have the direct and mediated forms of the relation according to whether we take as (relative) starting point a given language expression, i.e. something that can be uttered, written or communicated in some other detectable way, or what we have described as a state of the language user, i.e. that which cannot be detected directly and can only be reconstructed with more or less accuracy on the basis of the language user's outward behaviour.

However, on account of the principle that "si duo faciunt idem, non est idem", this schematic representation is no more than a simplification. The relations between the blocks are also therefore not determinist, but stochastic in character. This means that each block of a particular kind can have assigned to it another block of a different kind, with a certain probability. At the same time the distribution of probabilities will clearly depend on whether we are considering a single person or a somehow defined class of language users, or a given set of situations and tasks in which language is used, and so forth. Schematically the relations conceived in this way may be presented in the manner of the diagram in Figure 9.

It is assumed then that to a certain language expression is assignable a whole spectrum of blocks of another kind, in this case a spectrum of possible states of language users. This also means that terms such as 'express', 'state', 'signify' and others that play a similar role need not relate to any fixed relations, but invariably to a particular network of relations. The network may not of course have unlimited and randomly

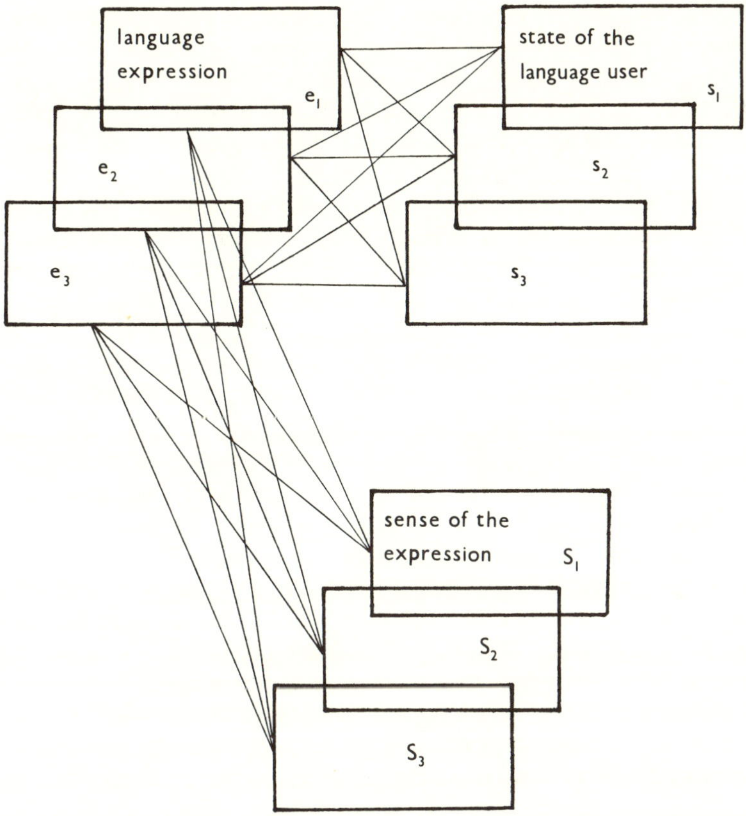

Fig. 9.

conceived possibilities of branching. If that were the case, mutual understanding would be impossible a language expression used in an attempt to communicate would simply not be able to transmit a sense capable of leading to the same, or at least similar, states in different users of the same language expression and the same language.

Let us now turn to a more detailed analysis of the relation denoted by the term 'signify'. We have already pointed out that signifying is not the same thing as being connected with a given experience or state of a language user. In other words, taking 'signify' as a two-place predicate, one of whose arguments is an expression of language, the other argument is not that experience or state of the language user that may be associated with the sense of that expression. If that were the case, it would be more appropriate to use instead the term 'express' or 'state'. Kotarbiński illustrates the difference by the following example: If person *X* says "Le soleil brille", that sentence, uttered by *X*, states something immediately about an experience of that person. If another person, say *Y*, elsewhere in Europe and in a different set of circumstances, says "Die Sonne scheint", then the sentence uttered by *Y* states something immediately about an experience of *Y*. Yet we can scarcely assume that the two experiences are identical, least of all if *X* lives in the warm climes of the Mediterranean shore and *Y* on the cool shores of the Baltic. While Kotarbiński says that both experiences are "the same in essence" [73, p. 15], that particular turn of phrase cannot resolve the whole problem but merely shifts it somewhat, since it raises the further issue of what 'the same in essence' is. Yet there is no doubt that the two sentences have the same sense, albeit the experiences of the users of the expressions, experiences that can be linked to the sense of the expressions, will differ in some degree.

Talking about the difference between what we call the 'sense' of an expression[3] and a state of the language user, e.g. an idea or experience that he has or any element of his 'subjective world', goes nowhere towards solving the problem of what 'signify' is or what the sense of a language expression is. In order to get at least somewhere nearer a solution of the problem, let us consider what we are doing when we put the question "what does ... signify?", when the dotted gap is an expression of language.[4]

There is obviously no universal schema by which to proceed in all situations. For situations where the question "what does ... signify" arises, the different kinds of procedures employed may be divided into two groups (the same going for other, analogous questions and problems):

The first group includes those procedures described by Russell as *ostensive definitions*. An ostensive definition is defined by Russell as "any process by which a person is taught to understand a word otherwise than by the use of other words" [136, p. 78]. He illustrates the concept of ostensive definition by a simple example: A man who speaks no French has been shipwrecked on the coast of France. He calls at a cottage and sees some bread on the table. Being hungry, he points to the bread in a way that shows that he does not know what to call the thing he wishes to ask for. If the cottager says "pain", the castaway will assume, at least preliminarily, that that is the French word for bread, and he will thus have acquired the sense of the term 'pain'.

This example obviously illustrates the procedure where the answer to "what does ... signify" is found indirectly, i.e. by pointing to a particular denotate. By this procedure a new and hitherto unknown expression has a sense assigned to it.

But the kind of procedure that starts from a known denotate to a new or known expression is not the only one possible. We may start from a known expression, usually one from a language we speak, and proceed from that expression to one that is unknown, e.g. from 'bread' to French 'pain', having explained that it has the same sense. The application of the two procedures obviously depends on circumstances, especially in the case of the acquisition of a foreign language. A child acquires the terms of its own language, or in a bilingual environment even the terms of a foreign language, by having as his main starting point the objects of his own specific environment, especially objects that are frequently repeated and are exciting to the emotions, and then, on the basis of ostensive definitions, he gradually acquires the sense of the terms and language expressions generally that apply in behaviour that relates to the particular objects.

If in connection with the acquisition of the sense of new and unknown terms attention is drawn to the role of association, the basis of which is the well-known process of conditioned reflex, then this is primarily to do with the process of ostensive definition. But this process undoubtedly has its limitations, thanks to the limitations of the empirical and behavioural starting point of the procedure. And because this starting point is limited, there are also limits to the familiar hypothesis, based on the conception of the second signalling system, which says that a word is a "signal of a signal". If the hypothesis was justified for all the expressions of a language, it would have to be able to cope with the problem of the sense of the logical constants, i.e. words such as 'and' 'or', 'all', 'none' etc. The limitations of ostensive definition, together with the limitations of associationalist approaches as means for acquiring sense, are also due to the fact that, generally speaking, only part of any universe is accessible empirically, although it seems unlikely that there will ever be any firm and universally valid limits to what can be acquired through ostensive definition.

Bertrand Russell points out that ostensive definition can lead to the acquisition of the sense of such words as 'baptism' or 'half-moon', but that it is impossible to acquire the sense of words such as 'kilogon'. It follows from Russell's account of ostensive definition that, overall, it is a procedure that is adequate to only a proportion of language expressions, notably those he characterises as 'names'. This does not mean, of course that ostensive definition can be used for all names, but the following categories are listed by Russell for where it can:

(a) generic names like ‘man’, ‘woman’, ‘cat’, ‘bread’, ‘tree’ etc.; in fact these are the names of classes.

(b) names of qualities, e.g. ‘red’, ‘hard’, ‘cool’ etc. (this category obviously cannot be limited just to empirical predicates, but may also include dispositional predicates and others);

(c) names of certain relations such as ‘up’, ‘down’, ‘on the right’ (Russell includes words like ‘fast’, ‘slow’ etc. in this category);

(d) the so-called egocentric words like ‘I’, ‘you’, ‘here’, ‘now’, etc.[5]

This categorisation of language expressions, in the context of the acquisition of their senses, is certainly neither complete nor unproblematical, especially in terms of the possibilities for handling it logically. It omits the names of some of the various types of behaviour and activity, such as ‘write’, ‘stroll’ or ‘rejoice’, though it is true that not all names of behaviour or activity can have their sense acquired through ostensive definition, as is the case of ‘decide’, ‘resolve’ and others. Nor can the categorisation be really warranted logically, for if we accept the name of a quality as a one-place predicate, the name of a relation can only differ in being a multi-place predicate. On the whole it holds that ostensive definition as a means for acquiring the sense of an expression of language, or as a procedure for answering the question ‘what does . . . signify’, is limited by a denotate’s being sufficiently discriminable for the language user, either in terms of the potential of the technical or empirical equipment, or in terms of the potential of his behaviour. At the same time this means that the procedure only concerns the ‘lower’ levels of abstraction, the term ‘lower’ being of course entirely relative and depending on the overall empirical and intellectual capabilities of the language user. (Hence we are in one situation in the case of a child and in other in that of a research worker.)

The second group of procedures in which we clarify for ourselves the sense of a language expression and therefore resolve a situation where the question ‘what does ... signify’, or some analogous question, has been put, includes procedures where, as opposed to definition by example as an aid to the acquisition of the sense of an expression of language, we use instead other language expressions.

In his analysis of the term ‘signify’ Kotarbiński [73, p. 16] stated that one of the simplest procedures that could be included in this group hings on the answer formula that has the shape ‘ ... signifies that ---’, in which the dotted expression is that whose sense is to be determined and the text in dashes is an expression that enables us to acquire the sense of....

This schema assumes that it is possible to make the sense of the two expressions identify. It is important to recall here that identity or similarity of sense does not, or need not, mean the same measure of identity or similarity between the entities that are denoted by the two expressions. Closer analysis of this problem therefore calls for a solution to the problems of synonymy, which will be put forward in a later section.

[*C*] *Denote*

With the analysis of the term 'denote' we enter the very heart of the matter of semantic problems. Even an intuitive approach to 'denote' and the difference between 'denote' and 'signify' suggests that the term 'denote' assumes a relation between an expression of language and something that has been denoted, marked out, an entity that is an element of a universe. Many problems present themselves here, and in order to have some idea of them before we turn to their actual solution, which will be presented separately in the proper context, we shall start with a simple form of the use of the term 'denote'.

In a language $\mathscr{L}$ and with respect to the semantic rules valid for $\mathscr{L}$, or to the customs accepted by the users of $\mathscr{L}$ with regard to a given expression, or to any other analogous convention, an expression denotes such and such an object. This formulation corresponds to similar formulations that operate with other terms, such as 'is a name for'.[6]

The very formulation hints at some of the problems that arise in connection with the term 'denote'. In general terms they amount to three kinds, although in different conceptions the three groups are closely linked and cannot be fully isolated.

(a) One group of problems may be described as ontological. In the expression '... denotes such and such an object' they relate to that object, and they include such questions as the nature of the object, how different kinds of objects can be differentiated and classified, how the object might be apprehended otherwise than in connection with the language expression that denotes it, and so forth. The group also includes the famous controversy between nominalism and platonism, the argument over abstract entities and their nature, and others.

(b) Then there are the problems connected with the denoting phrase, of which the most important one is the question of what language expressions can actually denote something. The term usually used for these expressions is 'name', hence the question may be rephrased as: What expressions in a given language may be described as 'names'?

(c) The third group of problems are the questions connected with the analysis of the 'denote' relation, i.e. semantic relations in the proper sense of the word.[7]

The chapter on sense and denotation will contain a closer analysis of the second and third groups of problems, especially a finer definition and differentiation of the concept of 'name', and analysis of the most important relations between names (in particular synonymy and homonymy), and an analysis of the problems of the naming relation. The ontological problems associated with the term 'denote' together with semantic relations generally will be treated in Chapter VIII.

[*D*] *Represent*

The interpretation of the term 'represent' may, as in the case of the previous terms, be ambiguous. Kotarbiński [73] narrows the interpretation in such a way that the term 'represent' relates to the interrelations between expressions of language.

Language expressions represent each other if they can be used instead of each other, a certain characteristic of the expressions being preserved. It is worthwhile determining what characteristic is preserved and how far it is a semantic characteristic of the expressions. Understandably that depends on the conditions under which the interchangeability of two or more expressions of language is admissible. If we admit that one semantic characterisation of the expressions is preserved, we are saying that the expressions correspond in meaning. In view of its usage in everyday speech this formulation is considerably vague and needs defining somewhat. This can be done in two ways, by defining exactly the conditions under which one expression may represent another, and by defining exactly what is preserved.[8]

If we treat the terms 'replace' and 'represent' as a relation of expressions of language, it is possible, with one expression replacing another, to study the conditions for that relation, for example the conditions needed for it to be a symmetrical and transitive relation. These conditions may be further specified by a statement of the language or languages in which interchangeability is admissible, the operations that admit of such interchangeability, and those that do not.[9]

Given that one expression may replace another and vice versa, we need to be more exact about what is preserved. In terms of the concepts we have already mentioned it is possible to distinguish between whether what remains unchanged is what the expressions signify or what they denote. The expressions 'Karel Čapek' and 'the author of "RUR"' may replace each other if it is sufficient that both expressions denote the same person. But their senses differ, hence the expression 'Karel Čapek' cannot replace 'the author of "RUR"' if what the expressions signify is required to remain unchanged. From this it is clear that the requirement of the preservation of sense is stricter than that of the invariability of what the expressions denote.[10]

To summarise this account of semantic terms and their common parlance affines, we may say that the terms 'signify' and 'denote' are the ones connected with semantic problems in the proper sense. They are also connected with the two basic modes of meaning,[11] i.e. sense and denotation, and it is in the analysis of sense and denotation that we shall return to these terms in greater detail.

2. SEMANTIC CONCEPTS IN FORMALISED LANGUAGES

[A] *The Concept of a Formalised Language*

The concept of a 'formalised language' and its all-round analysis fall mainly within the competence of logic.[12] The following account will therefore assume a knowledge of the basic principles of logic and of the devices used for building up a formalised language, including type theory.

Carnap [22] treats the concept 'formalised language' (he himself uses the term 'language system') as subordinate to the concept 'semantic system'. By a semantic system, or a system interpreted as such, Carnap understands a system of rules, formulated in a metalanguage and relating to an object language, that determine the truth conditions for each sentence of the object language, i.e. the necessary and sufficient truth conditions of all sentences of the object language. He then distinguishes two kinds of semantic systems: 'code systems' and 'language systems'. A code system is one that contains a finite number of sentences and supplies a complete review of the truth conditions for each sentence. This means that a code system translates each sentence of the object language into the metalanguage individually and so performs its interpretation. Thus a code system is a more primitive kind of semantic system (Carnap gives the examples of the cable code, the semaphore system and others). A language system, i.e. a formalised language, is assumed to contain an infinite number of sentences, hence the rules of a language system are assumed to be general and, as general rules, to determine the truth conditions for every sentence.

This distinction that Carnap draws between a semantic system and (formalised) language can scarcely be upheld today. It is fairly restrictive in its conception of a code, which is not at all that now current in information theory and the theory of communication in general. By this conception any communication, and therefore any language communication, assumes a mapping of objects, events or states in a system of constructive objects that may be realised by certain rules. Any such mapping is then described as coding, and the sum of rules as a code. The sum of constructive objects that may be used in coding is called an alphabet. In these terms any semantic system, irrespective of whether it makes possible a finite or infinite number of sentences, assumes some kind of coding and is therefore a 'code system'. Hence Carnap's suggested distinction between a code system and a formalised language is not tenable. This is by no means to say that there are not and cannot be differences in the richness or potential of different formalised languages.

In building up a formalised language, i.e. a particular object language, the existence of some other language, i.e. metalanguage, must always be assumed, without which it would not be possible to formulate the basic rules of the object language or to make the basic moves that building the language up involves.

The structure of a formalised language can be divided into two components: the syntactic, which concerns the build-up of the non-interpreted formalised language,[13] and the semantic, which concerns the interpretation of the formalised language.

For present purposes we shall use a simple formalised language corresponding to first order predicate calculus in syntactic terms. As signs of this object language $\mathscr{L}$ we shall use the letters of the Latin alphabet, and as signs of the metalanguage $\mathscr{M}\mathscr{L}$ relativised to $\mathscr{L}$ we shall use letters of the Greek alphabet.

The vocabulary of $\mathscr{L}$ contains as primitive signs:[14]

(1) individuals: (a) constants p, t, d, a with numerical indices as the need arises,
(b) variables x, y, z with numerical indices as the need arises;

(2) predicates: (a) constants A, B, C, P with numerical indices or numerical indices and superscript dashes, e.g. P''_1;
(b) variables P, Q with or without alphabetical indices.

(The number of arguments may be indicated by a number in brackets at top right-hand edge; e.g. $P_2^{(2)}$ is a two-argument predicate.)
(3) formulae: H, I, J;
(4) sentences: S, with or without numerical indices;

Type markers, insofar as they are used, will be placed before the sign at its upper left-hand edge, e.g. ${}^{(1)}P_1^{(3)}$ is a three-argument predicate of the first type.

As logical signs the vocabulary of $\mathscr{L}$ contains:

(5) negation: $\sim$, conjunction: $\cdot$, disjunction: $\vee$, equivalence: $\equiv$, implication: $\rightarrow$, incompatibility: $|$;
(6) general quantifier: $(\forall x)$, existential quantifier: $(\exists x)$, operator for abstraction: (λx) and the operator of individual description: (ιx);
(7) the sign $\in$ for membership in a class.

We shall also assume the use of brackets in the usual manner.

The vocabulary of $\mathscr{M}\mathscr{L}$ contains:

(1) individual variables (in $\mathscr{L}$): ξ;
(2) predicates (in $\mathscr{L}$): Π;
(3) formulae (in $\mathscr{L}$): Γ;
(4) sentences (in $\mathscr{L}$): Σ, sentences or classes of sentences (in $\mathscr{L}$): Φ, classes of sentences (in $\mathscr{L}$): Ω;
(5) the sign for the individual, predicate, sentence or class of sentences (in $\mathscr{L}$): Δ;[15]

Unless otherwise stated we shall distinguish constants and variables among metalanguage signs by numerical indices (for the constants) and letters of the alphabet (for the variables); for example Σ_1 is a sentence (in $\mathscr{L}$), Σ_i is a variable of a sentence (in $\mathscr{L}$).

As logical signs the vocabulary of $\mathscr{M}\mathscr{L}$ contains:

(6) *NON* for negation, *ET* for conjunction, *VEL* for disjunction, *AEQ* for equivalence, *SEQ* for implication, and *SHEFF* for incompatibility;
(7) *OMNIA* for the general quantifier, *EXIST* for the existential quantifier;
(8) a sign for concatenation: $\frown$.[16] This means, for example, that '$OMNIA \frown \Pi \frown \xi$' is the name[17] of a formula as recorded in the object language $(\forall x)(Px)$. In order to give a more accurate idea of the symbol $\frown$, it will be assumed that all the signs used will be used in the abstract sense.[18]

In order to simplify the notations in the metalanguage we shall not use different metalanguage symbols for the brackets in $\mathscr{L}$. Since brackets cannot, of course, be understood separate from other signs, the possible need for distinguishing them is obvious. As the need arises certain additional signs will be introduced, but they will only apply for the small section where they are brought in.

Looking over the signs that feature in $\mathscr{L}$, we see that they can be divided into logical and non-logical. This division, which we have already made in the inventory of the primitive signs of $\mathscr{L}$, is not, strictly speaking, feasible at the syntactic level. Only when we are in a position to give an interpretation of signs and therefore analyse them at the semantic level, can we separate them into logical and non-logical. This kind of division is usually made on the basis of certain semantic concepts, especially that of model.[19] It is customary to assume a universe that is the full set of all possible models of $\mathscr{L}$, or, to use Leibniz's original terminology, all the possible worlds about which statements can be made in $\mathscr{L}$. Then any sign which has the same interpretation in all possible worlds as it has in the actual world is a logical sign. Other signs whose interpretation may differ in various possible worlds from the interpretation they have in the actual world are non-logical. Put this way, the distinction between logical and non-logical signs shows that logical signs are not rid of all connection with the semantic level of analysis. On the contrary, this connection, as viewed by Kemeny [69] points to the existence of two kinds of semantic concepts:

(1) concepts defined in terms of interpretation with respect to the actual world,

(2) concepts defined in terms of all possible interpretations without regard to the one interpretation that is with respect to the actual world or any other intended interpretation.

A component of the vocabulary of $\mathscr{M}\mathscr{L}$ are, as is clear from the foregoing, semantic concepts, for $\mathscr{L}$ is not a semantically closed language.[20]

The formation rules of $\mathscr{L}$ correspond to the rules of first order predicate calculus. If each finite linear sequence of signs of $\mathscr{L}$ is a formula of $\mathscr{L}$, then the formation rules for introducing the concept of a well formed formula of $\mathscr{L}$ (abbreviated as wff) may be formulated as follows:

$\mathscr{F}$(1) a wff has one of these forms: (a) $\Pi^{(1)} \dot{\frown} \xi_1$, (b) $\Pi^{(2)} \dot{\frown} \xi_1 \dot{\frown} \xi_2$, (c) ... in general $\Pi^{(m)} \dot{\frown} \xi_1 \dot{\frown} \xi_2 \dot{\frown} \ldots \dot{\frown} \xi_m$

$\mathscr{F}$(2) if Γ_i is a wff, then $NON \frown \Gamma_i$ is also a wff;

$\mathscr{F}$(3) if Γ_i and Γ_j are wff, then $\Gamma_i \dot{\frown} ET \frown \Gamma_j$ is also a wff;

$\mathscr{F}$(4) if Γ_i and Γ_j are wff, then $\Gamma_i \dot{\frown} VEL \dot{\frown} \Gamma_j$ is also a wff;

$\mathscr{F}$(5) if Γ_i and Γ_j are wff, then $\Gamma_i \dot{\frown} AEQ \dot{\frown} \Gamma_j$ is also a wff;

$\mathscr{F}$(6) if Γ_i and Γ_j are wff, then $\Gamma_i \dot{\frown} SEQ \dot{\frown} \Gamma_j$ is also a wff;

$\mathscr{F}$(7) if Γ_i and Γ_j are wff, then $\Gamma_i \dot{\frown} SHEFF \dot{\frown} \Gamma_j$ is also a wff;

$\mathscr{F}(8)$ if Γ_i is a wff and ξ is a free individual variable contained by this formula, then $OMNIA\ \dot{\frown}\Gamma_i$ is also a wff;

$\mathscr{F}(9)$ if Γ_i is a wff and ξ is a free individual variable contained by this formula, then $EXIST\dot{\frown}\Gamma_i$ is also a wff.

Well formed formulae having one of the above forms and not containing free variables will be considered sentences.

Well formed formulae whose free individual variables are bound by the abstraction operator will be called abstracta. Well formed formulae whose free individual variables are bound by the individual description operator will be called individual descriptions.

As rules of inference of $\mathscr{L}$ we shall assume the modus ponens and substitution rules.

We shall leave to one side the problems of the axiomatic structure of $\mathscr{L}$, i.e. we shall not state which wff are regarded as axioms.[21] We shall likewise leave aside the problems of logical consequence, which fall within the purview of logic.[22]

[*B*] *The Interpretation of a Formalised Language*

In setting up an interpretation, i.e. in the semantic component of building up $\mathscr{L}$, we have to avail ourselves of a metalanguage as well. Thus we have to distinguish:

(1) a metalanguage that relates to a given $\mathscr{L}$ and contains the translation of the primitive signs of $\mathscr{L}$ to $\mathscr{M}\mathscr{L}$ in such a way that those steps that fall within the syntactic component of the build-up of $\mathscr{L}$ can be taken;

(2) a metalanguage that contains also semantic concepts and so makes possible statements about the relations of the expressions of $\mathscr{L}$ to a particular universe.

Thus it is essential to distinguish between the syntactic component of the metalanguage[23] and its semantic component. This distinction is of course relative, being relativised to $\mathscr{L}$.

Interpretation may be of two kinds (which corresponds, as has been suggested, to the two kinds of semantic concepts): interpretation with respect to all possible worlds,[24] and interpretation with respect to a particular or intended world.[25]

Analysis of the connections between the syntactic and semantic components of the structure of a formalised language and, simultaneously, of the connections between formalisation and interpretation, may serve two rather different aims. One is connected with the analysis of certain important properties of formalised systems generally, including, for example, properties such as indisputability and completeness. The other concerns the use of a formalised language in specific tasks, for example in the empirical sciences. The first aim is outside the competence of this work, so our remarks on it will be of the briefest. The central theme of this work is connected with the second aim hence primarily with questions in the methodology of science.

The analysis of the ways of formalisation and interpretation, and the analysis of semantic and syntactic aspects generally, is of importance in the solution of methodological problems such as the modelling of scientific procedures, the investigation of the relations of empirical procedures, the formation of concepts and generally the introduction of new items of scientific nomenclature.

This is by no means to suggest that the methodological problems of the empirical sciences are confined simply to those that are connected with the various stages in the syntactic and semantic build up of a formalised language. Nevertheless the analysis of the procedures of formalisation and interpretation, analysis of inference in science, of the procedures of measuring and processing of data, the creation of concepts and operations with them all undoubtedly constitute a major component in the range of methodological problems of science. The study of these methodological problems must, among other things, be based on a strict differentiation between the statements of a science about a given domain of objects and statements about those statements, or about the procedures that lead to what in the situation, and in terms of the tasks set, are considered the results of that science. Thus the scientific analysis of the methodological problems of science is largely metatheoretical in character.

The approach to the methodological problems of science that is based on analysis of the ways of formalisation and interpretation of the system of means of communication of science is, therefore, as Martin has stated so poignantly [89, p. 21], approximative and abstract. This itself means that no abstraction need be permanently and universally productive for decision-making in concrete methodological problems. The value of this or that abstraction is set in the same way as that of actual scientific activity. In certain circumstances scientific endeavour has found it more profitable to deal with the behaviour of an ideal gas than with that of the various empirically observable gases, as it has found it more profitable – but always with respect to a given set of aims – to experiment with the abstraction of a thermal machine than with a real thermal machine. There is no doubting the value of geometrical abstractions, the abstraction of multidimensional space and suchlike. Similarly in the study of methodological problems it may prove useful to deal not only with actual empirical and logical procedures, but also with abstract schemas of them. And it is here that we may put to good use the possibilities offered by logical semantics and syntactic and semantic analysis generally, along with the study of ways of formalisation and interpretation.

The system of science's means of communication that may also be studied in its abstract form may be described as the 'language of science'. This system contains, of course, more than just linguistic means in the traditional sense, since it also covers all the other devices that are capable of representing, and hence transmitting, scientifically relevant information,[26] data on empirical and experimental procedures and so forth. Nor can the study of this system be confined just to syntactic relations, as implied by, for example, the original programme of logical empiricism and its conception of the tasks of scientific methodology.

Abstract schemas of procedures is science – and this will also apply to those schemas based on the analysis of ways of formalisation and interpretation and generally on the possibilities offered by logical semantics – may of course in certain circumstances provide not only usefully simplified pictures of these procedures, but also pictures that may be distorted. This can lead to a measure of mistrust of any kind of abstraction and schematisation. Mistrust, as Carnap has shown [24, p. 210], leads to the insistence that the methodological analysis of science should concentrate on the actual behaviour of scientists, whether in the laboratory or at their typewriter. Carnap does not find this insistence fully warranted. And doubtless this viewpoint, if not debased by superficial empiricism and behaviourism, may be highly productive. Yet despite that, abstract schemas that neither forget the essential features of real procedures nor concentrate on non-essentials can throw more light on the procedures than any narrowly singular view. If we were to deprive methodological study of these possibilities and deprive metatheoretical analysis of science of the right to abstraction, we would be guilty of the same folly as if we tried to deprive scientific activity itself of the right to abstraction.

These features of logical semantics may well be useful also in the study of natural languages. Here too the abstract character of the schemas of logical semantics may be a starting point for the kind of approximation that in certain circumstances is capable of seeing some of the properties of natural language more clearly and with better definition than mere empirical study, and may well reveal certain properties that empirical study is just not fitted to discern. On the other hand formalised languages may have certain other properties that cannot be demonstrated for natural languages. Hence it would be unnatural to require complete sameness of approach to both formalised and natural language, just as much as it is impossible to carry over all the results of one domain into the other and vice versa.

[C] *The Introduction of Semantic Concepts by Definition*

In introducing semantic concepts sight must not be lost of the fact that semantic concepts that are relevant to a given $\mathscr{L}$ are always introduced in $\mathscr{ML}$. Two possibilities present themselves: one rests in the explicit definition of the concepts in $\mathscr{ML}$ using the potentials offered by $\mathscr{ML}$. This is the way used by Tarski and his adherents for introducing semantic concepts. A different method is based on the introduction of semantic concepts as undefined concepts of $\mathscr{ML}$ contained within axioms. Both methods can be shown to correspond to procedures for the axiomatisation of arithmetic. For example, Frege and Russell introduce arithmetical concepts in a way corresponding to the definitional introduction of semantic concepts, which enables them to apply a programme of reducing arithmetic to logic. On the other hand Peano's axiomatic method corresponds to the other way. Tarski also mentioned the possibilities of the second way [146, p. 401]. He argued that in this way semantics becomes an independent deductive theory, but he retained some misgivings as to the viability of the method, which seems at first sight straightforward enough. Tarski was convinced that the choice of axioms for an axiomatically built semantics tended to be somewhat random and dependent on non-essential factors. Another question that arises is the indubitability of axiomatically built semantics. Tarski's main objection to this way of introducing semantic concepts lay in certain psychological considerations by which it is natural to admit as primitive signs only those whose meaning is self-evident. Lastly, Tarski feels that the axiomatic approach to semantics and the introduction of semantic concepts is scarcely compatible with the conception of the unity of science in the form moulded by the Vienna Circle.[27]

Tarski's arguments are not entirely reliable. They are unreliable not only because they failed to deter certain logicians from taking the road that Tarski had thought hardly viable,[28] but also because they are subservient to certain opinions and tendencies dependent on the conceptions of the Lwow-Warsaw School and the Vienna Circle of the thirties.

The approach that introduces semantic concepts by means of explicit definitions using elements of a metalanguage is, according to Tarski, more suitable because in this way the concepts are, as he puts it, "reduced to purely logical concepts". In this way semantics "becomes a part of the morphology of language if the latter is understood in a sufficiently wide sense" [146, p. 406].

An attempt to incorporate semantics in the morphology of a language cannot of course be excluded, but in the end it leads to the priority of the syntactic level, which is scarcely to be accepted, and to attempts to explain the properties of a language purely from the language itself. Tarski did not reject the second method out of hand, leaving the question open.[29] He himself opted for the first method, i.e. the introduction of semantic concepts by definition. The basic features of his viewpoint are formulated in the conclusion: "It is possible to construct in the metalanguage methodologically correct and materially adequate definitions of the semantical concepts if and only if the metalanguage is equipped with variables of higher logical type than all the variables of the language which is the object of investigation" [146, p. 406]. Because the term 'methodologically correct' (or better 'logically correct') can be defined in harmony with the syntactic rules of $\mathcal{ML}$, there is still the *problem of the material adequacy* of the definitions of semantic concepts, or the problem of a criterion of adequacy. This is why the formulation of that criterion is one of the foremost problems in the definitional introduction of semantic concepts. The criterion is usually formulated in connection with the definition of the semantic concept of truth,[30] but it may also be formulated in connection with the definitions of other semantic concepts.

We shall try first to outline an intuitive sense of the requirement of material adequacy and then give certain definitions of a criterion of adequacy. We shall be starting out from a concept that Tarski considers one of the simplest, and hence fundamental, semantic concepts, namely *satisfaction*.

Assume three well formed formulae of $\mathcal{L}$:

$$wff(1)\ P_2^{(2)}x_1, x_2$$
$$wff(2)\ P_1x \cdot \sim P_1x$$
$$wff(3)\ P_1x \vee \sim P_1x$$

The domain of individuals X are all points or places on the surface of the Earth, 'be in Bohemia' is a metalanguage name for P_1, 'be more northerly than' a metalanguage name for $P_2^{(2)}$. Given that we know the meaning of the logical constants occurring in the formulae, we may say that wff(1) can be satisfied, wff(2) cannot be satisfied, and wff(3) can be generally satisfied. In other words, the ordered set $\{\xi_1, \xi_2\}$ satisfies wff(1) if and only it ξ_1 is more northerly than ξ_2, wff(2) is not satisfied by any ξ, and wff(3) is satisfied by any ξ.

As an example of satisfaction we may take the following: If 'Prague' and 'Tábor' are metalanguage names for p and t, then $\{p, t\}$ satisfies the formula $P_2^{(2)}x_1, x_2$ if an only if Prague is more northerly than Tábor. Meanwhile the requirement of material adequacy rests in the fact that the names that can be inserted in the well formed formula that is a component of the definiendum correspond in content to expressions of the metalanguage that are contained in the definiens. In other words, we want the expressions of $\mathcal{M}\mathcal{L}$ to be a translation of the relevant expressions of $\mathcal{L}$.

Failure to observe the requirement of material adequacy thus necessarily means opening the door to complete arbitrariness, which would render the given language useless as a means of reliable communication. But well formed formulae of $\mathcal{L}$ may, if it is allowed by the formation rules of $\mathcal{L}$, contain any finite number of free variables. Tarski therefore formulated his definition of satisfaction in such a way as to include also these cases.

In order to define the semantic concept 'satisfy' in general terms – vis-à-vis a given $\mathcal{L}$ – the formal structure of $\mathcal{L}$ has to be taken into account. Since it is possible, on the basis of syntactic rules, to put in a recursive definition of a well formed formula of $\mathcal{L}$, a recursive procedure is also possible in defining 'satisfy'. We can consider the simplest instance of a well formed formula, the so-called atomic formula, and various instances of the structure of more complex formulae, molecular formulae. In the case of $\mathcal{L}$ that we have introduced, the wff created by formation rule $\mathcal{F}(1)$ will be treated as atomic, the other wff as molecular formulae.

Since all the ways of constructing molecular formulae must be taken into account, the definition of the general concept of satisfaction ought to include all instances of part definitions. But these may run to an untidy mass. This leads Tarski, in his conception of a general definition, not to take into account just ordered n-tuples of individuals, but an infinite sequence of individuals. For the general definition of the concept of satisfaction we select from that infinite sequence only those individuals that can be inserted into the wff under consideration. Individual variables may also be ranged in sequence, and then the two sequences can be made mutually assignable by having matching indices to the individuals in one sequence and the variables in the other.

To give a general definition of satisfaction in $\mathscr{L}$, we have to introduce an interpretation of the non-logical constants in $\mathscr{L}$. If φ is a sequence of points (or places) on the Earth's surface and if we confine ourselves to the predicates P_1 and P_2 in the interpretation already given, we can define satisfaction in $\mathscr{L}$ as follows:

(D_1 – III 2 C) The sequence φ satisfies Γ in $\mathscr{L}$ if and only if φ is the sequence of points on the Earth's surface, Γ is a wff in $\mathscr{L}$ and if one of the following conditions applies:

(1a) Γ is $\Pi_1 \frown \xi_i$ for some positive number i and the i-th term of the sequence φ is a point on the Earth's surface that is in Bohemia.

(1b) Γ is $\Pi_2 \frown \xi_i \frown \xi_j$ for the positive numbers i and j and the i-th term of the sequence φ is a point on the Earth's surface that is more northerly than the point on the Earth's surface that is the j-th term of the sequence.

(2) There is a wff Γ_n such that Γ_n is $NON \frown \Gamma_m$ and φ does not satisfy Γ_m.

(3) There is a wff Γ_m such that Γ_m is $\Gamma_n \frown ET \frown \Gamma_o$ and φ satisfies both Γ_n and Γ_o.

(4) There is a wff Γ_m such that Γ_m is $\Gamma_n \frown VEL \frown \Gamma_o$ and φ satisfies Γ_n or φ satisfies Γ_o or satisfies both Γ_n and Γ_o.

(5) There is a wff Γ_m such that Γ_m is $\Gamma_n \frown AEQ \frown \Gamma_o$ and φ satisfies either both Γ_n and Γ_o or φ satisfies neither Γ_n nor Γ_o.

(6) There is a wff Γ_m such that Γ_m is $\Gamma_n \frown SEQ \frown \Gamma_o$ and φ either does not satisfy Γ_n or φ satisfies Γ_o or φ satisfies both Γ_n and Γ_o.

(7) There is a wff Γ_m such that Γ_m is $\Gamma_n \frown SHEFF \frown \Gamma_o$ and either φ does not satisfy Γ_n and satisfies Γ_o or φ satisfies Γ_n and does not satisfy Γ_o or φ satisfies neither Γ_n nor Γ_o.

(8) There is a wff Γ_m such that Γ_m is $OMNIA \frown \Gamma_n$ and it can be said of all points on the Earth's surface that they satisfy Γ_n.

(9) There is a wff Γ_m such that Γ_m is $EXIST \frown \Gamma_n$ and it can be said of at least one point on the Earth's surface that it satisfies Γ_n.

This definition of 'satisfaction' corresponds to Tarski's definition of the concept [147, p. 193], expect that Tarski's definition concerns a formalised language that is, in terms of its syntactic structure, a calculus of classes. What is important, however, is that for Tarski the definition of 'satisfaction' is the starting point for introducing other semantic concepts, while for Carnap the starting point was the concept 'true'. This is to be explained by the fact that Tarski and Carnap stressed rather different aspects in their conception of a formalised language. With Tarski the emphasis is on the role of variables, and because a formalised language in the general sense contains variables, Tarski's procedure is more appropriate.[31] Then the semantic concept 'true' features intuitively as a special instance of 'satisfaction'. The concept of 'true' and 'satisfaction' are identical when a wff is a sentence.

Carnap's approach takes as starting point much simpler formalised languages (his 'semantic systems'), which in the simplest form of all do not assume the occurrence of variables in the object language.[32] This is why Carnap first gives his characterisation of the concept 'true' before proceeding to deal with other concepts, although in fact he does not offer a definition of 'true', but merely formulates a criterion for the adequacy of such a definition in the shape of a convention.

Tarski's more general starting point enables him to move from the definition of 'satisfaction' to that of 'true'. If we continue to use the language that has served in the introduction of the concept of 'satisfaction', the definition of 'true' has the form:

(D_2 – III 2 C) Γ is a true sentence in $\mathscr{L}$ if and only if Γ is a sentence in $\mathscr{L}$ and if Γ is satisfied in every sequence of points on the face of the Earth[33]

Let us now return to the problem of the material adequacy of the above definitions of semantic concepts. Tarski's procedure assumes that the concept 'true' is definable in the metalanguage and relates to sentences ($S_1, S_2, \ldots$) of the object language $\mathscr{L}$. The expression '... is true if and only if---' therefore contains expression components '...' and '---', where '...' is occupied by S_1 and '---' is the translation of S_1 into $\mathscr{ML}$. This is the intuitive sense of the convention for a criterion of adequacy of the definition of the semantic concept 'true'.

Tarski summarises these intuitive requirements in convention **T**[147, pp. 187–188]:

A formally correct definition of the symbol ***Pr*** (where the symbol ***Pr*** denotes the class of all true sentences in $\mathscr{L}$), which is formulated in the metalanguage, is called an adequate definition of the concept 'true' if it has these consequences:

(α) all sentences that can be acquired from the expression '$S \in \boldsymbol{Pr}$ (i.e. S is a true sentence in $\mathscr{L}$) if and only if Σ' are acquired by replacing the sign S by the name of any sentence of the object language and the sign Σ by an expression that is the translation of that sentence into the metalanguage;

(β) the sentence: for any S, if $S \in \boldsymbol{Pr}$, then S is a sentence in $\mathscr{L}$.

Carnap formulated an essentially similar convention [22, p. 26–27].

If we wonder about the sense of such a criterion of material adequacy for the semantic concepts mentioned, and if we look both at Tarski's procedure which sets out from 'satisfy' and at Carnap's where 'satisfy' is left until after other semantic concepts have been introduced, we can point to a number of common features:

(1) Both cases assume a $\mathscr{L}$ and a $\mathscr{ML}$ into which $\mathscr{L}$ is translatable but which is 'richer' than $\mathscr{L}$.

(2) A convention is assumed as a valid criterion of material adequacy. This convention relates in essence to the possibility of translating $\mathscr{L}$ into $\mathscr{ML}$ in terms of content.

(3) The nature of this convention is not investigated any further. It is merely formulated as a convention, assumptions for its feasibility and tenability being beyond the framework of the analysis.

There can be no objections to the procedure if the tenability of the convention is taken as a matter of course. But it is obvious that this is possible only in the extreme case, i.e. in an analysis (of a formalised language) that serves purely logical purposes. However, if we wish to investigate the genuine tenability of the convention, we have to consider aspects that are actually pragmatic in character, i.e. we shall have to respect, for example, the users of the language, their capabilities (i.e. capacities, memories and other properties of the language users' channels), the aims associated with using the language, and so on.

The approach to definitions of semantic concepts outlined above has, as we have said, a built-in assumption that $\mathscr{ML}$ is richer than $\mathscr{L}$. Tarski, as we have also said before, pointed out that $\mathscr{ML}$ must be equipped with variables of a higher logical type than the variables of $\mathscr{L}$.

In the light of this we may confront the vocabulary of $\mathscr{L}$ and that of $\mathscr{ML}$. If we analyse the vocabulary of $\mathscr{ML}$ from the point of view of its correspondence with the vocabulary of $\mathscr{L}$, we may distinguish the following kinds of expressions:

(1) Expressions for the logical signs of $\mathscr{ML}$ that correspond to the logical signs of $\mathscr{L}$ (in the case we have been describing *NON* in $\mathscr{ML}$ corresponds to $\sim$ in $\mathscr{L}$, *ET* in $\mathscr{ML}$ to $\cdot$ in $\mathscr{L}$, etc.);

(2) Non-logical signs in $\mathscr{ML}$ that correspond to the non-logical signs in $\mathscr{L}$ (e.g. $\xi_1, \xi_2 \ldots$ in $\mathscr{ML}$ correspond to $x_1, x_2, \ldots$ in $\mathscr{L}$, Π_1, Π_2 in $\mathscr{ML}$ correspond to P_1, P_2 in $\mathscr{L}$);

(3) In $\mathscr{ML}$ there have to be variables available for all the expressions that may occur in $\mathscr{L}$(Σ for sentences in $\mathscr{L}$, Φ for sentences or classes of

sentences in $\mathscr{L}$, Ω for classes of sentences in $\mathscr{L}$, the sign Δ for individuals, predicates, sentences of sentences in $\mathscr{L}$, and the sign Γ for formulae in $\mathscr{L}$). It is here that the need for a higher type of variables in $\mathscr{M}\mathscr{L}$ becomes apparent.[34]

(4) Finally $\mathscr{M}\mathscr{L}$ is the place for semantic terms such as 'satisfy', 'true' and any others that can be defined on the basis of them.

The characteristic features of the definitional introduction of semantic concepts is the reduction of semantic concepts to non-semantic concepts (i.e. reduction of (4) to (1), (2) and (3) relativised to $\mathscr{L}$, but only on the assumption that $\mathscr{M}\mathscr{L}$ is richer than $\mathscr{L}$ in the type given). This matches the two main results of Tarski's work [147], which Tarski himself denoted as results **A** and **B**, or **A**′ and **B**′. The formulations of **A** and **B** concern the introduction of the semantic concept 'true', while that of **A**′ and **B**′ concerns the introduction of semantic concepts generally. Since we are presently interested primarily in the principles of introducing semantic concepts as such, and not the finer points of procedure connected with any one of them, we shall give here only formulations **A**′ and **B**′:

> **A**′: "The semantics of any formalized language of finite order can be built up as a part of the morphology of language, based on correspondingly constructed definitions.
>
> **B**′: It is impossible to establish the semantics of the formalized languages of infinite order in this way"
>
> [147, p. 266].

Conclusion **B**′ is clearly not categorical. It merely points to the impossibility of introducing semantic concepts in one particular way, i.e. by means of definitions employing the means of expression of $\mathscr{M}\mathscr{L}$, which contains variables of a higher logical type. This was what led Tarski to formulate a third conclusion, which is not in fact a conclusion to his investigations, but a hypothesis based on the limitations of his approach:

> **C**′: "... the semantics of any formalized language of infinite order can be extablished as an independent science based upon its own primitive concepts and its own axioms, possessing as its logical foundation a system of the morphology of language"
>
> [ibid.].

This means that semantic concepts can be included in the vocabulary of the metalanguage as primitive signs and that the basic properties of those concepts have to be determined axiomatically.

[*D*] *The Axiomatic Introduction of Semantic Concepts*

As already demonstrated, Tarski's chief merit is in having given some precision to the definitional introduction of semantic concepts. Tarski also showed the limitations of this approach to semantics. However, there is one approach to semantics that does not require the metalanguage to be richer than the object language in the sense that the metalanguage contains variables of a higher logical type than the object language. This approach has it that semantic constants, that are not defined, are introduced into the metalanguage and their use is determined axiomatically. By contrast to the definitional introduction of semantic concepts, this approach may be described as their axiomatic introduction.

The method was worked out in detail by R.M. Martin [89], who has as his fundamental semantic primitive term "*multiple denotation*". It must be noted now that the concept of 'multiple denotation' is connected in Martin's work with only one particular conception of language: language is built nominalistically, hence the special conception of 'multiple denotation'.

Martin distinguishes three different ways of using the term 'denotation', each way tied to a different conception. (1) We are speaking about individuals, classes or relations that denote the relevant objects. (2) We may be speaking about expressions for a class or property that denote the members of the relevant class. (3) Finally we may be speaking about a conception of denotation in the broadest sense whereby, for example, sentences denote propositions or truth values.[35]

In the language $\mathscr{L}$ that we have introduced we have so far adhered to a conception that might be described as 'semantically neutral'. This means, for instance, that we have not explicitly prescribed that predicates of $\mathscr{L}$ be interpreted as classes or properties. In the given concrete interpretation in the case of the predicate 'be in Bohemia' we have not stated explicitly whether it relates to the class of all points on the Earth's surface that are in Bohemia, or to the property of each separate element of that class that is in Bohemia. By contrast, Martin, as he himself emphasises, keeps strictly to the second of these uses of the term denotation. This means that in Martin's conception, which may be described as strictly nominalist, language expressions that are capable of denoting 'do not denote' classes, but separate individuals. Thus the predicate 'be in Bohemia' denotes each separate point on the Earth's surface that is in Bohemia. Similarly 'dog', for instance, does not denote the class of dogs or the property 'be a dog', but Fido, Lassie, Spot etc. So Martin's conception of 'multiple denotation' must be understood in this nominalist spirit.

The term 'multiple denotation' ('*Den*') is introduced as the sole primitive semantic concept of the metalanguage. It may be treated as a two-place predicate, while it is desirable that one of its arguments should always be an individual.[36] It must of course be firmly established what the other argument of this predicate may be. For Martin the basic kind of denoting expressions ('*Den*') are abstracta that contain no free variables, which leads him to introduce the operation of abstraction, symbolised by the operator of (nominalist) abstraction $\ni$.[37]

The operation of abstraction works by binding the free variables that are marked with the operator in any wff to which the operator is applied. For example, from the wff $P_1 x$ we can create the abstractum $x \ni P_1 x$. Similarly from the wff $P_2 x_1 x_2$ we can create an abstractum by binding either x_1 or x_2 or both variables by the operator of abstraction. If we apply the operation of abstraction to a wff H in which there are no free variables, $x \ni H$ is an empty abstractum. Hence non-empty abstracta correspond to one- or many-place predicates, provided that the individual free variables can be bound by the operator of abstraction. It is therefore possible to speak, as in the case of predicates, of one- or many-place abstracta.

For the axioms connected with the use of '*Den*' Martin puts forward an axiom schema, being forced to assume an infinite number of axioms. If Δ is the name of an abstractum that arises from the wff H that contains one free variable (i.e. if Δ is the name of $x \ni H$), the schema of axioms has the form:

$$(A_1 - \text{III 2 D})(\forall x)(\Delta \textit{Den}\, x \text{ if and only if } \ldots x \ldots)$$

in which Δ is either the name of the abstractum $x \ni H$ and '... x...' is the translation of H into $\mathscr{ML}$, or it is the name of a one-place predicate and '... x ...' is the translation of that one-place predicate with argument x into $\mathscr{ML}$.

This axiom schema matches the nominalist nature of the entire conception: only separate individuals can be denoted ('*Den*') and only one-place predicates or one-place abstracta can denote. The universe is therefore a domain of individuals, which also corresponds to the nominalist conception of a language of the first degree.

With '*Den*' having been introduced like this, other concepts can now be brought in by definition, which Martin does in the case of 'satisfy', 'semantic product', 'semantic sum' and others. The semantic concept 'true' may also be introduced on the basis of '*Den*'. It is useful to recall in this connection that definitions of 'true', whether made on the basis of the introduction of certain semantic concepts as primitive terms of a metalanguage, or by means of explicit definitions, obviously do not provide a concrete criterion by which to decide whether this or that sentence is true. These definitions of the concept 'true' give only the abstract framework for analysis of that which means that a sentence is true. It provides therefore the conditions under which sentences can be assigned the attribute 'true', while any actual decision may depend also on other than semantic means, such as results of experiments, observations etc.

The axiomatic method of introducing semantic concepts, which treats one of the concepts as a primitive term of $\mathscr{ML}$, clearly has certain advantages over the definitional introduction of semantic concepts. In the definitional method we do not leave the sphere of language, assuming merely that we have at our disposal $\mathscr{ML}$ that is richer then $\mathscr{L}$ in that it contains variables of a higher type. In the axiomatic introduction of semantic concepts we also need an $\mathscr{ML}$ that is richer than $\mathscr{L}$, albeit in a rather different respect. The vocabulary of $\mathscr{ML}$ itself contains '*Den*' or perhaps one of the other semantic concepts which is treated as primitive.

The introduction of semantic concepts as conceived here makes it possible on the whole to process the entire semantic problems of formalised languages. Even here, however, it could after all be objected that we remain in the sphere of language and that this approach to semantics is also typified by the principle with which Wittgenstein ends his *Tractatus* [165]: "Wovon man nicht sprechen kann, darüber muss man schweigen". While it is doubtless true that it is impossible to talk about any kind of objects otherwise than by using a particular language, the same applying to relations between those objects and expressions of language and to the expressions themselves, the introduction of semantic concepts as primitive terms sui generis nevertheless indicates that the very nature of these concepts is not within the competence of the *internal* structure of language – the competence of what Tarski called the morphology of language.

Of course, there are here certain disadvantages that may be substantive under certain circumstances. The axiomatic method of elaborating a theory – in our case a theory of semantics – may be of use if it helps not only to be specific about certain procedures but also to make them more economical. This in turn requires that the number of axioms that have to be respected in those procedures should not be too high. For the semantics of a relatively simple formalised language, where the universe is accessible to complete and exhaustive processing, this is a suitable approach.

To put it crudely, this is the intuitive reason why this method of introducing semantic concepts is associated with the tendencies of finitistically conceived constructivism. Martin summarises these tendencies in a methodological principle: Wherever finitisation is possible, it is to be recommended [89, p. 269]. This principle need not conflict with the conception of an 'evolving language' which expands gradually as its universe expands, parallel to which further axioms are added using semantic concepts as primitive terms of the metalanguage.

However, if we assume that we shall have to reckon with an infinity of axioms, objections may be raised against the feasibility of the axiomatic method of introducing semantic concepts and the advantages of the definitional approach may become apparent.

In conclusion there is no escaping the question of whether the two basic methods of introducing semantic concepts in formalised languages have anything in common with the semantics of natural languages, or the semantics of real processes of communication. It would appear that the purely philological approach, which disregards extralinguistic connections, is closer to the definitional method of introducing semantic concepts whereby semantics is reduced to the morphology of a language. But when it comes to carrying the approach out in practice, regard to extralinguistic entities and connections is inevitable, whether these are connections with a given universe, or social or psychological connections, or whatever. Therefore, in our view, the axiomatic method of introducing semantic concepts is closer to the semantics of real processes of communication: while it fixes certain semantic rules in the form of axioms, without going into their origins, character or possible changes, it does not derive those rules from language itself. It has to be added that the rules of the semantics of real processes of communication are by natural empirical, they are conditional on the social and cultural level of the users of the language, they are connected with certain real tasks, and so forth. At the same time any real intercommunication assumes a measure of stability in these rules. And in science the procedure whereby we take the relatively stable as absolutely stable (in a given theory) and conceive axiomatically that which is based on experience, is very often the productive one.

IV. THE SEMANTICS OF LOGICAL CONCEPTS

In the preceding chapter we touched briefly on the problem of the two kinds of concepts: concepts that relate to a particular envisaged world, e.g. the actual or empirically accessible world, and concepts that relate to all possible worlds. It would be wrong to think that the competence of semantics can be confined to the first kind only. On the contrary it is to the credit of logical semantics that it has shown ways of semantically analysing the second kind of concepts. This kind of problem area may be described as the semantics of logical concepts, or as logical semantics (L-semantics for short) in the proper sense of the word.

Within this problem area a special place is occupied by the semantic analysis of what are called logical concepts (L-concepts), such as L-true, L-untrue or contradictory. This group of logical concepts includes many others, such as L-implication, L-equivalence, L-disjunction etc. Many of these concepts have had, and still have, great significance for methodology. For example, the concept of L-truth and the associated concept of analyticity have played a major role in modern scientific thought ever since the days of Leibniz, for it has been important to know and prove that this or that sentence holds 'for purely logical reasons', irrespective of any accessible or even only possible empirical support. Similarly L-implication has an important position in the explanation of such phrases as 'this sentence follows from that sentence for purely logical reasons'. Many of these L-concepts feature in scientific thinking and the language of science in less obvious guise and often in a different terminology. For example, we are unlikely to find the concept of L-truth in the natural sciences, yet it is intuitively obvious that contradictory statements about one and the same universe are inadmissible.

It is to the credit of logical semantics that it has put forward several conceptions that offer a systematic interpretation of these concepts, whose use in the language of science is frequently associated with an intuitive understanding. It is still not possible to speak of a united and universally accepted conception of the logical world as distinct from the non-logical world that may be treated as sufficiently general and universally applicable.[1] What is significant is that all the conceptions that have attempted to solve these problems have, in one way or another, referred across to problems of semantics and appealed to certain semantic concepts.

The pages that follow will be devoted to analysing primarily two fundamental conceptions: (1) the conception of L-semantics elaborated chiefly by Carnap and offering a semantics of L-concepts in relation to sentences or classes of sentences; (2) a conception of interpretation drafted by J.G. Kemeny and offering a semantics of L-concepts on the basis of the semantic concept of 'model'.

1. PROBLEMS OF L-SEMANTICS

In the previous chapter we spoke of the possibility of introducing certain basic semantic concepts such as 'true' and 'satisfy'. If we take as a basis the concept 'true' and operate with sentences or classes of sentences, we can define a number of other important concepts that concern relations between these expressions of language. Carnap describes these as radical concepts, as opposed to the concepts prefixed by L- or F-, i.e. what he calls logical and factual concepts [22, p. 33].

In order to introduce some important definitions of these concepts we have to use once more the metalanguage signs already introduced. With these aids and on the basis of the semantic concept 'true', Carnap defines several radical concepts concerning the relations between sentences or classes of sentences.

(D_1 – IV 1) Ω_i is *true* (in $\mathscr{L}$) if and only if each sentence in Ω_i is true.

(D_2 – IV 1) Φ_i is *untrue* (in $\mathscr{L}$) if and only if Φ_i belongs to $\mathscr{L}$ and if it is not true in $\mathscr{L}$.

(D_3 – IV 1) Φ_i *implies* Φ_j (in $\mathscr{L}$) (to put it another way:[2] $\Phi_i \rightarrow \Phi_j$) if and only if both Φ_i and Φ_j belong to $\mathscr{L}$ and if either Φ_i is untrue or Φ_j is true or both Φ_i and Φ_j are true

(D_4 – IV 1) Φ_i *is equivalent* to Φ_j (to put it another way: $\Phi_i \equiv \Phi_j$) (in $\mathscr{L}$) if and only if both Φ_i and Φ_j belong to $\mathscr{L}$ and if either both are true or both are untrue.

(D_5 – IV 1) Φ_i *is disjunctive* with Φ_j (to put it another way: $\Phi_i \vee \Phi_j$) (in $\mathscr{L}$) if and only if at least one of the two is true.

(D_6 – IV 1) Φ_i *is incompatible* with Φ_j (to put it another way: $\Phi_i | \Phi_j$) (in $\mathscr{L}$) if and only if at least one of the two is untrue (i.e. if they are not both true).

Of the other important radical concepts that can be similarly introduced by definition we shall mention the concept of an *empty class of sentences* in $\mathscr{L}(\wedge_L)$ and a *universal class of sentences* in $\mathscr{L}(\vee_L)$. On the basis of these definitions Carnap formulates a series of theorems which are fairly easy to derive and will therefore be omitted here. Now let us note some of the assumptions behind these and other possible definitions of radical concepts. In the case of some of them Carnap also notes them, but others are left implicit.

(1) The procedure above only concerns sentences and classes of sentences. It is tacitly assumed these are conjunctive classes of sentences, because the truth value of the class of sentences $\Sigma_1, \Sigma_2, \ldots \Sigma_n$ is the same as the truth value of the conjunction whose elements are $\Sigma_1, \Sigma_2, \ldots \Sigma_n$.[3]

(2) It is further assumed that the semantic characteristic (or meaning) of a sentence is the same thing as the truth value of the sentence. This standpoint, which was first given a more exact form by Frege, doubtless possesses many advantages. But at the same time it narrows the view of the semantic problems of language not only because of the difficulty without which it could not be extended to take in other expressions of language, but also because it is only suited fully to extensional languages.[4]

The above definitions of certain radical concepts express in effect the familiar matrix of truth values for negation, implication, equivalence, disjunction and incompatibility. It should be noted, however, that these definitions and their associated theorems are not themselves the proper subject of L-semantics, being more of an elementary prerequisite. The actual aim of L-semantics is to justify how the switch can be made from these concepts to the L-concepts that correspond to these (and possibly other) radical concepts.

In *Meaning and Necessity* [23, p. 10] Carnap formulates the following convention:

"A sentence Σ_i is L-true in a semantical system S (our language $\mathscr{L}$, LT) if and only if Σ_i is true in S in such a way that its truth can be established on the basis of the semantical rules of the system S alone, without any reference to (extra-linguistic) facts."

Thus the proper task of L-semantics rests in the analysis of what is meant by 'on the basis of the semantical rules of the system S alone'. This of course does not apply just to the concept of L-true, but to many other concepts as well.

Carnap puts forward to different ways of building up L-semantics: A. The introduction of L-concepts (in addition to the above-mentioned semantic concepts) as non-defined concepts by means of several postulates from which the properties of the concepts can be stated formally; **B**. The introduction of L-concepts on the basis of an idea of logical range and the concept of 'state description'. With the subsequent appearance of certain difficulties connected with method **B** he added a number of complementary constraints that are usually described as 'meaning postulates' [25].

[*A*] *The Introduction of L-Concepts as Non-Defined Terms*

From the semantic point of view this method is the least interesting since it involves semantic relations only in a very narrow sense. Semantic relations proper only enter the conception as determined by the relations between L-concepts and the above-mentioned radical concepts.

For the constitution of L-semantics Carnap formulates 15 postulates in all containing as primitive terms: (1) the concept 'true' as a radical concept (the other radical concepts are assumed to be definable on the basis of the above definitions), and (2) L-concepts: L-true, L-untrue, L-implication, L-equivalent, L-disjunctive.

The postulates that contain these primitive terms concern the following relations:
(1) postulates $\mathbf{P}_1$ to $\mathbf{P}_4$ concern relations between L-concepts and the corresponding radical concepts (essentially this amounts to the relation of inclusion, which it is important to realise given that this is a transitive, but not symmetrical relation); (2) postulates $\mathbf{P}_5$ to $\mathbf{P}_{10}$ concern properties of L-concepts; (3) postulates $\mathbf{P}_{11}$ to $\mathbf{P}_{13}$ concern relations between sentences and classes of sentences, and (4) $\mathbf{P}_{14}$ and $\mathbf{P}_{15}$ concern important properties of the concepts L-true and L-untrue.

In formulating them we shall use the previously employed metalinguistic signs for sentences and class of sentences in $\mathscr{L}$.

($\mathbf{P}_1$ – IV 1 A) If Φ_i is *L-true*, it is (also) true.
($\mathbf{P}_2$ – IV 1 A) If Φ_i is *L-untrue*, it is untrue.
($\mathbf{P}_3$ – IV 1 A) If Φ_i *L-implies* Φ_j (to put it another way: if $\Phi_i \underset{\mathrm{L}}{\longrightarrow} \Phi_j$), *then* $\Phi_i \longrightarrow \Phi_j$.
($\mathbf{P}_4$ – IV 1 A) If Φ_i and Φ_j are *L-disjunctive*, they are also disjunctive.

The sense and purpose of the postulates is on the whole quite evident and they have to do with the relation of inclusion between L-concepts and the corresponding radical concepts. Another postulate introduces the important property of L-implication – transitivity. It should be noted that in Carnap's system the concept of L-implication plays a particularly important role. It is a concept that represents an elucidation of the intuitive concept of logical consequence. This property also corresponds to the intuitive sense of the turn of phrase that says that something follows logically from something else.

($\mathbf{P}_5$ – IV 1 A) L-implication is transitive, i.e. if $\Phi_i \underset{L}{\longrightarrow} \Phi_j$ and $\Phi_j \underset{L}{\longrightarrow} \Phi_k$, then $\Phi_i \underset{L}{\longrightarrow} \Phi_k$.

The next three postulates also concern properties of the concept of L-implication:

($\mathbf{P}_6$ – IV 1 A) If $\Phi_i \underset{L}{\longrightarrow} \Phi_j$ and Φ_i are L-true, then Φ_j is also L-true.

($\mathbf{P}_7$ – IV 1 A) If $\Phi_i \underset{L}{\longrightarrow} \Phi_j$ and Φ_j are L-untrue, then Φ_i is also L-untrue.

($\mathbf{P}_8$ – IV 1 A) Any sentence Σ_i L-implies itself, i.e. $\Sigma_i \underset{L}{\longrightarrow} \Sigma_i$.

The next two postulates concern properties of L-equivalence and L-disjunction:

($\mathbf{P}_9$ – IV 1 A) Φ_i is L-equivalent to Φ_j if and only if $\Phi_i \underset{L}{\longrightarrow} \Phi_j$ and $\Phi_j \underset{L}{\longrightarrow} \Phi_i$.

($\mathbf{P}_{10}$ – IV 1 A) If Φ_i is L-true, Φ_i is L-disjunctive with any Φ_j.

The next three postulates characterise the relations between sentences and class of sentences on the assumption that the sentences are elements of those classes:

($\mathbf{P}_{11}$ – IV 1 A) If Σ_j is an element of Ω_i, then $\Omega_i \underset{L}{\longrightarrow} \Sigma_j$.

($\mathbf{P}_{12}$ – IV 1 A) If Φ_i L-implies each element of Ω_j, then $\Phi_i \underset{L}{\longrightarrow} \Omega_j$.

($\mathbf{P}_{13}$ – IV 1 A) If each element of Ω_i is L-true, then Ω_i is L-true.

Finally the last two postulates characterise other properties of L-implication:

($\mathbf{P}_{14}$ – IV 1 A) If Φ_j is L-true, then for any Φ_i it holds that $\Phi_i \underset{L}{\longrightarrow} \Phi_j$.

($\mathbf{P}_{15}$ – IV 1 A) If Φ_i is L-untrue, then for any Φ_j it holds that $\Phi_i \underset{L}{\longrightarrow} \Phi_j$.

(Also these last two postulates correspond to the properties of logical consequence, i.e. that a logically true sentence can follow from anything, and a logically untrue sentence may have any sentence as its consequence.)

On the basis of these postulates Carnap formulates a number of theorems of L-semantics. By means of definitions he then introduces some other logical concepts. In comparing these L-concepts with their corresponding radical concepts what is crucial is that the radical concepts are definable on the basis of the concept 'true'. The same approach is not possible, however, in L-semantics: L-untrue and L-implication are not definable on the basis of L-true. This is the real reason why L-concepts are introduced as non-defined primitive terms and why their properties and relations are set up by means of the above postulates.

A major disadvantage of this method of bringing in L-concepts, and hence of this kind of differentiation between the spheres of the 'logical' and the 'non-logical', is that the logical semantics of the object language as formulated in the metalanguage itself already contains the concept 'logical' and its counterpart 'non-logical'. And yet the concept of 'non-logical' may be either definable on the basis of the radical semantic concepts, e.g. 'true', or introduced as another primitive term of the metalanguage.

[*B*] *The Concepts of 'Logical Range' and 'State Description'*

The above method of introducing L-concepts as non-defined terms by means of postulates is not really a semantic method in the proper sense of the word. It assumes no universe with respect to the object language, but only a metalanguage the postulates of which introduce some of the properties of logical concepts and so too the differentiation of the logical and non-logical spheres.

By contrast, the second way of constituting L-semantics assumes, at least at first sight, the existence of an extralinguistic sphere to which language expressions are assigned.[5]

From the intuitive point of view, this second method is a kind of reconstruction of 'possible events' corresponding to all the sentences that can be formulated in $\mathscr{L}$. This reconstruction is, however, only formal in the sense that the 'possible world' that corresponds to any atomic sentence in $\mathscr{L}$ is reduced to a positive or negative decision about the assignment. Hence the matrix of truth values in logic is an application of this semantic viewpoint. In other words, at the root of this method lies the assumption, formulated early by Frege, that the denotation of a sentence is identical with its truth value. In addition, this approach rests, as has been mentioned, on some of the ideas of Leibniz, who also considered 'possible worlds' as opposed to the actual world.

The 'possible world' with respect to a given language is to be understood as the complete stock of states that may contain objects that make up the universe of the given language. It is assumed simultaneously that these states can be described by means of expressions of that language. Another attempt to formulate the same ideas is Wittgenstein's concept of 'logical space' (or more exactly 'logical free play' – *logischer Spielraum*).[6]

Carnap takes up Wittgenstein's concept of 'logical space', calling it 'logical range', and links it with the concept of 'state description'(*Zustandbeschreibung*). He then uses the concepts for several distinct purposes: as a starting point for building up L-semantics [22], as the basis of his conception of 'logical probability' and the logical foundations of probability theory [24], and as the starting point for the semantic theory of information [6]. Reference to these broader connections is very important, because information about these three different groups of problems in the literature treats them as if they were quite separate matters. In fact they are three different formulations of substantially the same problem, or at least conceptions whose scientific importance *and* limitations cannot be appreciated without clarification of their common origins. Moreover, it must be stressed that methods based on the concepts of 'logical range' and 'state description' are limited by the implicit assumption that they can be related only to languages that contain no more than a denumerable infinity of signs and so too no more than a denumerably infinite set of individuals.[7]

Intuitively speaking, the concept of 'logical range' represents the stock of 'possible' worlds in respect of a given language or given expressions of that language. Then the concept of 'logical state' corresponds to a quite definite state of affairs, i.e. one realisation out of the stock of possible worlds. A sentence or class of sentences (again this is a conjunctive class of sentences) which exactly describes or, preferably, unambiguously defines such a state is called a state description. It must be added that a state description leaves nothing unclear: it gives an unambiguous pronouncement on all the objects, properties and relations about which statements can be made by the means of the given language. This is why state descriptions have been said to be the 'most powerful' statements of fact that can be made by means of a given language with respect to a given universe. For example, if the universe contains n individuals to which it is possible to assign or not assign one property (i.e. if with this group of individuals we are operating with just one one-place predicate), the state description is the (conjunctive) class of n atomic sentences. If we operate with two one-place predicates, the state description is a class of $2n$ atomic sentences, and so forth. If we operate with two-, three- down to k-place predicates, we have to consider ordered pairs, triples and k-ples of individuals, and so it is a simple matter of combination.

For the intuitive view of the concept of 'state description' as Carnap works with it in his logical semantics, the following circumstances are also important:

(1) No consideration is paid to whether all logical states are really possible (and hence whether all state descriptions have a sense). This shortcoming, which was pointed out by critics of Carnap's L-semantics [22], Carnap attempted to remove in part by means of the conception of meaning postulates [25].

(2) Nor is there any consideration of the measure of likelihood of individual logical states. It is this circumstance that provides one logical starting point for the concept of 'logical probability' [24].

(3) If it is possible to connect individual possible states, and hence also the *content* of individual statements, with a concept that has the property of additivity, a formulation can be reached that is analogous to the elementary form of Shannon's formula for a measure of the quantity of information. This is also the main result of Carnap and Bar-Hillel's theory of semantic information [6].

To understand these circumstances and their interconnections it is useful to look at the connection between the concept of 'state description' and certain of the ideas of Boltzmann. The possibility of this kind of statistical-mechanical analogy was pointed out by Cherry [33, p. 238]. We shall here attempt to take Cherry's point further. Let us assume a universe consisting of n individuals. To these individuals correspond n individual constants in $\mathscr{L}$. The universe can be *decomposed* according to whether certain properties, relations etc. appertain to the separate individuals. This can be expressed by using one- or multi-place predicates as the names of the properties, relations etc. Atomic sentences of $\mathscr{L}$ with n primitive individual constants and π primitive predicate constants (i.e. sentences of a language which we shall call $\mathscr{L}_n^\pi$) are formed by assigning or not assigning the various individual constants to the various predicates – in the case of one-place predicates, or ordered pairs of individual constants to the predicates – in the case of two-place predicates, and so on.[8] It is possible also to imagine that the set of individuals can be decomposed into subsets of individuals (or subsets of ordered pairs, triples etc., according to the nature of the relevant predicates). Given this decomposition, which splits the whole set of individuals into at least two subsets, we can distinguish two kinds of statements:

(1) *structure descriptions*, which are statements about *how many* individuals (ordered pairs, triplets etc., according to case) belong in each subset;

(2) *state descriptions*, which are statements about *which* individuals (pairs, triplets etc.) belong in each subset.

The analogy between this analysis of $\mathscr{L}_n^\pi$ and some of the ideas of classical statistical physics rests in the parallels between structure description and 'macrostate' on the one hand, and state description and 'microstate' on the other.

We shall take as an example of this kind of analysis of a system of individuals the system consisting of the individual points on the Earth's surface. Assuming that we are able, within a given task, at a given time, and with a given (limited) system of means for drawing distinctions, to distinguish a finite number of points on the Earth's surface, we may denote them $a_1, a_2, \ldots a_n$. For simplicity's sake let us further assume that we can decompose the system on the basis of two properties, corresponding to two one-place predicates: "P_1" for the property 'be in Bohemia' and "P_2" for the property 'be in a town'. The whole system may then be decomposed as in Figure 10. It is obvious from the diagram that part II contains those points that have both properties "P_1" and "P_2", part I those that have "P_1" but not "P_2", part IV those that have "P_2" but not "P_1", and part III those that have neither "P_1" nor "P_2". For the structure description of the system, i.e. for the macroscopic approach to the system,

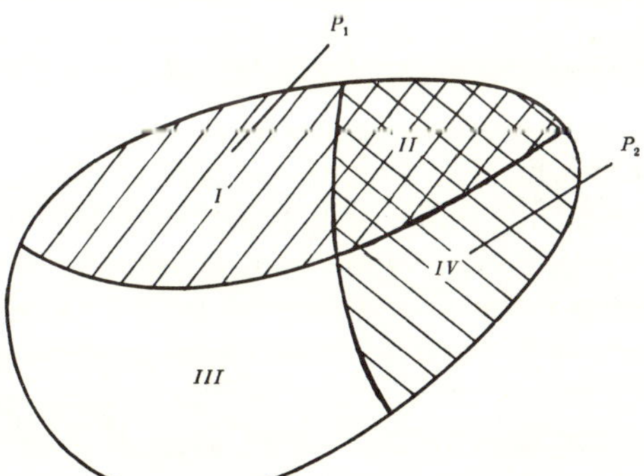

Fig. 10.

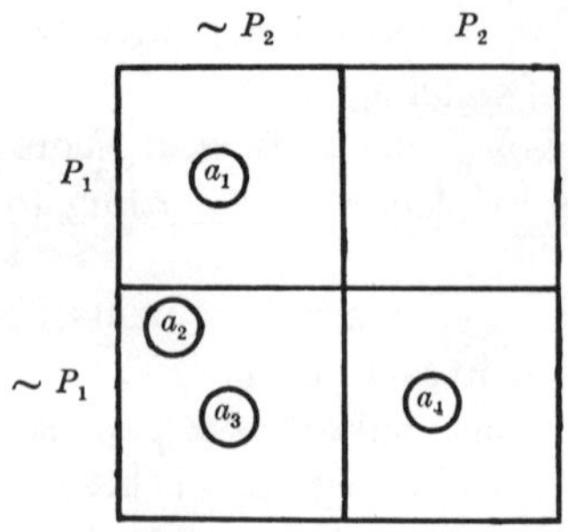

Fig. 11.

it is enough to ascertain how many points are contained in parts I, II, III and IV. By contrast, the state description, i.e. the microstate of the system, requires us to ascertain which individual points occupy I, II, III and IV. One of the possible state descriptions may be represented diagrammatically as follows in Figure 11 (for reasons of the practical limitations we shall only distinguish four points: a_1, a_2, a_3, a_4). The situation illustrated by the diagram expresses the state description: $P_1a_1 \cdot \sim P_2a_1 \cdot \sim P_1a_2 \cdot \sim P_2a_2 \cdot \sim P_1a_3 \cdot \sim P_2a_3 \cdot \sim P_1a_4 \cdot P_2a_4$.

All the possible situations in this system with n points are expressed by the disjunction of all the possible state descriptions:

$$(P_1a_1 \cdot P_2a_1 \cdot P_1a_2 \cdot P_2a_2 \cdot \quad \ldots P_1a_n \cdot P_2a_n) \vee$$
$$\vee (P_1a_1 \cdot P_2a_1 \cdot P_1a_2 \cdot P_2a_2 \cdot \quad \ldots P_1a_n \cdot \sim P_2a_n) \vee$$
$$\vdots$$
$$\vee (\sim P_1a_1 \cdot \sim P_2a_1 \cdot \sim P_1a_2 \cdot \sim P_2a_2 \cdot \ldots \sim P_1a_n \cdot \sim P_2a_n)$$

This decomposition of the system is made on the basis of a two-value decision process by which we either assign or do not assign a particular property to each separate individual. There is no difficulty in envisaging a three-value decision process (with the principle of quartum non datur) or a n-value decision process (with the principle of '$n+1$-um' non datur). This means then that in the case of a k-value decision process about a system that can be described by means of the language $\mathscr{L}_n^\pi$, there are in all $k^{n\pi}$ possible distinct state descriptions (disregarding the practicability or genuine admissibility of those states). We have already stressed that this method is applicable only to languages that do not contain more than a denumerably infinite set of individuals. There are yet other problems that arise when we consider the method's application in an empirical context:

(a) The state description method does not take into account the *problems of the distinguishability* of individuals and the general problems of the 'individualisation' of properties and relations. This distinguishability is not absolute, but relative, depending on the nature of the objective system and on its interaction with the system of the language user (who may be equipped with a particular observational and experimental technique, certain a priori knowledge or set of aims etc.).

(b) We have said that the method of state description assumes a decomposition of the given universe. The criteria for this decomposition are of course taken 'dogmatically', being set up in the 'vocabulary' and remaining invariable in the course of the 'use of the language'. A real communicational process, and especially processes of communication, control and decision (all of which assume a certain 'use of language'), in fact change these criteria. We are often forced to change them in the course of solving a single task or set of tasks of the same type. Hence the statistical-mechanical analogy as a kind of instrument for understanding the semantics of a language gives only a rough and simplified view of things. We might ask whether it is possible to extend and enrich the analogy by a kind of statistical-dynamic or rather *statistical-decision* analogy. (Some of the possibilities for applying that kind of approach will be demonstrated in later chapters, especially that on the criterion for sense and on reduction and constitution.)

(c) The method of state description, as is by now obvious, is incapable of taking account of the interdependencies of objects, properties, relations and so forth. Therefore, while we were looking at the possibilities for decomposing a given universe by more than one criterion (in our example this was decomposition on the basis of predicates P_1 and P_2) we ignored the possibility that the criteria might be dependent on each other. Yet it is natural that the fact of statistical independence of the occurrence of certain objects, or properties etc., can only be assumed in the extreme case. It thus transpires that an important instrument in the semantic analysis of a language used in an empirical context is the *concept of dependence*, and with it the means for quantifying the measure of dependence.

After this digression on 'state description', which has partly shown the difficulties and limitations of the concept, let us return to the concept of logical range. Intuitively speaking the logical range of a sentence is the class of state descriptions in which the sentence is true. This is one starting point both for a solution to some of the problems of logical semantics and for consideration of the logical foundations of probability and the foundations of so-called inductive logic. Let it be also noted that the foregoing critical comments on 'state description' concern both, and possibly yet other, applications of a method based on the concept of 'state description'.

We assume that for each state description there is always a corresponding logical state (***LS***), i.e. a possible situation of the given universe. Carnap suggests two basic properties of a logical state: (a) ***LS*** is logically possible, i.e. its corresponding state description cannot be an L-untrue sentence; (b) ***LS*** must be totally unambiguously definite, i.e. its state description must not leave anything unclear or leave any open questions. In line with these intuitive considerations Carnap formulates two basic postulates which concern the concept of logical range (***LR***):[9]

($\mathscr{P}_1$ – IV 1 B) If $\Phi_i \longrightarrow \Phi_j$, then $\boldsymbol{LR}(\Phi_i) \subset \boldsymbol{LR}(\Phi_j)$
($\mathscr{P}_2$ – IV 1 B) If $\boldsymbol{LR}(\Phi_i) \subset \boldsymbol{LR}(\Phi_j)$, then $\Phi_i \longrightarrow \Phi_j$

On the basis of the concept of 'logical range' (***LR***) and that of 'state description' (***SD***) other logical concepts can be brought in by definition. In addition to these two basic concepts, which suffice to define the concepts of logic, another semantic concept that is important is that of 'real state' (***RS***). This is a logical state that is established by a true state description. It is a concept that may be treated as fundamental in relation to 'factual', as opposed to logical, concepts.

On the basis of '***LR***' and '***LS***,' and also '***RS***' as the case demands, we can introduce yet further concepts. These include above all 'universal logical range' (as the class of all possible worlds that can be described by a given $\mathscr{L}$, i.e. the class of all possible 'logical states', symbolised as $\vee_{\boldsymbol{LS}}$), and the complementary concept of 'empty logical range' (as the empty class of logical states, symbolised as $\wedge_{\boldsymbol{LS}}$). It is obvious that $\vee_{\boldsymbol{LS}}$ and $\wedge_{\boldsymbol{LS}}$ can be defined on the basis of the ***LR*** and ***LS***, given a knowledge of $\mathscr{L}$. This means that these are logical concepts and that the semantic properties of the language expressions that correspond to them can be defined on the basis of the semantic rules of that language. Unlike '***LR***', '***LS***', '$\vee_{\boldsymbol{LS}}$' and $\wedge_{\boldsymbol{LS}}$', however, ***RS*** is to not a logical concept. This means that the semantic properties of the language expressions that correspond to a ***RS*** cannot be determined on the basis of the semantic rules of the language alone.

On the basis of these concepts it is possible to define the most important logical concepts. We shall enter only some definitions. At the same time it is possible to express some important properties of logical concepts in the shape of theorems of the L-semantics built up on the concepts '***LR***' and '***LS***'.

(D_1 – IV 1 B) $\vee_{LS}$ (in $(\mathscr{L})^{10}$ = df class of all ***LS*** (in $\mathscr{L}$))
(D_2 – IV 1 B) $\wedge_{LS}$ = df empty class of ***LS***
(D_3 – IV 1 B) Φ_i is *L-true* = df $\boldsymbol{LR}(\Phi_i) = \vee_{LS}$
(D_4 – IV 1 B) Φ_i is *L-untrue* = df $\boldsymbol{LR}(\Phi_i) = \wedge_{LS}$
(D_5 – IV 1 B) Φ_i *L-implies* $\Phi_j (\Phi_i \underset{L}{\longrightarrow} \Phi_j)$ = df $\boldsymbol{LR}(\Phi_i) \subset \boldsymbol{LR}(\Phi_j)$

This definition is at the same time a summation of [$\mathscr{P}_1$] and [$\mathscr{P}_2$].

(D_6 – IV 1 B) Φ_i is *L-equivalent* to Φ_j = df $\boldsymbol{LR}(\Phi_i) = \boldsymbol{LR}(\Phi_j)$
(D_7 – IV 1 B) Φ_i is *L-disjunctive* with Φ_j = df $\boldsymbol{LR}(\Phi_i) + \boldsymbol{LR}(\Phi_j) = \vee_{LS}$
(D_8 – IV 1 B) Φ_i is *L-incompatible* with Φ_j = df $\boldsymbol{LR}(\Phi_i) \times \boldsymbol{LR}(\Phi_j) = \wedge_{LS}$.

In L-semantics based on ***LR*** and ***LS***, it is possible to define a number of other logical concepts. What is important, however, is to note some of the properties of logical concepts. It can be proved, for example, that L-equivalence is reflexive, symmetrical and transitive. On the other hand, L-implication is merely reflexive and transitive.

If to the bases of L-semantics we also attach ***RS***, we can define some other semantic concepts that belong among the radical concepts:

(D_9 – IV 1 B) Φ_i is *true* = df ***RS*** is an element of $\boldsymbol{LR}(\Phi_i)$
(D_{10} – IV 1 B) Φ_i is *untrue* = df Φ_i is not true
(D_{11} – IV 1 B) Φ_i *implies* Φ_j ($\Phi_i \longrightarrow \Phi_j$) = df Φ_i is untrue or Φ_j is true or both Φ_i and Φ_j are true, etc.

On the same bases it is also possible to introduce some general logical concepts. These include in particular that of L-determinacy:

(D_{12} – IV 1 B) Φ_i is *L-determinate* = df Φ_i is L-true or L-untrue.

This concept may also serve the defining of systems for which the semantic rules of the systems themselves are enough for characterising them semantically. This kind of system may then be described as L-determinate systems:

(D_{13} – IV 1 B) A system $\mathscr{L}$ is L-determinate = df all the sentences of $\mathscr{L}$ are L-determinate.

It is to be understood that systems of this kind are not language systems in the true sense of the word and represent only the extreme case of language systems. The language systems used in real processes of communication, systems that are the basis of scientific observations, experiments and scientific cognition generally, contain first and foremost sentences of the kind usually described as synthetic. L-determinate systems, on the other hand, contain only L-determinate sentences, i.e. using traditional terminology, only analytical sentences or contradictions.

Synthetic sentences, i.e. sentences whose semantic properties cannot be determined by the semantic rules of the language itself, therefore require reference to some 'actual world'.[11] Carnap calls this kind of sentence *factual*, or F-determinate, sentences. This corresponds to Carnap's definition:

Φ_i is F-determinate (or factual) = df Φ_i is not L-determinate [22, p. 141]. This definition and Carnap's accompanying viewpoint lead however, as we shall be seeing, to considerable difficulties. The definition cannot therefore be accepted as correct.[12]

Identifying the concepts 'factual' (synthetic) and 'not L-determinate' assumes that all logical states are potentially realisable, i.e. that any ***LS*** may become an ***RS***. This assumption is unacceptable not only for natural languages, but also for many formalised languages that operate with non-logical terms between which there are relations of dependence. If Carnap's attitude to F-concepts may be treated as controversial or valid only in special cases, then it must be considered problematic, especially with respect to the relations of dependence, to identify analyticity with validity in all ***LS***. (These problems will be dealt with in more detail in the chapter on analytical and synthetic aspects in science.)[13]

[C] *The Concepts of 'Logical Content' and 'Semantic Information'*

The concepts of 'logical range' and 'logical state' are linked by Carnap to a further concept, which later played a major role in the formulation of the Carnap and Bar-Hillel theory of semantic information: the concept of 'logical content'. We will give first some of the intuitive ideas that gave rise to the introduction and more exact formulation of this concept.

If, with respect to a given object language $\mathscr{L}$, a universe is specified as a set of possible worlds, it is obvious that a given language expression in $\mathscr{L}$ is the 'richer in content' the more 'possible worlds' it excludes. The 'contentual power' of a sentence in $\mathscr{L}$ resides in its excluding certain possible worlds; this power increases as the range of the possible worlds excluded by the sentence increases.

If we have, for example, Φ_i and Φ_j and Φ_j can be deduced logically from Φ_i, then the logical content of Φ_j must be contained in the logical content of Φ_i. (The relation is then the exact converse of that of 'logical range'.) If the logical content of Φ_j is contained in that of Φ_i, then Φ_j cannot be a statement of something of which Φ_i could not be a statement. Therefore it is impossible for Φ_i to be true and Φ_j untrue. Since this impossibility is based on logical reasoning, Φ_j follows logically from Φ_i. These intuitive ideas may be formulated in two postulates:

($\mathscr{P}_1$ – IV 1 C) If $\Phi_i \underset{L}{\longrightarrow} \Phi_j$, then $\boldsymbol{LC}(\Phi_j) \subset \boldsymbol{LC}(\Phi_i)$

($\mathscr{P}_2$ – IV 1 C) If $\boldsymbol{LC}(\Phi_j) \subset \boldsymbol{LC}(\Phi_i)$, then $\Phi_i \underset{L}{\longrightarrow} \Phi_j$

We shall now note some of the important properties of the concept '***LC***'. As with ***LR***, so with ***LC*** we shall use operations with classes.

The logical content of a sentence or class of sentences in $\mathscr{L}$ is equal to the logical content of another sentence or class or class of sentences in $\mathscr{L}$ if and only if both sentences or classes of sentences are L-equivalent. In such a case the logical ranges of both sentences or classes of sentences are the same:

(T_1 – IV 1 C) $\boldsymbol{LC}(\Phi_i) = \boldsymbol{LC}(\Phi_j)$ if and only if Φ_i and Φ_j are equivalent.
(T_2 – IV 1 C) $\boldsymbol{LC}(\Phi_i) = \boldsymbol{LC}(\Phi_j)$ if and only if $\boldsymbol{LR}(\Phi_i) = \boldsymbol{LR}(\Phi_j)$.

One vital question for assessing the role of logical content from the point of view of L-semantics is the question whether it is possible to introduce '***LC***' by means of definition on the basis of those concepts that are treated as the starting point for L-semantics, i.e. ***LR*** and ***LS***. Since $\boldsymbol{LC}(\Phi_j) \subset \boldsymbol{LC}(\Phi_i)$ if and only if $\Phi_i \underset{L}{\longrightarrow} \Phi_j$, and since from both postulates for ***LR*** it is possible to deduce that $\boldsymbol{LR}(\Phi_i) \subset \boldsymbol{LR}(\Phi_j)$ if and only if $\Phi_i \underset{L}{\longrightarrow} \Phi_j$, it holds that

(T_3 – IV 1 C) $\boldsymbol{LC}(\Phi_j) \subset \boldsymbol{LC}(\Phi_i)$ if and only if $\boldsymbol{LR}(\Phi_i) \subset \boldsymbol{LR}(\Phi_j)$.

Then it becomes possible to define $\boldsymbol{LC}(\Phi_i)$ as a class of ***LS*** that do not logically imply Φ_i, i.e. a class of ***LS*** that do not belong to $\boldsymbol{LR}(\Phi_i)$. This can be expressed in a definition by using the concept of complement:

(D_1 – IV 1 C) $\boldsymbol{LC}(\Phi_i) = \mathrm{df}\ \overline{\boldsymbol{LR}(\Phi_i)}$

This definition also expresses the motives given in the intuitive ideas at the start of this section: the 'contentual power' of a sentence increases as the range of the ***LS*** excluded by the sentence increases.

In addition to the relations between ***LC*** and ***LR*** and ***LS*** we may also consider those between ***LC*** and ***RS***. Assuming that we have at our disposal a ***RS*** that is one of the ***LS*** of a given $\mathscr{L}$, we may also define such radical concepts as 'true'. For example, Φ_i is true if and only if its corresponding ***RS*** is not one of the ***LS*** excluded by Φ_i.

Since the concept of 'logical content' is introduced on the basis of 'logical range' and 'logical state', the same critical reservations that we expressed with respect to these concepts and the method of state description will also apply to it. This means that the circumstances that restrict the utility of the method of state description in semantics also restrict the utility of the concept of 'logical content'. It must be emphasised that the concepts 'logical state', 'logical range' and therefore also 'logical content' actually derive from the structure of the language itself. Therefore they are essentially syntactic concepts, and not semantic concepts in the true sense of the word.

The concept '***LC***(Φ_i)' i.e. the concept of the logical content of the sentence or class of sentences Φ_i, was taken by Carnap and Bar-Hillel as the 'explicans'[14] for the concept of the "information mediated by the statement 'Φ_i'", taken in its semantic sense [6]. Since ***LC***(Φ_i) is equal to the complement of ***LR***(Φ_i), the concept of semantic information may also be based on the concepts '***LR***' and '***LS***' and hence on the method of 'state description'.

Carnap and Bar-Hillel introduce the additional concept of 'content element'. If a state description is the most powerful factual statement that may be made using the means of a given language, content element is, on the contrary, the weakest factual statement that can be made using the means of the given language. Content element is the disjunction of all the atomic sentences of the given language, or of their negations. For example, in the language used above, which contained two one-place predicates P_1 and P_2 and four individual constants a_1, a_2, a_3 and a_4 and in which atomic sentences could be formed by a concatenation of any (one) predicate and any individual constant, one of the 256 content elements may have the form:

$$P_1a_1 \vee P_2a_1 \vee P_1a_2 \vee P_2a_2 \vee P_1a_3 \vee P_2a_3 \vee \sim P_1a_4 \vee \sim P_2a_4$$

The class of all the content elements that are L-implied by a sentence Φ_i is ***LC***(Φ_i). If Φ_i is L-true, it contains no element; if it is L-untrue, it contains all the elements. (This also matches the intuition that the contentual power of a tautology which contains all possible worlds, therefore not excluding any possible world, is zero. Under contradiction the reverse is the case.)

In order to use the method of state description and its associated concept of 'logical content' as the explicans for the concept of information taken in the semantic sense, account has to be taken not only of L-determined sentences, but also of factual sentences. What is also needed is some measure for quantifying ***LC***.

This measure can be derived if we know the like measure for the corresponding ***LR***. Accordingly what is taken as starting point is Carnap's system of inductive logic, which introduces this measure.

It has to be stressed that this system of inductive logic abides fully by the method of 'state description'. At the same time, the concept of 'state description', in the sense in which we have described it, is taken as the basis for making the concept of a sentence's meaning more precise. The method of 'state description' consists in interpreting a sentence (or class of sentences) Φ_i by ascertaining the state descriptions in which Φ_i applies and those in which it does not. The concept of 'applies' in the expression 'the sentence Φ_i applies in state description $\boldsymbol{SD}_h$' can be explained by a recursive procedure for all the sentences in $\mathscr{L}$. (At the same time an atomic sentence or its negation belongs with all the ***SD*** of $\mathscr{L}$ in which it occurs as conjunctive element.) The sentence Φ_i applies in $\boldsymbol{SD}_h$ if and only if one of the following conditions is satisfied:

(1) Φ_i is an atomic sentence and applies in a ***SD***;
(2) Φ_i is L-true (and therefore applies in any ***SD***);
(3) Φ_i has the form $NON \frown \Phi_j$ and Φ_j does not apply in $\boldsymbol{SD}_h$;
(4) Φ_i has the form $\Phi_j \frown ET \frown \Phi_k$ and both components apply in $\boldsymbol{SD}_h$;
(5) Φ_i has the form $\Phi_j \frown VEL \frown \Phi_k$ and at least one of the components applies in $\boldsymbol{SD}_h$;
(6) Φ_i has the form $\Phi_j \frown AEQ \frown \Phi_k$ and either both components apply in $\boldsymbol{SD}_h$ or they both do not apply in $\boldsymbol{SD}_h$;
(7) Φ_i has the form $\Phi_j \frown SEQ \frown \Phi_k$ and either Φ_j does not apply in $\boldsymbol{SD}_h$, or Φ_k applies in $\boldsymbol{SD}_h$, or both Φ_j and Φ_k apply in $\boldsymbol{SD}_h$;
(8) Φ_i has the form $\Phi_j \frown SHEFF \frown \Phi_k$ and either Φ_j does not apply in $\boldsymbol{SD}_h$ and Φ_k does apply in $\boldsymbol{SD}_h$, or Φ_j applies in $\boldsymbol{SD}_h$ and Φ_k does not apply in $\boldsymbol{SD}_h$, or neither Φ_j nor Φ_k apply in $\boldsymbol{SD}_h$;
(9) Φ_i has the form $OMNIA \frown \Sigma$ and every Σ applies in $\boldsymbol{SD}_h$;
(10) Φ_i has the form $EXIST \frown \Sigma$ and at least one Σ applies in $\boldsymbol{SD}_h$.

Clearly the simplest situation would be that where the individual ***LS*** would be equally probable. But in any event it is necessary to bring in a measure that can be expressed quantitatively. Carnap's system of inductive logic sets out one such pattern of probability statements about sentences.[15] This is also why probability statements belong to the semantic part of the metalanguage. The concept of inductive probability, or degree of confirmation, is taken as a relative concept that depends on both a hypothesis **h** and the established datum **e**. Since we assume that these are sentences, i.e. Σ_h and Σ_e, the probability statement is a function of the two sentences and takes the form $\mathbf{c}(\Sigma_h, \Sigma_e)$, in which Σ_h and Σ_e are metalanguage signs for sentences. (In abbreviated form this probability statement will be expressed as $\mathbf{c}(\mathbf{h}, \mathbf{e})$.) In order to work with the quantitative concept of inductive probability, we also need to be able to record quantitatively the meaning of Σ_h and Σ_e with respect to the universe of the given $\mathscr{L}$. Hence Carnap regards the relation between Σ_h and Σ_e, which is studied by inductive logic, as a semantic relation.

Thus the concept of inductive probability is taken as relative, i.e. it concerns the inductive probability of Σ_h vis-à-vis Σ_e. At the same time another relativity is assumed, namely that to the given $\mathscr{L}$ and its universe. The quantification of the function $\mathbf{c}(\mathbf{h}, \mathbf{e})$ means that $\mathbf{c}(\mathbf{h}, \mathbf{e})$ is equal to a definite number, i.e. $\mathbf{c}(\mathbf{h}, \mathbf{e}) = \mathbf{r}$ (in the given $\mathscr{L}$).

The concept of inductive probability, or the degree of confirmation, is treated by Carnap as the basic concept of inductive logic in much the same way as L-implication is the basic concept of deductive logic. We have already seen how it is possible to make an interpretation of 'L-implication' on the basis of 'logical range'. Σ_e L-implies Σ_h if and only if the logical range of Σ_e is contained in the logical range of Σ_h. Analogously it is possible to interpret the concept of inductive probability: The probability statement $\mathbf{c}(\Sigma_h, \Sigma_e) = \frac{1}{2}$ shows that not the whole logical range of Σ_e is contained in that of Σ_h, but that only $\frac{1}{2}\boldsymbol{LR}(\Sigma_e)$ is contained in $\boldsymbol{LR}(\Sigma_h)$.

Schematically this analogy of deductive and inductive logic may be illustrated as in Figure 12.[16]

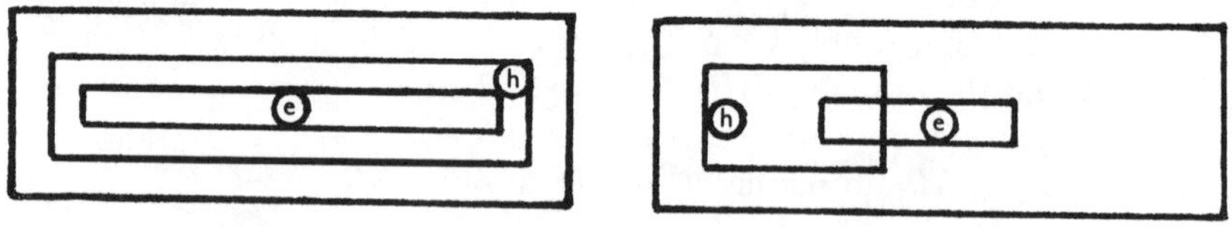

Fig. 12.

The properties of the function $\mathbf{c}(\Sigma_h, \Sigma_e)$ were originally presented by Carnap in the form of a number of basic conventions.[17]

In the formulations of these conventions, we shall, for reasons of concision, give only the indices of sentences, i.e. $\mathbf{c}(\mathbf{h}, \mathbf{e})$ instead of $\mathbf{c}(\Sigma_h, \Sigma_e)$. Instead of an L-true sentence in $\mathscr{L}$ (i.e. a tautology) we shall use the sign **t**. Furthermore, to shorten the notation of the logical constants we shall record the conjunction of the sentences Σ_h and Σ_e by means of $\mathbf{h}\cdot\mathbf{e}$ and their disjunction by means of $\mathbf{h} \vee \mathbf{e}$.

The basic conventions for **c**-functions have the form:

(C_1 – IV 1 C) If **e** and **e'** are L-equivalent, then $\mathbf{c}(\mathbf{h}, \mathbf{e}) = \mathbf{c}(\mathbf{h}, \mathbf{e}')$
(C_2 – IV 1 C) If **h** and **h'** are L-equivalent, then $\mathbf{c}(\mathbf{h}, \mathbf{e}) = \mathbf{c}(\mathbf{h}', \mathbf{e})$
(C_3 – IV 1 C) $\mathbf{c}(\mathbf{h}\cdot\mathbf{j}, \mathbf{e}) = \mathbf{e}(\mathbf{h}, \mathbf{e}) \times \mathbf{c}(\mathbf{j}, \mathbf{e}\cdot\mathbf{h})$
(C_4 – IV 1 C) If $\mathbf{e}\cdot\mathbf{h}\cdot\mathbf{j}$ is L-true, then $\mathbf{c}(\mathbf{h} \vee \mathbf{j}, \mathbf{e}) = \mathbf{c}(\mathbf{h}, \mathbf{e}) + \mathbf{c}(\mathbf{j}, \mathbf{e})$
(C_5 – IV 1 C) $\mathbf{c}(\mathbf{t}, \mathbf{e}) = 1$ for any **e** that is not L-true
(C_6 – IV 1 C) For any $\boldsymbol{SD}_i$ in $\mathscr{L}$ it holds that $\mathbf{c}(\boldsymbol{SD}_i, \mathbf{t}) > 0$

The last convention has as the second argument of the **c**-function an L-true sentence. Carnap calls functions of this kind (abbreviated to $\mathbf{c}_t$-functions, so that instead of $\mathbf{c}(\boldsymbol{SD}, \mathbf{t})$ we may write $\mathbf{c}_t(\boldsymbol{SD})$) regular **c**-functions if the following conditions are met:

(1) For any $\boldsymbol{SD}_i$ in $\mathscr{L}$ it holds that $\mathbf{c}_t(\boldsymbol{SD}) > 0$

(2) The sum of all $\mathbf{c}_t$-values for all $\boldsymbol{SD}$ in $\mathscr{L}$ is equal to 1.

The above-listed properties of the concept of inductive probability show that a system of inductive logic can be built up on the basis of the method of 'state description'. This means that the snares and snags of the method, of which we have already spoken, also concern these foundations of inductive logic. What is particularly evident is the fact that for each ***SD*** there is a corresponding possible world which may be characterised by non-zero probability. This is of course only the extreme possible case, assuming that each logical state is also 'really possible'. We apparently have here two different attributes: 'possible', which is taken as an attribute of a world that is possible with respect to a given $\mathscr{L}$, irrespective of any constraints;[18] and 'really possible', which is taken as an attribute of of a world which, with respect to certain conditions (e.g. observational or experimental facilities, time of observation, 'memories', 'capacities' and 'delays' at our disposal), has a non-zero probability of occurrence, or, which is no more than a refinement of these conditions, has a probability of occurrence that differs from zero by at least some ε that we are capable of discerning.

The actual build-up of the system of inductive logic is based on the concept of regular **m**-function. The starting point is the concept of the regular function of measure **m** for the individual state descriptions; the concept corresponds to the following conditions:

(D_2 – IV 1 C) **m** is the regular function of measure for ***SD*** in $\mathscr{L}$ = df it satisfies the conditions:
(a) for each ***SD*** in $\mathscr{L}$, **m**(***SD***) is a positive real number;
(b) the sum of the values of **m** for all ***SD*** in $\mathscr{L}$ is 1.

The concept **m** can also be extended to sentences in $\mathscr{L}$:

(D_3 – IV 1 C) (a) For each L-untrue sentence **j** in $\mathscr{L}$ it always holds that **m**(**j**) = 0;
(b) For a sentence **j** in $\mathscr{L}$ which is not L-untrue **m**(**j**) is equal to the sum of the values of **m** for all ***SD*** that make up the logical range of **j**.

(D_4 – IV 1 C) **c** is the regular function of confirmation for $\mathscr{L}$ = df **c** satisfies for **m** and any sentences **h** and **e** in $\mathscr{L}$ both the following conditions:
(a) if **m**(**e**) is other than zero, then $\mathbf{c}(\mathbf{h}, \mathbf{e}) = \dfrac{\mathbf{m}(\mathbf{e}\cdot\mathbf{h})}{\mathbf{m}(\mathbf{e})}$;
(b) if for some **e** **m**(**e**) = 0, then **c**(**h**, **e**) contains no value.

From these definitions it can also be deduced that $0 < \mathbf{m}(SD) < 1$ insofar as **m** is the regular function of measure for ***SD*** in $\mathscr{L}$. From (D_4 – IV 1 C) it is also easy to see the difference between inductive and deductive logic that we intimated earlier. It is equally obvious that the $\mathbf{c}_t$-function and the **m**-function coincide.[19]

On the basis of (D_1 – IV 1 C) it is also possible to establish a suitable function of measure for the logical content of a sentence Φ_i, which we may briefly call the content measure of Φ_i. If we know **m** for Φ_i, then the content measure (designated **cont**) of the sentence Φ_i, i.e. $\mathbf{cont}(\mathbf{i}) = \mathbf{m}(\sim \mathbf{i}) = 1 - \mathbf{m}(\mathbf{i})$. Since $\mathbf{m}(\mathbf{i}) + \mathbf{m}(\sim \mathbf{i}) = 1$[20] and since $0 \leqslant \mathbf{m}(\mathbf{i}) \leqslant 1$, it holds that $0 \leqslant \mathbf{cont}(\mathbf{i}) \leqslant 1$.

It is obvious that the concept of 'content measure'[21] corresponds to the **m**-functions of sentences in $\mathscr{L}$ in the same respect in which the concept '***LR***' of a sentence in $\mathscr{L}$ corresponds to that of '***LC***' of a sentence in $\mathscr{L}$. Since **m**(**i**) corresponds to $\mathbf{c}_t$-functions or **c**-functions of which the second argument is a tautology, we are offered the possibility of formulating the corresponding conditional content measure, i.e. the content measure of **j** with respect to **i** (symbolically: **cont(j|i)**). This conditional content measure essentially shows how much more information is transmitted by one sentence with respect to another sentence. Hence Carnap and Bar-Hillel define **cont(j|i)**:

$$\mathbf{cont}(\mathbf{j}|\mathbf{i}) = \mathrm{df}\,\mathbf{cont}(\mathbf{i}\cdot\mathbf{j}) - \mathbf{cont}(\mathbf{i})$$

The maximum value of **cont(j|i)** is when **cont(j|i)** = **cont(j)**. At the same time **cont(j|i)** = **cont(j)** if and only if $\mid\vdash \mathbf{i} \vee \mathbf{j}$. By contrast if $\mathbf{i} \underset{L}{\rightarrow} \mathbf{j}$, then **cont(j|i)** = 0.

Since **cont(i)** may only have values between 0 and 1, Carnap and Bar-Hillel considered an alternative solution, although using the same starting point. This alternative solution assumes a measure of information that may acquire values from zero to infinity. This measure (of the quantity) of information is designated **inf(i)** and defined as follows:

$$\mathbf{inf}(\mathbf{i}) = \mathrm{df}\,\mathbf{Log}\frac{1}{1 - \mathbf{cont}(\mathbf{i})}$$

Since **cont(i)** = 1 − **m(i)**, it holds that

$$\mathbf{inf}(\mathbf{i}) - \mathbf{Log}\,\frac{1}{\mathbf{m}(\mathbf{i})} = -\,\mathbf{Log}\,\mathbf{m}(\mathbf{i})$$

As will be apparent, this is a formula that is analogous to the elementary form of Shannon's well known formula for the measure of the quantity of information in information theory. This conception of the measure of information, taken in the semantic sense, is only a semantic theory of information insofar as its starting point, i.e. the method of state description, may also be considered a semantic method in the true sense of the word. Therefore the reservations expressed above against state description also relate to this theory.

Nevertheless, the theory can be elaborated in a number of respects. In his system of inductive logic, Carnap introduces estimate functions based on **c**-functions. In a function **c(h, e)** we assume that we know the values of the arguments. But in practice there are certain important situations where we have to consider a series of possible arguments. One instance of this is when we are to consider the possibility of some future state. With the object of capturing these situations and others like them, Carnap introduces a special estimate function, based on **c**-functions, and also uses this function (denoted **est**) to express the measure of information provided by a series of hypotheses with respect to a known reality that is expressed by a particular sentence.

Another possibility, but not one suggested in [7], is to expand this approach to the problems of semantic information by exploiting the concepts of structural description and the distribution of individuals. We have already pointed out that the concepts of both 'state description' and 'structural description' assume a particular decomposition of the universe, structural descriptions saying how many individual objects belong in individual parts of the universe, and state descriptions saying which objects belong in each part.

We may demonstrate the meaning of the concept of 'structural description' on the same example used to demonstrate 'state description'. If we take a language $\mathscr{L}_n^2$ which contains two one-place predicates and n individual constants, we can use the predicates P_1 and P_2, which we shall call primitive predicates, to form molecular predicates. It is particularly useful to create the molecular predicates in such a way that the individual constants are distributed so that each belongs only to one class. In other words, we are making the requirement that concatenation of each constant with only one molecular predicate is all that is admissible. Some of these classes may, understandably, be empty. The molecular predicates so formed may also be said to constitute the decomposition of the universe of the n individuals.

Thus in the case of language $\mathscr{L}_n^2$ the molecular predicates give rise to classes I, II, III and IV. This kind of molecular predicate, which fulfils the stated conditions, is called by Carnap a Q-predicate. By a Q-predicate is meant a molecular predicate that has the form $P_1 \cdot P_2 \ldots P_\pi$, or the form that differs from the foregoing in that one of the primitive predicates is replaced by its negation. All in all, if the primitive predicates are one-place predicates and if we assume two-value decision-making, there are 2^π such Q-predicates. If, for example, $\pi = 3$, there will be 8 Q-predicates:

$$
\begin{array}{ll}
Q_1 : P_1 \cdot P_2 \cdot P_3 & Q_5 : \sim P_1 \cdot P_2 \cdot P_3 \\
Q_2 : P_1 \cdot P_2 \cdot \sim P_3 & Q_6 : \sim P_1 \cdot P_2 \cdot \sim P_3 \\
Q_3 : P_1 \cdot \sim P_2 \cdot P_3 & Q_7 : \sim P_1 \cdot P_2 \cdot P_3 \\
Q_4 : P_1 \cdot \sim P_2 \cdot \sim P_3 & Q_8 : \sim P_1 \cdot \sim P_2 \cdot \sim P_3
\end{array}
$$

It is easy to prove that Q-predicates are L-incompatible and that the sum of all the Q-predicates is L-disjunctive.

On the basis of the decomposition of the universe characterised like this we may consider the distribution of the individuals with respect to the molecular predicates $Q_1, Q_2, \ldots, Q_\alpha$ (where α is 2^π under the assumption that there are π primitive predicates in $\mathscr{L}$). This makes it possible to state that n_1 individuals have the property Q_1, n_2 individuals the property $Q_2, \ldots$ and n_α individuals the property Q_α. (It is obvious that $n_1 + n_2 + \ldots + n_\alpha = n$.)

This way of analysing the universe of a given $\mathscr{L}$ is clearly in a sense 'weaker' than the way represented by the method of 'state description'. But this way is the more suitable if the individuals are more or less homogeneous and if the attention is focused not on distinguishing the individuals, but on distinguishing their properties.

Since a structural description is a disjunction of all the isomorphous state descriptions, it is possible to disregard the differences between individual state descriptions, except for the difference that state descriptions concern the same number of individuals. Where there is nothing to be gained from distinguishing the individuals, and if on the contrary the number of individuals is the main thing to be taken into account, it is sufficient for the semantic analysis of $\mathscr{L}$ to be based just on the concept of 'structural description'.

This approach is suited to the 'symmetrical **m**-functions', which may be defined as follows:

> **m** is a symmetrical function of measure for ***SD*** in $\mathscr{L}$ = df it is a regular function of measure for ***SD*** in $\mathscr{L}$ and it has the same value for all isomorphous ***SD***.

The concept of a symmetrical function of measure may also serve as the point of departure for introducing the concept of the measure of information **inf(i)**, which is definable in the same way as the case of the regular function of measure, i.e.

$$\mathbf{inf}(\mathbf{i}) = \mathrm{df}\ \mathbf{Log}\frac{1}{\mathbf{m}(\mathbf{i})}$$

However, since the starting point here is only structural description, or the macro-approach to the universe of a given $\mathscr{L}$, this conception of measure may be said not to differ in principle from the conception quoted in elementary accounts of information theory.

[*D*] *Meaning Postulates*

The method of 'state description' based on the concepts of 'logical range' and 'logical state' leads to difficulties because, other reasons apart, it also fails, in its original form, to respect the relations of the non-logical constants of the given language. But since the non-logical constants are usually not independent of each other, but are ascertainably interdependent in some way or another, the method of 'state description' cannot be used in its original form. This method not being based on the semantics of non-logical constants, it really only takes into account the semantics of logical concepts. Hence, as we have already stressed, 'logical range' and 'logical state' are not semantic concepts in the proper sense of the world.

Therefore the method of 'state description' can only be applied as a semantic method if we are concerned with some L-determined system (in the sense given previously) or a system with non-logical signs whose meaning is not taken into account. The method of 'state description' is, as already suggested, influenced by some of the ideas of statistical mechanics, and has been used primarily in simple logical systems where the meaning of the non-logical signs was irrelevant.

In a language $\mathscr{L}_n^\pi$, the vocabulary of which contains π predicates and n individuals, it is possible to formulate n atomic sentences and $2^{n\pi}$ different state descriptions (given two-value decisions). But any two differing ***SD*** are L-incompatible. Also any two differing primitive predicates must be logically independent of each other. But in many cases this means surrendering any claim to semantic analysis, for example in the sense that we would be assigning to the predicates individual properties, relations or classes among which certain dependences could be demonstrated. These shortcomings of the method of 'state description' have been pointed out especially by J.G. Kemeny [68] and Y. Bar-Hillel [5].

These difficulties with the method of 'state description' can be demonstrated by the example of the very simple language that may be used in traffic control. Imagine a small wayside station with a single track. The station approaches are both protected by a signal. For simplicity's sake let us assume that both signals may show or not show only two different lights, green and red. To make things even more simple, i.e. for traffic handling through the station to be controllable by just the two signals, let us assume that movement through the station goes only one way. Diagrammatically this would appear as in Figure 13.

If we wished to express all the possible situations at this station, the vocabulary of $\mathscr{L}$ would have to contain two individuals a_1 and a_2 and two one-place predicates P_1 and P_2.

Disregarding any contentual interdependence of P_1 and P_2, we have (as in any $\mathscr{L}_2^2$) 16 possible state descriptions in all.

Fig. 13.

P_1a_1	P_2a_1	P_1a_2	P_2a_2	SD
1	1	1	1	1
			0	2
		0	1	3
			0	4
	0	1	1	5
			0	6
		0	1	7
			0	8
0	1	1	1	9
			0	10
		0	1	11
			0	12
	0	1	1	13
			0	14
		0	1	15
			0	16

Fig. 14.

Expressing the truth value of the atomic sentences in columns by using 1 (for Σ_i) and 0 (for $\sim \Sigma_i$), we arrive at Figure 14.

However, the assumption of logical independence is hardly suited to any intelligent use of the given language system. Certain constraints need to be formulated, and these may be quite diverse. Two possible solutions are suggested here:

(a) It could be required that each signal show one particular colour; then it is not necessary for signal a_1 and signal a_2 to show the same colour. This may be formulated in the following postulates ($\mathscr{P}_a$):

$$(\mathscr{P}_a) \qquad P_1a_1 \longrightarrow \sim P_2a_1 \qquad \sim P_1a_1 \longrightarrow P_2a_1$$
$$P_1a_2 \longrightarrow \sim P_2a_2 \qquad \sim P_1a_2 \longrightarrow P_2a_2$$

It holds that

$$[(P_1a_1 \longrightarrow \sim P_2a_1)\cdot(\sim P_1a_1 \longrightarrow P_2a_1)] \longrightarrow (P_1a_1 \equiv \sim P_2a_1)$$

and analogously that $P_1a_2 \equiv \sim P_2a_2$.

Clearly, by recognising these limiting conditions as formulated in the postulates, many state descriptions are eliminated. Only $\boldsymbol{SD}_6$, $\boldsymbol{SD}_7$, $\boldsymbol{SD}_{10}$ and $\boldsymbol{SD}_{11}$ can be considered descriptions of admissible *models* of the given language when also containing the postulates.

(b) The above constraints may of course prove impractical. In effect they mean that a single light suffices, since the second light is the negation of the first. Thus the language contains superfluous signs, or there are other than linguistic reasons for the use of certain signs.

It is still possible to state somewhat weaker constraints. It might be required, for instance, that the possibility of one signal showing both lights at the same time be eliminated. At the same time it is not impossible for both lights not to show. Conceding that traffic is allowed only in one direction, it is useful to require that one of the lights on the second signal should always be on if a light is showing on the first signal. But it is not always necessary for the first signal to be showing a light if there is a light on on the second signal, for instance if a train is standing in the station. These requirements may be formulated in the following postulates:

$$(\mathscr{P}_b) \qquad \begin{array}{l} P_1a_1 \longrightarrow \sim P_2a_1, \text{ i.e. } P_1a_1 | P_2a_1 \\ P_1a_2 \longrightarrow \sim P_2a_2, \text{ i.e. } P_1a_2 | P_2a_2 \\ (P_1a_1 \vee P_2a_1) \longrightarrow (P_1a_2 \vee P_2a_2) \end{array}$$

Since it is impossible for one and the same signal to show green and red simultaneously, we can eliminate as contradictory those $\boldsymbol{SD}$ in which there is 1 in both the first and second or the third and fourth columns simultaneously, i.e. $\boldsymbol{SD}_1$, $\boldsymbol{SD}_2$, $\boldsymbol{SD}_3$, $\boldsymbol{SD}_4$, $\boldsymbol{SD}_5$, $\boldsymbol{SD}_9$ and $\boldsymbol{SD}_{13}$. But since if the first signal is on, the second must also be showing a light (though not vice versa), we can eliminate $\boldsymbol{SD}_8$, and $\boldsymbol{SD}_{12}$. Therefore in addition to $\boldsymbol{SD}_6$, $\boldsymbol{SD}_7$, $\boldsymbol{SD}_{10}$ and $\boldsymbol{SD}_{11}$ it is possible to treat $\boldsymbol{SD}_{14}$, $\boldsymbol{SD}_{15}$ and $\boldsymbol{SD}_{16}$ as admissible.

This kind of constraint concerning the interdependences of non-logical constants in a given language are usually called *meaning postulates* (see especially [25], [144, p. 303ff.] and [32, p. 233ff.]).

The conception of meaning postulates as suggested in [25] was in pursuit of two objectives: one was to give some precision to the concept of analyticity in languages in which a major role is played by the meaning of non-logical terms and their interdependence; the other was to rectify some of the shortcomings of the original version of the method of 'state description'. For the former, the recognition of the importance of meaning postulates means that two concepts of analyticity have to be distinguished: analyticity only with respect to the semantics of logical terms, and analyticity which respects in addition the semantics of non-logical terms. For the second, Carnap [25] has put forward a fairly simple solution which assumes that with respect to the meaning of the non-logical terms it is possible to formulate certain relations between them and express those relations by means of the devices of logic. The solution in question concerns not only the relations between one-place predicates, but also those between multi-place predicates.

In a language which has built into it certain meaning postulates the concept of analyticity has to be changed. If all that is respected is the method of state description, to which we have denied the status of a semantic method in the proper sense of the word, analyticity can be defined in the way mentioned above. The concept of analyticity in $\mathscr{L}$ can, as Carnap has shown, be explicated by means of the concept of L-truth, of which the definition has also been given previously. On the basis of the method of 'state description' it can be said that a sentence Φ_i is L-true if and only if $\boldsymbol{LR}(\Phi_i)$ is the universal class of all the possible state descriptions. But if we respect meaning postulates, this definition is no longer suitable. If $\mathscr{P}$ is a (conjunctive) class of all the meaning postulates accepted in a given language, then with respect to the class $\mathscr{P}$ the concept of analyticity must be changed in the following way:

(D_1 – IV 1 D) Φ_i is L-true in $\mathscr{L}$ with respect to $\mathscr{P}$ = df $\mathscr{P}$ L-implies Φ_i (i.e. $\mathscr{P} \underset{L}{\longrightarrow} \Phi_i$).

In reality it is impossible to relate both concepts of analyticity to the same language. There are really two languages: the language $\mathscr{L}$, in which no meaning postulates are set up, and the language $\mathscr{L}'$, which arises out of the former by the addition of meaning postulates to the semantics of $\mathscr{L}$.

The stipulation of meaning postulates does not modify just the concept of analyticity, but other logical concepts as well. Since the major logical concepts include not only that of L-truth as a device for the explication of analyticity, but also L-implication as a device for the explication of logical consequence, we must give the modification of the definition of L-implication entailed by the switch from $\mathscr{L}$ to $\mathscr{L}'$:

(D_2 – IV 1 D) Φ_i L-implies Φ_j (i.e. $\Phi_i \underset{L}{\longrightarrow} \Phi_j$) (in $\mathscr{L}'$) = df $\mathscr{L} \underset{L}{\longrightarrow} (\Phi_i \longrightarrow \Phi_j)$

The definiens in (D_2 – IV 1 D) may also be expressed in another, equivalent way: $\mathscr{P} \longrightarrow (\Phi_i \longrightarrow \Phi_j)$ is L-true, or $\mathscr{P} \cdot \Phi_i \underset{L}{\longrightarrow} \Phi_j$ or $\mathscr{P} \cdot \Phi_i \longrightarrow \Phi_j$ is L-true.

If we switch from $\mathscr{L}$ to $\mathscr{L}'$, it is also necessary to change the concept of 'state description'. From the example of the language $\mathscr{L}_2^2$ in which we introduced the meaning postulates ($\mathscr{P}_b$) it is visibly necessary to eliminate a number of ***SD*** because as a consequence of ($\mathscr{P}_b$) they become contradictory. It is obviously unthinkable for a state description to be L-untrue.

Given that $\mathscr{L}'$ arises out of $\mathscr{L}$ by the semantics of $\mathscr{L}$ being extended by meaning postulates $\mathscr{P}$, the concept of 'state description' in $\mathscr{L}'$ may be defined as follows:

(D_3 – IV 1 D) ***SD*** in $\mathscr{L}' =$ df any ***SD*** in $\mathscr{L}$ in which $\mathscr{P}$ applies.

After this modification of the concept of 'state description' it becomes possible to formulate an alternative definition of L-truth in $\mathscr{L}'$ on the basis of the concept of ***SD*** in $\mathscr{L}'$ (or ***LS*** in $\mathscr{L}'$): Φ_i is L-true in $\mathscr{L}'$ if and only if ***LR***(Φ_i) is the universal class of all the logical states that are admissible in $\mathscr{L}'$ with respect to $\mathscr{P}$. Hence also:

(D_4 – IV 1 D) Φ_i is L-true in $\mathscr{L}' =$ df Φ_i holds in any ***SD*** in $\mathscr{L}'$.

The use of meaning postulates is also important for languages of the $\mathscr{L}_n^\pi$ type, where n rises above all the limits that contain the variables and quantifiers. We may return for an example to the language that we have already described: the vocabulary of the language contains a denumerably infinite set of individual constants denoting the individual points on the surface of the Earth, P_1 for 'be in Bohemia', P_2 for 'be more northerly than', P'_1 for 'be in Europe', and P''_1 for 'be in Africa'.

Since the meaning of these predicates lies in certain relations of dependence, the meaning postulates have to be formulated in such a way as to capture the dependence. Firstly, it is obvious that any point that lies in Bohemia is also in Europe. Therefore:

($\mathscr{P}$1) $(\forall x)(P_1 x \longrightarrow P'_1 x)$

Furthermore it is necessary to express the relations between P_1, P'_1 and P''_1. Any point that is in Bohemia cannot be in Africa. If any point is in Europe, it cannot be in Africa. Therefore:

($\mathscr{P}$2a) $(\forall x)(P_1 x \longrightarrow \sim P''_1 x)$, i.e. $(\forall x)(P_1 x | P''_1 x)$

($\mathscr{P}$2*b*) $(\forall x)(P'_1 x \longrightarrow \sim P''_1 x)$, *i.e.* $(\forall x)(P'_1 x | P''_1 x)$

Other meaning postulates relate to the relation 'be more northerly than'. This relation is (a) irreflexive, (b) asymmetrical, and (c) transitive, and we can therefore formulate further meaning postulates:

($\mathscr{P}$3a) $(\forall x) \sim P_2 x, x$

($\mathscr{P}$3b) $(\forall x)(\forall y)(P_2 x, y \longrightarrow \sim P_2 y, x)$

($\mathscr{P}$3c) $(\forall x)(\forall y)(\forall z)(P_2 x, y \cdot P_2 y, z \longrightarrow P_2 x, z)$

But it is also necessary to express the dependency relation between P'_1 and P''_1 on the one hand and P_2 on the other. It is obvious that any point that is in Bohemia is more northerly than any point in Africa. Therefore we have to formulate another meaning postulate:

($\mathscr{P}$4) $(\forall x)(\forall y)(P_1 x \cdot P''_1 y \longrightarrow P_2 x, y)$

There is also the possibility of expressing the dependency relation between P'_1 and P''_2 on the one hand and P_2 on the other. Europe as a whole (or as a class of points with the property P'_1) lies more northerly than Africa (i.e. the class of points with the property P''_2). Nevertheless the formula

$$(\forall x)(\forall y)(P'_1 x \cdot P''_1 y \longrightarrow P_2 x, y)$$

cannot be recognised as a meaning postulate for the given language since it is well known that there are points in Europe that are not more northerly than the northernmost points in Africa. On the basis of this empirical finding it holds that

$$(\exists x)(\forall y)(P'_1 x \cdot P''_1 y \longrightarrow \sim P_2 x, y)$$

is untrue, making it impossible to recognise the previous formula as a meaning postulate.

The role of meaning postulates can, as has been seen, be taken over by sentences capable of expressing the *dependences* between the non-logical terms of a language. So if we introduce a new non-logical term into a language (a procedure usually described as *constitution*), the class of meaning postulates which have to be respected in the language must be extended by other meaning postulates concerning the dependences between the newly constituted term and those already introduced. In the analysis of constitution procedures (and their converse, so-called reduction procedures) 'reduction sentences' are used to express these dependences. Therefore reduction sentences and the formalisms that are capable of taking over their function may be treated as meaning postulates or an element of the class of meaning postulates.[22]

The conception of meaning postulates may also be applied in the system of inductive logic, of which the basis is, as shown above, the regular function of measure **m** and in which **c** is the function of confirmation, whereby $\mathbf{c}(\mathbf{h}, \mathbf{e}) = \frac{\mathbf{m}(\mathbf{e} \cdot \mathbf{h})}{\mathbf{m}(\mathbf{e})}$. If we extend semantics to take in the meaning postulates $\mathscr{P}$ (i.e. if we switch from $\mathscr{L}$ to $\mathscr{L}'$), we may also transfer to a new function of measure for state descriptions in $\mathscr{L}'$, which we shall denote **m'**. If **m'** is the function of measure for ***SD*** in $\mathscr{L}'$, it must satisfy all three conditions:

(1) $\mathbf{m}'(\boldsymbol{SD}_i) = 0$ for any $\boldsymbol{SD}_i$ in $\mathscr{L}$ in which $\mathscr{P}$ does not apply;

(2) For any $\boldsymbol{SD}_i$ in $\mathscr{L}$ in which $\mathscr{P}$ applies, $\mathbf{m}'(\boldsymbol{SD}_i)$ is directly proportional to $\mathbf{m}(\boldsymbol{SD}_i)$, i.e. $\mathbf{m}'(\boldsymbol{SD}_i) = \mathbf{K}.\, \mathbf{m}(\boldsymbol{SD}_i)$;

(3) The sum of all $\mathbf{m}'(\boldsymbol{SD})$ is 1.

Since the class of meaning postulates is L-true (in $\mathscr{L}'$), it holds that

$$\mathbf{m}'(\mathscr{P}) = \mathbf{K}.\mathbf{m}(\mathscr{P}) = 1.$$

Therefore $\mathbf{K} = \frac{1}{\mathbf{m}(\mathscr{P})}$. On the basis of this Carnap introduces **m'** as the regular function of measure for $\mathscr{L}'$ (i.e. for $\mathscr{L}$ where the class of meaning postulates $\mathscr{P}$ applies).

From this it can be easily deduced that

(4) $\mathbf{m}'(\boldsymbol{SD}_i)$ in $\mathscr{L}' = \frac{\mathbf{m}(\boldsymbol{SD}_i)}{\mathbf{m}(\mathscr{P})}$;

(5) If Φ_i is a sentence or class of sentences in $\mathscr{L}'$, then $\mathbf{m}'(\Phi_i) = \frac{\mathbf{m}(\mathscr{P} \cdot \Phi_i)}{(\mathscr{P})} = \mathbf{c}(\Phi_i, \mathscr{P})$.

(6) If Φ_h and Φ_e are sentences in $\mathscr{L}'$, and if Φ_e is not L-untrue, it holds that $\mathbf{c}'(\Phi_h, \Phi_e) = \frac{\mathbf{m}'(\Phi_e \cdot \Phi_h)}{\mathbf{m}'(\Phi_e)} = \frac{\mathbf{m}(\mathscr{P} \cdot \Phi_e \cdot \Phi_h)}{\mathbf{m}(\mathscr{P} \cdot \Phi_e)}$ $= \mathbf{c}(\Phi_h, \mathscr{P} \cdot \Phi_e)$.

It is also easy to suggest how meaning postulates may be respected in defining content measure for sentences in $\mathscr{L}'$.[23]

Since **cont(i)** = 1 – **m(i)**, it is possible to introduce **cont'(i)** = 1 – **m'(i)** = $1-\dfrac{\mathbf{m(i)}}{\mathbf{m}(\mathscr{P})}$. Likewise **inf'(i)** = $\mathbf{Log}\dfrac{1}{\mathbf{m'(i)}} = \mathbf{Log}\dfrac{\mathbf{m}(\mathscr{P})}{\mathbf{m(i)}} = \mathbf{Log\ m}(\mathscr{P}) - \mathbf{Log\ m(i)}$.

This procedure has certain associated problems and difficulties relating to content. For one thing, there is no doubt that expressing the dependences of non-logical terms merely by using devices of logic offers relatively restricted facilities. Another problem resides in the fact that these dependences must be required to be unchanging. But this kind of constancy need not be absolute, but merely relative, for instance to a given type of tasks.

In the context of the application of meaning postulates along with the method of state description, there are two distinct situations that may be treated as generally possible:

(a) In a given $\mathscr{L}$ the meaning postulates are set up beforehand in such a way that the semantically relevant relations of the non-logical terms are exactly defined. This is clearly the simpler solution and only this simpler solution can be resolved by the procedure suggested by Carnap in [25].

(b) Yet allowance must be made for the situation where the meaning postulates are known beforehand only in part. If the meaning postulates were completely unknown; the application of the method of state description would be a purely formal matter and it would be impossible to deduce from it any consequences for the semantics of the non-logical terms used in the given language. In using natural languages and in the assimilation of their semantics this is clearly the more fundamental situation. This also applies to the methodological problems of science and the use of language in the activities of science. Accordingly there are in this situation new, quite different problems from those for which we have suggested solutions in the foregoing account.

2. THE SEMANTICS OF LOGICAL CONCEPTS ON THE BASIS OF THE CONCEPT OF INTERPRETATION

As we have already seen, the method of state description operated with a conception of 'possible worlds', the definition of a possible world being based on means of a syntactic nature. This standpoint has two major consequences:

(a) From the philosophical point of view it leads to the assumption (which incidentally coincides with the views of Wittgenstein) that the 'limits of my world are set by the limits of my language'. Although the philosophical analysis of this assumption lies outside the purview of this study, it can at least be pointed out that all the limits existing in nature and ascertained by the natural sciences, e.g. the limits set by the second principle of thermodynamics, the theory of relativity, the Heisenberg relation, Shannon's theorems etc., are not limits of our language, but limits of naturé itself as established by science. This is not to say that language or language use has no limits, although they too are important in a way that is analogous to the limits of nature in that they are due to restrictions of capacity, or the memory and delay facilities of the language users, and so forth.

(b) From the semantic point of view proper, it has to be noted that the method of state description means that the semantics, or at least the logical semantics, of a language is built up on the basis of the language's syntax. Moreover, as we have already shown, the original form of the method of state description, used as the starting point for the semantics of logical concepts, does not take into account the semantics of non-logical terms. Thus although it facilitates the analysis of the semantically important relations of logical terms, it simultaneously impoverishes that analysis by disregarding the semantically important relations of non-logical terms.

It is true that the introduction of meaning postulates removes some of the difficulties under (b) above, but in principle their introduction does nothing to change the consequences of the method of state description given under (a).

For these reasons the method of state description, especially in the form evolved by Carnap, has been the subject of criticism. While the construction of fundamental semantic concepts, i.e. concepts corresponding to Carnap's radical concepts, may not provoke particularly serious objections (whether the semantic concepts are introduced by definition, as in Tarski's case, or by axioms, as suggested by Martin), the construction of concepts of logical semantics on the basis of the method of state description has given rise to serious objections. Of the logicians it was Quine in particular who accepted definitions for the basic semantic concepts while rejecting definitions of concepts of logical semantics based on the method of state description.

J. G. Kemeny [69] pointed out that Carnap's definitions of concepts of logical semantics were marred by three fundamental flaws: (1) their starting point, i.e. the concepts of 'logical range' and 'state description' lead on to serious difficulties; (2) these definitions concern only fairly simple languages; (3) the concept of 'analyticity', which Carnap identifies with L-truth, is not entirely satisfactory, assuming as it does validity in all possible worlds, though the concept of a 'possible world' is determined with respect to a given language by the syntax of that language.

Kemeny advanced a conception for a 'new approach' to semantics based on the concept of 'model' and facilitating the definition of both kinds of concept, i.e. some fundamental semantic concepts and concepts of logical semantics. However, the concept of model can only play its part satisfactorily in the case of a formalised language which is complete. In the case of languages of a higher degree than the first, problems are encountered, making it necessary to determine what models are taken into account. Kemeny calls these models interpretations. Therefore a metalanguage relative to a given object language must provide not only the relevant data required for syntactic purposes, but also certain extra data, e.g. as to the model (Kemeny's 'intended models') that are to be treated as interpretations. Then it can be said that the metalanguage is semantically adequate. Kemeny maintains that an interpretation is completely determined if the following data are given:

(1) translation of the object language into the metalanguage;
(2) division of the primitive constants into logical and non-logical.

[*A*] *The Concepts of Model and Interpretation*

Kemeny has worked out a system of logical semantics and a system of definitions for the basic concepts of semantics using four fundamental concepts: 'value-assignment', 'semi-model', 'model' and 'interpretation'. Moreover he defines these concepts with respect to a very generally conceived formalised language, i.e. he does not restrict himself just to languages of the first degree.

Value-assignment is understood to mean the assignment of individuals from the domain of the universe of $\mathscr{L}$ to each of the individual variables of $\mathscr{L}$. Two value-assignments differ if and only if on the basis of one value-assignment an individual is assigned to at least one individual variable while not being assigned to the same variable on the basis of the other value-assignment.

For the concept of 'semi-model' to be introduced, one needs to know the types of different expressions. It is further necessary to know what expressions of one type are assignable to other expressions of another type. This kind of assignment is set out by what Kemeny describes as a 'semi-model'. This means, for example, that each one-place predicate has assigned to it by the semi-model a class of individuals, each n-place predicate a class of ordered n-tuples of individuals, and so on. Clearly then, not just any assignment is admissible; an accepted categorisation and hierarchy of types has to be respected.

After defining 'semi-model' Kemeny defines the concept of 'holding' of wff in $\mathscr{L}$ in a given semi-model with respect to a given value-assignment. In its formal aspect this definition is recursive and essentially analogous to the definition we have already given in connection with the concept 'satisfy'. The role of 'holding' here is analogous to that of 'satisfy' in Tarski's approach.

The concept 'holding' must further be distinguished from the concept 'valid', the difference between them corresponding to the difference between 'satisfies' and 'always satisfies'. This means that if the value of a wff with respect to each value-assignment is always 1, we shall say that the wff is valid in the given semi-model. If it is always 0, it is invalid.

In this connection it is worthwhile pointing out some of the advantages of this approach over the method of state description in the form evolved by Carnap. Carnap was always forced to assume that the syntax of a language under consideration (or, to put it another way, the logical system that was the basis for the relevant formalised language) was known in its entirety. Only with this assumption was he able to build up the semantics of the language. On the other hand, the 'new approach' suggested by Kemeny is not necessarily so constrained, which is an unquestionable advantage.

At this stage it becomes possible to define the concept of 'model'. A 'model' of the object language $\mathscr{L}$ may be defined as a semi-model $\mathbf{M}$ if

(1) each axiom of $\mathscr{L}$ is valid in $\mathbf{M}$;

(2) if inference rules allow us to infer ${}^{(1)}H$ from ${}^{(1)}I$, ${}^{(1)}J$, ... and if ${}^{(1)}I$, ${}^{(1)}J$, ... are valid in $\mathbf{M}$, then ${}^{(1)}H$ is valid in $\mathbf{M}$;

(3) not all well formed formulae are valid in $\mathbf{M}$.

This may also be summarised by taking the concept 'meaning postulates', but giving it a rather broader sense than hitherto, i.e. having it include not only the non-logical terms, but also all the axioms and definitions of $\mathscr{L}$. We can then say that a model of the object language $\mathscr{L}$ is a semi-model in which 'meaning postulates' (in the sense just given) are valid.

From the intuitive point of view, as Kemeny also shows, the concept of 'semi-model' delineates everything that remains open or undecided in a given interpretation of language. The value of a particular formula is essentially therefore its interpretation. If a formula contains free variables, its interpretation depends on the interpretation of those free variables.

The concept of a model of language $\mathscr{L}$ then delimits the possible world with respect to $\mathscr{L}$. Thus it may appear to be a good place to start for defining some of the important concepts of logical semantics. For example, the concept of the 'analyticity' of a sentence (in $\mathscr{L}$) might seem to express that the sentence (in $\mathscr{L}$) is valid in all models of $\mathscr{L}$. But this view of analyticity as something valid in all models is only admissible in the case of language systems with a fairly simple logical base or, to put it more exactly, in cases where it is not necessary to take into account the incompleteness of systems. However, some important results of logic, notably Gödel's theorems of the incompleteness of some systems, show that with systems that are not demonstrably complete the 'model' approach to semantics is unsuitable. Moreover, the borderline between the 'analytical' and the 'synthetic' in these systems is problematic, i.e. they may contain sentences that are analytical but unprovable. This is also the main reason why Kemeny recommends that 'analyticity' be defined not in terms of validity in *all* models of $\mathscr{L}$, but only in certain models that he describes as 'intended models' or '*interpretations*'.

Before proceeding to suggest a more accurate definition of the concept of interpretation, we have to stress that the concept is introduced in order to facilitate the more exact definition of 'analyticity' and 'analytical truth' of a sentence. Intuitively speaking a sentence is analytically true if and only if its truth can be established on the basis of the logic of the given $\mathscr{L}$, without regard to any 'factual data'. But there is a problem here: are 'factual data' the same thing as 'extralinguistic data'? If so, then of course even the data in $\mathscr{L}$ are 'extralinguistic' in the sense that they assume a metalanguage. If not, the concept 'factual data' calls for some kind of qualification. The use of the concept of 'interpretation' as a starting point for logical semantics assumes this second possibility. Kemeny therefore tries to unite two essential features: (a) preservation of the tradition, evolved preeminently in philosophy, whereby the concept 'analytical' corresponds to Leibniz's term 'necessary', and 'synthetic' to Leibniz's 'contingent',[24] or, to put it another way, 'analytical' and 'synthetic' correspond to 'true in all possible worlds' and 'true in an actual world' respectively; (b) respect for the results of the theorems on incompleteness.

Thus 'interpretation' relates to possible worlds, and the concept of 'possible world' is relativised to $\mathscr{L}$ not only in the sense suggested by the method of state description, i.e. syntactically, but also semantically. Any 'world' is therefore a possible world with respect to $\mathscr{L}$ insofar as it does not upset the logical basis of $\mathscr{L}$ and insofar as it is suited to the accepted usage of terms in $\mathscr{L}$. Leibniz's concept of 'possible world' or the concept of 'description of a possible world' then corresponds in this sense to the concept of interpretation.

Yet it remains a matter of course that terms may have more than one possible interpretation. And the starting point of this interpretation, as follows from Kemeny's conception, is *interpretation in an actual world.* The same goes for the differentiation of logical and non-logical terms in $\mathscr{L}$. Terms that have the same interpretation in all possible worlds, that interpretation being the same as in an actual world, can be described as *logical constants.* Other constants are then non-logical.

In this connection we may also refine the previously mentioned differentiation of semantic concepts. The conception of the first group of semantic concepts remains unchanged – that is those that can be defined in terms of an interpretation that describes the actual world. The second group of semantic concepts is definable in terms of all possible interpretations, without it being necessary to pay particular regard to any one of them.

At the same time, however, it is assumed that this corpus of all possible interpretations may, with respect to the logical concepts, realise the same kind of assignment as an interpretation that describes the actual world. In this respect (and in this respect only) it is not therefore necessary to heed the interpretation vis-à-vis the actual world.

This line of thought is the basis for the refinement of the concept of 'interpretation'. The procedure that leads to the concept of 'interpretation' is roughly as follows: A model **M** is selected which ensures an assignment of meanings to terms in $\mathscr{L}$, which meanings we intend to consider their 'official' meanings. Kemeny denotes such a model **M***. This model has a decisive role: it is the interpretation that relates to the actual world. Interpretation is then understood to mean a model that assigns to $\mathscr{L}$ the same domain as **M*** and differs from **M*** only in its different assignment of meaning to those constants of $\mathscr{L}$ that are not logical constants. In other words, any interpretation whatsoever differs from **M*** only in the assignment of meaning to non-logical constants; it does not differ in the assignment of meaning to logical constants.

[B] *The Most Important Definitions of Logical Semantic Concepts on the Basis of 'Interpretation'*

After this clarification of the concept of 'interpretation' we may proceed to the definition of the most important concepts of logical semantics. As noted previously, the introduction of 'interpretation' is connected with efforts to uphold Leibniz's traditional conception while at the same time avoiding the difficulties that might arise as consequences of the theorems on incompleteness. Hence one of the first steps is to define the concept of 'analyticity' (also 'analytical truth', A-truth for short).[25]

(D_1 – IV 2 B) A wff (in $\mathscr{L}$) is *A-true* if it is valid in all interpretations of $\mathscr{L}$.

(It has to be added that this definition is intuitively obvious only if the well formed formula contains no free variables. The same applies to the definitions that follow.)

(D_2 – IV 2 B) A wff (in $\mathscr{L}$) is *contradictory* if it is invalid in all interpretations of $\mathscr{L}$.

(D_3 – IV 2 B) A wff (in $\mathscr{L}$) is *analytical* if it is either A-true or contradictory.

(D_4 – IV 2 B) A wff (in $\mathscr{L}$) is *synthetic* if it is valid in some interpretations and invalid in others.

On the basis of these definitions it is possible to use theorems to express some of the properties and relations of the terms given. It is obvious that no well formed formula can be both analytical and synthetic. A well formed formula is contradictory if and only if its negation is A-true, and so forth.

Other definitions of concepts that fall within the same category as A-truth, analyticity and contradiction concerns the concepts of (logical) implication and (logical) equivalence. Essentially these are concepts corresponding to L implication and L equivalence in the same sense in which A-truth corresponds to L-truth:

(D_5 – IV 2 B) Σ_i is (*logically*) *equivalent* to Σ_j if both Σ_i and Σ_j are valid in the same interpretations.

(D_6 – IV 2 B) A sentence Σ_i (*logically*) *implies* the sentence Σ_j if Σ_j is valid in all the interpretations in which Σ_i is valid.

A different category covers those concepts that are not defined in terms of all the interpretations of $\mathscr{L}$, but in terms of relations to the actual world, i.e. in terms of relations to **M***, where **M*** is understood, as we have said, as a model that assigns to the terms of $\mathscr{L}$ interpretations that we consider to be their 'official interpretations', i.e. interpretations that we wish to take into consideration. In this context the fact should not be overlooked that these are by no means only the kind of interpretations that must have an empirical basis in all circumstances. Thus the concept of 'actual world' cannot be identified invariably with that of 'empirical world'.[26] Also it must be emphasised that the decision as to the 'official interpretation' lies outside the competence of the semantic level of the analysis of language, being within the competence of the pragmatic level of analysis. The definitions which are based on $\mathbf{M}^*$ are as follows:

(D_7 IV 2 B) A wff (in $\mathscr{L}$) is *true* if it is valid in $\mathbf{M}^*$.
(D_8 IV 2 B) A wff (in $\mathscr{L}$) is *untrue* if it is invalid in $\mathbf{M}^*$.

Of the other definitions given by Kemeny, that for the synonymity of constants is important and interesting:

(D_9 IV 2 B) Two constants are synonymous if, whatever the interpretation, the same element is assigned to both constants.[27]

So far we have been thinking of $\mathbf{M}^*$ as dependent on some kind of choice, which has not yet been further specified. This means that the notation for 'models' and 'interpretations' has also not been specified. Kemeny puts forward a narrowing of the choice in the shape of two constraints:

1st constraint: All interpretations work with a type-homogeneous set of objects ${}^{(\iota)}X$ that are being assigned.

2nd constraint: A model that differs from an interpretation only in the assignment of non-logical constants (and therefore does not differ in the assignment of logical constants) is also an interpretation.

Kemeny ties the first constraint to the assumption that a formalised language is based on an abstract hierarchically built structure whose foundation is a given set of individuals. This kind of assumption is compatible not only with formal sciences, but also with the empirical sciences. Kemeny is also quite explicit about the unnecessity of requiring that ${}^{(\iota)}X$ be determined 'purely logically'.

From the methodological point of view and in connection with the empirical sciences, there are a few remarks that might be added to this viewpoint. Basically they have to do with the ontological starting point of any language:

(a) In science, in defining tasks of a certain type, it is always possible to state more or less exactly what corresponds to the range of entities that may be taken as 'basic' from the point of view of those tasks.

(b) This 'basic' range of entities cannot of course be seen as definitive, given once and for all. It may be broadened or narrowed (the procedures of constitution and reduction at the same level). But elements of that range may also be considered entities of a 'higher' or 'lower' level with respect to another range of entities.

(c) The 'basic' range may also be the starting point for ascertaining the hierarchical structure of the entities of 'higher' levels. One of the fundamental general tasks of the empirical sciences overall is to discover the nature of the interdependence between entities of different levels and the corresponding possibilities for constitution and reduction within the different levels.[28]

These remarks have also shown that the methodologically significant use of language in the context of the empirical sciences corresponds to operations with logical systems that are *incomplete*.

The second of the above constraints assumes a differentiation of two types of constants. It is usually maintained that the rules of the syntactic stage in the build-up of a formalised language, including axioms, cover all the properties of the logical constants, while the semantic stage covers those of the non-logical constants. This assumption leads, however, to the difficulties associated with the theorems about the incompleteness of certain formalised systems. This leads Kemeny to stress that in consequence of the syntactic incompleteness of certain systems it cannot be assumed that in delineating the syntactic component we are saying nothing about the non-logical constants. At the very least we are saying that they are subject to the rules and axioms of $\mathscr{L}$. If, in a given $\mathscr{L}$, meaning postulates have been brought in, the use of the non-logical constants is also subject to them. (There is no need in principle to exclude meaning postulates from the corpus of rules and axioms, which is why Kemeny does not believe in the necessity of distinguishing meaning postulates from the other axioms.)

From the two constraints it is obvious just how decisive is the role played by **M*** in this approach to logical semantics. Interpretations are therefore those models that assign to the non-logical constants the same element as **M***. At the same time, those models that differ from **M*** only in assigning other elements to just the non-logical constants while assigning the same element to the logical constants, may also be described as interpretations.

With the role of **M*** thus characterised, it is possible to show what the prerequisites are for a given object language to be fully determined semantically. In this connection Kemeny uses the term 'semantic determinateness', which he defines as follows:

(D_{10} – IV 2 B) A language $\mathscr{L}$ is *semantically determined* if in addition to the formal characteristics[29] of $\mathscr{L}$ there is given (1) one model qualified as **M*** which is so denoted for the purposes of translation from $\mathscr{L}$ into $\mathscr{M}\mathscr{L}$, and (2) if it has been established which constants of $\mathscr{L}$ are non-logical constants.

This approach to semantics makes it possible to define a number of other concepts. Its undoubted merits include the following facts: Kemeny's approach respects the results of the theorems about incompleteness and can therefore be applied even to formalised languages of higher degrees; also it makes it possible to define both kinds of semantic concepts, i.e. concepts like A-truth, contradiction, analyticity, syntheticity on the one hand, and also concepts such as truth, untruth etc.[30]

[C] *The Concept of 'Translation'*

The approach to semantics that is based on the concept of 'interpretation' enables us to give precision to certain important concepts, that of '*translation*' in particular. What is referred to here is not the fairly vast and not always entirely straightforward philological problem area of translation from one natural language into another. The problems concern all the basic levels of language analysis, and therefore also has its specific pragmatic aspects. In the following we shall suggest just a very narrow approach to the concept of 'translation', based solely on criteria of logical semantics. In order to define the concept certain other concepts have to be brought in. Above all there is the concept of 'phrase'. Kemeny recommends that 'phrase' be used to denote those formulae of $\mathscr{L}$ that do not contain free variables. Moreover both constants and sentences may be phrases. A constant is a phrase consisting of a single symbol.

Here we shall give some definitions that have a bearing on the concept of 'phrase':

(D_1 – IV 2 C) A phrase is a *logical phrase* if it has the same value in all interpretations.

This means, for example, that a constant is a logical phrase if and only if it is a logical constant. A sentence is a logical phrase if and only if it is an analytical sentence.

(D_2 – IV 2 C) A phrase is a *factual phrase* if it is not a logical phrase.

We also need to bring in synonymity with respect to phrases:

(D_3 – IV 2 C) Two phrases are *synonymous* if they have the same value in all interpretations.

This means that two constants are synonymous phrases if and only if they are synonymous constants. Two sentences are synonymous phrases if they are equivalent.[31]

From the intuitive point of view it is desirable for any semantic relation (i.e. a relation expressed by semantic concepts of either kind, hence both concepts from the theory of meaning and those from the theory of reference, to use Quine's terminology) to remain invariant under translation.

Kemeny distinguishes two kinds of translation: *free translation* and *literal translation*. He only defines the former. First and foremost it is to be expected that if H_1 in $\mathscr{L}_1$ is a translation of H_2 in $\mathscr{L}_2$, H_1 and H_2 are synonymous. But since in general $\mathscr{L}_1$ and $\mathscr{L}_2$ are different languages, it is necessary to assume a metalanguage $\mathscr{ML}$ that is the metalanguage vis-à-vis both $\mathscr{L}_1$ and $\mathscr{L}_2$. That being so, H_1 in $\mathscr{L}_1$ is the translation of H_2 in $\mathscr{L}_2$ under the assumptions that $\mathscr{L}_1$ and $\mathscr{L}_2$ have a common $\mathscr{ML}$ and that the translations of H_1 and H_2 into $\mathscr{ML}$ are synonymous. This situation corresponds to the definition:

(D_4 – IV 2 C) A phrase of $\mathscr{L}_1$ is a *free translation* of a phrase in $\mathscr{L}_2$ if the translations of the two phrases into $\mathscr{ML}$ are synonymous.

It might well be objected that this definition of a free translation is based on (D_3 – IV 2 C), which offers only a thoroughly weak and limited criterion of synonymity. This conception also corresponds to the purely extensional view of language. By contrast, the concept of translation as we operate with it in both natural languages and in those formalised languages that also contain non-logical constants the relations among which have to be respected, i.e. an intuitive sense of the concept, points more in the direction of the intensional view of language. To this also corresponds the concept of literal translation.

Kemeny gives no definition of literal translation, for the means afforded by his approach to semantics are inadequate to such a definition. The concept of literal translation assumes not only identity of value, i.e. identity in terms of denotation, but also identity of sense. This means that the concept of literal translation can only be brought in against the background of the intensional view of language. In other words, here too semantic analysis has to take the line suggested by Frege through his distinction of 'sense' and 'denotation'.

For reasons that are understandable Kemeny's approach to semantics cannot resolve all the problems of semantics. The problems discussed above and connected with the concept of 'sense' remain quite apart. But left to one side with them are certain other problems, whether those of the theory of meaning (in the Quine sense) or those of the theory of reference. Nevertheless the importance of an approach to semantics which is based on the concept of interpretation and so too on that which corresponds to **M***, i.e. the idea of an interpretation that is privileged from the point of the view of a given task, cannot be denied, especially in connection with the use of a formalised language in a given empirical context.

V. SENSE AND DENOTATION

1. FREGE'S CONCEPTION OF SENSE AND DENOTATION

One of the basic concepts of both natural and formalised languages is that of '*name*'. It is of course a term that has manifold uses not only in natural language, but also in linguistics, logic and mathematics. From the point of view of semantics the most essential question is that of the entities to which a name can be related. The mathematician G. Frege cannot be credited with being the first to tackle this kind of problem, since questions along these lines can be encountered throughout the evolution of logic and, naturally, linguistics as well. But Frege was the first to come up with a sufficiently general solution, partly be generalising the concept of 'name' and partly by not confining his solution to the problems and needs of linguistics and the analysis of natural language, setting out instead from the needs of mathematical and logical thinking.[1]

[*A*] *Sense, Denotation and the Identity of Names*

The immediate inspiration for differentiating between the two aspects of meaning is the problem of the identity of names. The problem of identity has always played a major role in the evolution of logical and mathematical thought. One of the seriously controversial questions is how the relation of identity is to be understood and to what entities it is to be bound. In principle we can separate the problems of identity of the objects of a universe and problems of the identification of those objects, i.e. matters of an ontological nature, from the problems of the identity of the names of those objects, or their properties, classes etc. In the former case identity is a relation of extralinguistic objects, while in the latter we are dealing with relations between expressions within a particular language. Again it is important whether we are talking about the identity of names of individuals, classes or other language expressions.[2] Almost all post-medieval conceptions of identity take up the basic ideas of Leibniz, but it should be noted that Leibniz's ideas can be related both to the problems of the identity of extralinguistic objects and to problems of the identity of one kind of language expression.

If we stress the identity of objects, irrespective of whether these are 'things' or objects of some other nature, e.g. constructive objects, we are usually operating with the concept of a 'property' of the objects. This is how, for example, Tarski formulated his basic law of the theory of identity [145, p. 55], which gives three formulations:

(1) $x \underset{\text{id}}{=\!=} y$ if and only if x has every property that y has and y has every property that x has.

(2) $x \underset{\text{id}}{=\!=} y$ if and only if x and y have all properties in common.

Tarski considers his third formulation less correct:

(3) $x \underset{\text{id}}{=\!=} y$ if and only if anything that can be said of either x or y can also be said of the other.[3]

If we stress the identity of names, the identity concerns relations within language, and it is a problem of a semantic, not ontological nature. This semantic problem was also touched on by Leibniz. The criterion of identity is then the possibility of mutual substitutability with preservation of the truth value of the sentence in which substitution can be made. It is the 'salva veritate' test. Clearly the spirit of Leibniz's philosophy is more in tune with this second approach to the problems of identity, although there is no doubt that Leibniz held the semantic problems of identity to be independent and derived with respect to the ontological problems. Frege too, who was strongly subject to Platonist influences in his mathematics and logic, clung basically to this second way of dealing with identity.[4]

If we adhere to the second account of identity, certain difficulties arise, and it was a solution to these difficulties that was attempted by Frege. It we consider the nature of the expression $a \underset{\text{id}}{=\!=} a$, it is clearly a different kind of statement about identity from a statement of the type $a_1 \underset{\text{id}}{=\!=} a_2$. In the first case there is no need to take any account of the meaning of a, while in the second meaning has to be respected. At the same time it must be realised what aspect of meaning is concerned. It has not always been easy to demonstrate the identity of two names denoting the same object. It was the ancient Babylonians who first recognised that the clear star above the evening horizon (the evening star) was the same one that shone with equal strength before dawn (the morning star). To this day there are still natural phenomena of which it is difficult to say whether they are connected with one and the same object or not (because, for example, they occur at different times).[5] In general it is safe to say that the problem of the identification of elements of the input space in the 'observer channel' or 'language-user channel' on the basis of data at the output is not always a trivially simple task.

Frege therefore starts with some questions that may be formulated as follows: Are the names 'Morgenstern' and 'Abendstern' identical? If so, in what respect? The answer is simple: It is expedient to link the name not only with its assignment to the object denoted, which can be described as the name's *denotation*,[6] but also with its *sense*.[7] In the case of 'morning star' and 'evening star' the denotation is identical, they denote the same thing, but they are not identical in sense.

The discovery that the names 'morning star' and 'evening star' are identical in terms of denotation does not amount to the trivial statement that 'morning star $\underset{\text{id}}{=}$ evening star' – it is not an expression of the kind $a \underset{\text{id}}{=} a$. This kind of discovery always requires some *knowledge*. Hence a decision on the identity of names is not without assumptions, but depends on the possibility of assigning to names not just denotation, but also a sense.

The role of this kind of knowledge can be demonstrated on two examples, one taken from Frege, the other from Russell. If we have a triangle we can draw three connecting lines from the vertices to join each vertex to the mid-point of the opposite side of the triangle. We shall call the connecting lines α, β and γ. Anyone without a knowledge of elementary geometry will be unable to decide without looking whether the names 'the intersection of α and β' and 'the intersection of β and γ' denote the same point and therefore have the same denotation. But at the same time it is clear that the two names have a different sense, which has to be expressed by different signs or in different words. In order to decide that 'the intersection of α and β' and 'the intersection of β and γ' have the same denotation, certain geometrical knowledge has to be available.

A similar case is the classic and oft-repeated example given by Russell [131]. It is well know that Walter Scott left his famous novel *Waverley* unsigned out of fear of the critics. For some time after the success of this anonymously published work Scott signed subsequent works as 'the author of Waverley'. It was not until later that he shed his anonymity. Russell formulated the example as follows: King George IV wished to know whether Walter Scott was the author of Waverley (and he really was). So what did Goerge IV want to know? If the names 'Walter Scott' and 'the author of Waverley' can replace each other (by the test of 'salva veritate'), then George IV wanted to know whether Walter Scott was Walter Scott! While it might be unwise to overestimate the English king's knowledge, one may reasonably doubt that he would merely have asked if Scott was Scott.

Ideas such as these inevitably lead to the conclusion (in which Frege and Russell fully concur) that the use of Leibniz's 'salva veritate' test in relation to names, or 'denoting phrases' to use Russell's term, must respect the two different aspects of meaning:[8] sense and denotation.

Frege identified the concept of 'denotation' with the denoted object or class of objects. A denotation is then that which is denoted. By contrast, sense is that which the relevant name signifies. What is important here is to note that the concept of 'sense' in particular is prone to various interpretations. Frege, in the spirit of his rejectionist attitude to psychologising in logic, emphasised that the concept of the 'sense' of a word should not be identified with an associative image, or some purely subjective phenomenon. Accordingly he treated 'sense' as a category that is neither purely subjective, nor purely objective.

Church, who is an adherent of Frege's conception, describes that conception by saying that the sense that is connected with a word represents certain information about the object denoted. Sense characterises the object unambiguously, irrespective of whether the object is accessible or not, irrespective of whether it actually exists (or has existed), irrespective of whether it can be visualised or not, and so forth. This can be expressed in another way by saying that for the class of users of $\mathscr{L}$ (assuming that all its users have a command of the entire verbal wealth of $\mathscr{L}$, i.e. are capable of understanding every name in $\mathscr{L}$) the same sense of any name is always accessible. In other words, all the users of a language are capable of assigning the same sense to a name without having to call up the same ideas or associative images.

This standpoint of Frege's is not entirely unproblematical. For Frege the mathematician and logician the effort to depsychologise the concept of 'sense' is on the whole perfectly understandable. It is obviously tied to the tasks envisaged. If we were to treat sense as a purely psychological category, conceived subjectively moreover, we would fall foul of a measure of arbitrariness. This would be particularly so if the concept of 'sense' were to be connected with subjective ideas or associative images. In forming such images not only is it impossible to preclude arbitrariness, but there are many other cultural, social, psychological and linguistic factors that can influence their formation, even in the case of two people using the selfsame expression in like circumstances, for: *si duo faciunt idem, non est idem.* It is in order to avoid this situation that Frege does not conceive 'sense' as a psychological category but as a constructed abstract object that is relatively stable, while an associative image is much more variable. Sense corresponds to what the language user assimilates when he understands a name. It is always possible to understand the sense of a name without knowing anything about the object denoted. For example, we can understand the name 'the intersection of α and β' or 'the intersection of β and γ' without necessarily knowing that both names denote one and the same point. Church points out that 'sense' is a postulated abstract object with certain postulated properties [39, p. 343]. Yet it is undeniable that in the understanding of a sign, collection of signs or expression of language a certain role is always played, where we use natural language, by what we have called the 'subjective world' of the language user. While it is true that we cannot understand the sense of a language expression without knowing what kind of objects it may relate to, it remains the case that what we call 'understanding' has two aspects: the objective and the subjective. The objective aspect rests in the language user's assimilation of the sense of a sign, word, complex of signs. The subjective aspect rests in the fact that in an act of understanding the language user also operates with certain asssociative images, ideas etc. of his own. Moreover these elements of the subjective may have a greater or lesser influence on the objective aspect. This situation may also be expressed by saying that the pragmatic aspects of language using may modify the semantic aspects. The words 'a house is on fire' have a sense that we are capable of understanding without knowing what kind of house or fire is meant. Nevertheless matters differ considerably if the words are heard from the stage in the course of a theatrical production, or if they are heard by a duty fireman.

We may say in conclusion that the differentiation of the two aspects of meaning, sense and denotation, only takes into account the semantic level of language analysis. If in addition we consider the pragmatic aspect, there can be no clear division between what Frege described as 'sense' and what he described as 'subjective image'.

[B] *The Concept of 'Name'*

The possibility of having sense and denotation assigned to it is a property of the category we have described as '*name*'. In semantic terms there is no substantial difference between what is usually called a 'proper name' and what we describe simply as a 'name'. On the contrary, the general concept of 'name' is characterised as if it were a 'proper name'.

Frege does not provide any sure definition of 'name', giving merely a kind of explication from which it is evident just what may (with greater or lesser accuracy) be described as a 'name'. A name may be any expression that has the function of a proper name, or denotes any object (in the abstract sense). The basis for explication of the concept 'name' is therefore the relation which Carnap has called the 'name relation'. Moreover it does not matter whether it is a unique object or an object that occurs in different forms in different situations, or whether it is a group of objects, and so forth. Thus a name may be any expression of the language $\mathscr{L}$ which is well formed[9] and has the function of a proper name.

While it is the 'name relation' that is fundamental to the constitution of the 'sense' of a name, this is not to say that every name must have some quite definite denotation. In other words, Frege admits the possibility of cases where a name relates to an imaginary object or one that cannot exist. The expression 'the body that lies at the greatest distance from the Earth' is a name and has a sense, although it is impossible to imagine such an object or even determine whether any such object exists.[10]

The name relation need not necessarily relate only to objects (in the abstract sense) of an extralinguistic nature. Names may also be used for 'names', one may operate in a meaningful way with signs of 'signs', and so on. In this connection Frege notes in particular indirect speech and introduces the concepts of indirect sense and indirect denotation. In reality the use of expressions that are the names of 'names' need not be tied just to indirect speech; its significance is more far-reaching and actually concerns the internal hierarchy of a language's expressions. It is of course difficult to establish the layers of this kind of hierarchy in natural language. In formalised languages the role of hierarchy may be taken over by the differentiation of types.

Another of Frege's remarks which is of more general impact concerns the connecting or concatenation of names. The sense of a new name which is born of the concatenation of two or more names depends on more than just the denotation of the names that are the components of the complex name. In this respect Frege only spoke in negative terms, putting forward no positive solution to the problem. A solution was presented by Church [39, p. 20, and also note 30 on p. 349]. Church summarised the relations between the sense of a complex name and the sense of its components (provided these components are names) in the following points: (1) the sense of a compound name does not change if one (or more) of its components (if these are themselves names) is replaced by another component having the same sense. (2) The denotation of a compound name does not change if one (or more) of its components is replaced by another component (again assuming that these are names) having the same denotation. (It is of course accepted here that the sense of the compound name may change.)

However, these principles have certain limitations. Firstly it should be noted that they apply only in the case of what Frege called 'direct sense' and 'direct denotation'. In cases of the 'indirect use' of a name the principles do not hold. This also means that replacement is feasible if the expressions are of the same type. For example, it is possible to compare the names:

(A) 'one third of a prime number greater than two and divisible by two';

(B) 'folk-tales about fairies and water-sprites'.

Both are instances of compound names in the sense above. In case (A) principle (2) is broken, so the entire name has no denotation, because the name component 'a prime number greater than two and divisible by two' has no denotation. In case (B) it is not possible to apply principle (2). The names 'fairies' and 'water-sprites' naturally have a sense, but they need not have denotation.[11] If we concede that these names have no denotation, it does not follow that the entire compound name has no denotation. Basically, we have in case (B) a situation which corresponds to indirect speech and therefore we are dealing with indirect denotation. What we have then is a class (or property) of classes (or properties), if we may use these words to express the type difference of the two components.

Another remark has still to be made which restricts the use of principles (1) and (2). This restriction is due to various forms of what was already characterised in medieval logic as *suppositio*. It was usually the practice in scholastic logic to distinguish between the use of a name in relation to the name itself (suppositio materialis) and its use in relation to some other object.[12] In this connection Carnap speaks of the 'autonymous' use of names (words, expressions of language). Let us compare the expressions:

(1) Prague is the capital of Czechoslovakia.

(2) The Prague is one of the best restaurants in Moscow.

(3) 'Prague' contains six letters.

(4) Prague won the match with Brno.

In each of these sentences the name 'Prague' is clearly being used in a different way, indicating that there are many ways in which one and the same name may be used. Two basic patterns are usually distinguished: the 'autonymous' use ([20], § 42, [23], p.4) and the use in the proper sense of name relation. A fairly detailed argument for the differentiation has been worked out by Quine [116] [122], who distinguishes between cases where names are *used* and those where a name is *mentioned* or taken into consideration. The second case may be understood also in the sense that we may be taking into account its exterior form, e.g. phonetically or graphically; we are considering a particular way in which the sense of a name has been encoded. In this respect case (3) of the examples above is quite different in principle. By contrast, cases (1), (2) and (4) represent uses of the name in the proper sense. Nothing is altered by the fact that these are three distinct names: the name of a city, the name of a restaurant in Moscow and the name or abbreviated appellation of a sporting club from Prague.

If we now return to the principles of combining names (as formulated by Church), it still has to be added that on top of the stated restrictions to the principles, the form in which a name is used must also be respected. If the form in which the individual name components are used is changed under substitution, the principles need not continue to apply.

[C] *Frege's Extension of the Semantic Characterisation of Names*

The great benefit of Frege's conception resides in the fact that he extends his method of semantic analysis of names to other categories of language. In particular this applied to *sentences*, which Frege treated as names sui generis. These ideas of Frege's were an exceptional influence on the future evolution of contemporary logical thought. Although they have been rendered more exact or otherwise modified in a number of respects, they do constitute a productive starting point for the semantic and logico-semantic analysis of other language categories.

Frege set out to show that sentences can be approached in the same way as names in the narrow sense of the word. He therefore graduated through the investigation of that which may be the sense and that which may be the denotation of different kinds of sentence. But he makes no issue of the problem of differentiating sentences, adopting the basic traditional linguistic classification of sentences into declarative, imperative etc., and speaking of compound sentences and the like.

Frege first looked at the assignment of sense and denotation in the case of declarative sentences. In the analysis of sentence the very term 'sentence' must be understood in the linguistic sense, i.e. as an expression of language. 'Sentence' must be distinguished from what traditional logic called a 'thought' and what is in modern logic called a 'proposition'.[13] Frege stresses that 'proposition' must not be identified with the subjective activity of thinking, the sequence of psychical states that may accompany the utterance, writing down or registering of a sentence.

One possible subject of investigation is whether a proposition may be the denotation of a sentence. In looking for a solution to this problem we might set out from the above-mentioned principles for compound names. One might, for example, replace one of the components in a sentence (provided it is a name) by another name component in such a way that their denotation is the same. But if we replace the name 'evening star' in the sentence 'the evening star is a body illuminated by the sun' by the name 'morning star' we obviously arrive at a different proposition. Hence a proposition cannot be a sentence's denotation. A proposition is, however, the sense of a sentence.

The question of what a sentence's denotation is is tackled by Frege in a consideration of what remains preserved when the substitution is made. That which remains unchanged is the truth value of the sentence. This idea of the preservation of the truth value is not Frege's own original idea. It is an application of Leibniz's criterion of the identity of interchangeable names *salva veritate*. If we apply principle (2) for the denotation of compound names, then with the replacement of one name component by another having the same denotation the denotation of the compound name remains the same. At the same time and by Leibniz's criterion the truth value remains preserved. Thus we are offered a possible solution: we could treat truth value as the denotation of a sentence.[14]

The conception of truth value as the denotation of a sentence is clearly to do with Frege's overall logical and philosophical conception.[15] By truth value Frege meant that a sentence is either true or untrue. Unlike more recent many-valued logics he recognised no other truth values. Moreover truth and untruth (das Wahre und das Falsche) are treated as objects sui generis. For the platonising conception of Frege these objects are objective entities that are free of any kind of psychological overlay. B.V. Birjukov, who has made a detailed interpretation of Frege's conception of sense [172], rightly points out that regardless of the possible philosophical reservations, Frege's conception of the denotation of a sentence has been of immense importance to the development of modern logic. Frege's conception can of course be freed of some of its elements of objective idealism in the manner of those logicians who have linked it to the conceptualist and nominalist traditions. The truth values 'be true' and 'be untrue' can be treated as two objects that are abstract in themselves but capture some of the semantic properties of a sentence, namely the nature of the relations with extralinguistic entities. While the platonist Frege assumes the existence of these objects, Church, for example, formulates his conception by postulating them as abstract objects. The latter conception does have certain advantages. A conception of truth values that are derived from the relation between sentences and the extralinguistic world accords with the philosophical view that truth (or untruth) is a property that does not depend on the language user, but on the relation between an expression and an objective world, whether in the sense of an 'actual world' or in the sense of 'all possible worlds'.

Frege's analysis of the sense and denotation of sentences is not just confined to declarative sentences. Separate note is taken of imperatives and interrogatives. The semantic problems of these types of sentences, along with their other problems, have only been worked over in a small measure. Frege believes that imperatives have no denotation, but that they do have sense. Therefore if we accept Frege's viewpoint, there is no point in considering the truth values of imperative sentences. A similar situation applies in the case of interrogatives. These sections of Frege's semantic analysis are not very convincing and cannot be taken as satisfactory. The fact of the matter is that he confuses certain grammatical points of view – without investigating or critically analysing the criteria for differentiating sentences, but merely accepting dogmatically their traditional classification – with points of view of logic and logical semantics.

As against Frege's fairly simple solution there are other conceptions that could be stated. One is based on the concept of context, the other disregards context.

(1) In considering imperatives and interrogatives we may take into account the fact that someone is uttering a command, wish, request, question etc. The first possible semantic analysis of sentences with this kind of content is based on the fact that we recognise the context,[16] although it need not be expressed explicitly and may be simply suppressed. In this case, sentences of this kind cannot be assessed semantically in isolation from other possible language expressions, and their semantics cannot be adjudged separate from the stated or merely hypothetically assumed context, There may be a case for constituting such a context abstractly, in the form of a statement about an actual or possible being (or other system capable of imitating this kind of behaviour in this respect) who can utter imperatives or questions.

(2) The second possibility is to make the semantic analysis of sentences in isolation, disregarding any actual or possible, stated or merely assumed context. In this case two ways are open to us: either we may give up the assignment of denotation to sentences and be content with their sense alone (the way taken by Frege himself), or we may postulate special abstract objects that would play for commands and questions a role analogous to that of denotation in declarative sentences.[17]

In deciding on the justification of both possibilities, which are equally feasible, there are certain other factors to be respected. These we believe to be primarily circumstances of a pragmatic nature, i.e. facts connected with the purpose or aim of the use of the given language. What is of primary importance is that which might be called the *basic context*. By basic context in this connection we mean the determination of the sentences and the kind of sentences that govern the nature of the given use of language. We may compare two examples:

(a) In the narration of a sequence of events certain imperatives and questions may occur. The sentences that express them figure in context with the basic sequence of sentences that are indicative in character. In this respect the semantic analysis of these sentences will benefit more if we select possibility (1), i.e. if we choose to respect the nature of the context, or reconstruct the context from the circumstances given.

(b) It may, however, be expedient to build up sentences of some particular type, e.g. imperatives, into a system. One such system is that of traffic regulations.[18] This kind of system, consisting of injunctions and prohibitions, with the possible addition of recommendations, can be formalised and also interpreted. In the semantic analysis of the various sentences (injunctions, prohibitions etc.) there is an advantage in constructing abstract objects of the type 'is required' or 'is forbidden', which may be treated as the denotation of the sentences and which may play a similar role to that of truth values.

Frege also paid some attention to 'subordinate clauses', most particularly in the context of indirect speech, indirect sense and indirect denotation. In his analysis of these problems he may be said to have run into some of the subsidiary situations that were later to be solved more generally in the theory of extensional and intensional context.

We have seen that Frege's conception of sense and denotation is far from having solved all the questions of semantics, especially those that appear on the scene along with the more complex formations of language, such as the problems of sentence context. In the formalised languages of the kind customarily used by logic and mathematics, this more complex sentence context has never come really to the fore. It is to the credit of cybernetics and its associated theoretical and technical facilities, as in connection with machine translation, informational languages, the automation of documentation and reference, that questions of the more complex sentence context have once more come into the limelight. While there is no question of the huge role played here by syntactic aspects, which have been paid a great deal of attention (in the works of Chomsky and others), there can be no overlooking the semantic aspects as well.[19]

2. THE THEORY OF DESCRIPTIONS

Frege's conception of sense and denotation is based on the semantic analysis of names, and the name relation, which is derived from proper names, is extended to all language expressions to which semantic analysis can be applied. This method assumes, at least for the more complex expressions of language (e.g. sentences, sentence contexts etc.), the postulation or construction of special abstract objects such that these objects correspond to the real objects that are the denotates of names in the narrower sense of the world. A kind of restriction may be detected here, for example because all the entities that are an object of naming many appear to be subsumed in a single group. This means that the grounds for accepting or rejecting a particular semantic conception are ontological in nature. To be more precise, it is a question of whether only certain objects should be granted a kind of privileged position in the hierarchical system of objects, the reasons for establishing any such position of privilege being of various kinds.

The most widespread reasons are those which are based on a certain conception of empiricism. This is particularly the case of Russell's theory of descriptions, which admits only a fairly narrow conception of names; they are only the names of objects that are immediately accessible through sensory data. Unlike Frege's conception, which analyses as names (in the broad sense of the world) all expressions of language that may have sense and denotation, Russell distinguishes (1) *names*, which are simple signs denoting individuals, and (2) *descriptions*, which may consist of a number of signs whose meaning is fixed; the 'meaning'[20] of the whole description comes as a result of the combination of those signs. In contrast, a name has meaning independently of the meaning of other words. Another important characteristic of Russell's theory of descriptions is the fact that both names and descriptions are seen merely as fragments that can only achieve validity as components of a sentence. In an appraisal of this theory it must always be borne in mind that:

(a) there is the tacit assumption of that privileged class of objects (individuals) which may have a name (Russell moreover never quite excludes the possibility of expressing all human empirical knowledge in a language that contains no names proper);

(b) it is taken as a matter of course that the basic language expression by which any communication can be made is the sentence.

In natural language it is understandably difficult to draw a clear boundary between names in the proper sense of the world (proper names, individual names etc.) and the names of classes of objects, expressing a general or common property of all the objects of a given class. Hence it is also difficult to divide things unequivocally into names and descriptions in the conception put forward by Russell.

[*A*] *The Concept of 'Denoting Phrase'*

Russell formulated his theory of descriptions primarily in connection with attempts to give greater precision to the foundations of his system of mathematical logic. But the origins of the theory of descriptions predate the writing of his fundamental work, the *Principia Mathematica*. The most important ideas of the theory of descriptions are to be found in the earlier study *On Denoting* [131] of 1905, although the terminology used there is rather different. It was here that Russell analysed the role of what he called the 'denoting phrase'.

The basic idea of Russell's semantics, and one which also marks the theory of descriptions, is the notion that a denoting phrase does not have sense in its own right, but only as a component of the sentence in which it figures. The denoting phrase then has a substantial share in determining the meaning of the sentence. In deciding on the sense and denotation of a denoting phrase it is therefore necessary to take into account the fact that it relates to something.

There are three cases to be distinguished: (1) a phrase is a denoting phrase but does not denote anything (e.g. 'a city in Africa that lies to the north of Prague'); (2) a phrase denotes a certain unique object ('the highest place in Czechoslovakia'); (3) a phrase denotes an unspecified object out of a defined class of objects (e.g. 'a town in Czechoslovakia').

Russell analyses Frege's distinction of sense and denotation. He applies the distinction to the relation between a denoting phrase and the objects which together with the denoting phrase may be components of a sentence. The benefits of the distinction of sense and denotation are demonstrated by Russell on three problems: (a) the problem of identity (of individuals), (b) the problem of tertium non datur, and (c) the problem of the difference of two or more individuals.

(a) If x is identical with y, then anything true that can be said of x may be said with the same truth of y. Therefore x may be replaced by y *salva veritate*. But if (the example is Russell's) George IV wanted to know whether Scott was the author of Waverley, and Scott *was* the author of Waverley, was the king asking whether Scott was Scott? In this connection Russell makes the ironic remark that such interest in the principle of identity can scarcely be ascribed to "the first gentleman in Europe".

(b) On the basis of the principle of tertium non datur, the formula 'either x is P or x is not P' is true. But if for x we insert 'the present king of France' and for P 'bald', it is either true that 'the present king of France is bald' or it is true that 'the present king of France is not bald'. If we then proceed to name all the objects that are bald and all those that are not bald, we will find in neither category the present king of France.

(c) Both cases can be resolved by distinguishing sense and denotation. In an analysis of (b) we have to take zero denotation into account. The same applies in the problem of the difference of two or more individuals. Take for example the sentence 'x is different from y'. If the sentence is true we can say 'there is a difference between x and y'. If on the contrary the sentence is not true, we could conclude that 'there is no difference between *x* and *y*'. If we were not to distinguish sense and denotation, and if we were to admit that anything that may be the object of a denoting phrase is an object,[21] we would then have to admit the existence of non-existent objects.

Therefore it becomes a matter of principle to distinguish situations where a denoting phrase denotes an empty class (or has zero denotation), as in the case of the expression 'the present king of France', from situations where the denotation is non-zero. It is these latter situations, where a denoting phrase relates to a 'real object' or to some possible object from a class of real objects,[22] that are the actual basis of Russell's theory of descriptions.

[*B*] *The Differentiation of Descriptions*

Russell evolved his theory of descriptions in the proper sense of the word in connection with the analysis of the concept of statement function.[23] A description is therefore a denoting phrase which is employed as a constituent of a statement function. Russell distinguishes two kinds of descriptions: *definite descriptions* and *indefinite* or *ambiguous descriptions*. It is interesting to note that although the theory of descriptions proper was applied primarily in a system of mathematical logic, it was undoubtedly motivated by certain factors taken over from natural language. These include first and foremost the differentiation of the definite article ('the' in English, 'der', 'die', 'das' in German, etc.) and the indefinite article (English 'a' German 'ein', 'eine' etc.). A definite description corresponds to the use of a denoting phrase that is associated with the definite article, in those languages that have a definite article, and an indefinite description corresponds to uses of a denoting phrase with an indefinite article.[24]

The concept of 'description' in Russell's conception is the basis for the constitution of the concept of statement function. Formally speaking, a statement function arises from the concatenation of a description and a variable. The difference between a definite description and an indefinite description resides in the manner in which the variables are bound in statements made up of statement functions. According to Russell's conception a statement containing a definite description is always associated with uniqueness. The uniqueness of an object is therefore the only substantial difference between an indefinite and definite description. A statement about 'the so-and-so' always evokes the idea of the corresponding statement about 'a so-and-so', with the additional notion that there is only one 'so-and-so'. In the statement 'x is the son of XY', the phrase 'is the son of XY' is a definite description. If we concatenate an individual name with this definite description, we make the definite description apply to a unique individual, but only, of course, if the statement so formed is true. Thus we are entitled to call the theory of descriptions conceived in this way a theory of individual descriptions. If we take Russell's example, 'Walter Scott is the author of Waverley', we must first assume that the book 'Waverley' was in fact written. If 'Waverley' had never been written, the statement connot be true. If the statement is true, then the statement function 'x is the author of Waverley' can only be satisfied by a single individual.

[*C*] *Individual Descriptions and Statement Functions*

A description (individual description) assumes therefore a statement function φx[25] which is satisfied by only one value of x and by no other values. Then we can define 'a term which satisfies φx exists'. The definition, if expressed by means of natural language, runs:

(D_1 – V 2 C) 'A term which satisfies the function φx exists' means 'There exists a term a such that φx is always equivalent to "x is a"'.[26]

In the definiendum of this definition we have the expression 'exists'. Since this existence must be assumed to be demonstrable, confirmable, it is also possible to express the definition in more general terms:

(D_2 – V 2 C) 'A term which satisfies the function φx satisfies ψx' means: 'There exists a a such that (1) φx is always equivalent to "x is a", and (2) φx is true'.

Russell does not define the expression '$(\iota x)(\varphi x)$', i.e. the individual description; what he is actually defining is a statement in which the individual description occurs.

For an appreciation of the role of an individual description the interpretation of the concept of existence, expressed in Russell by the symbol 'E!' is important. If '$(\iota x)(\varphi x)$' does not exist, it is not possible to assign to it any property meaningfully. The expression 'the present king of France' is a 'denoting phrase', but it denotes no one, hence it is not possible to assign to this (non-existent) object either the property 'be bald' or the property 'not be bald'. So if $(\iota x)(\psi x)$' has any property, or, to put it another way, if '$\psi(\iota x)(\varphi x)$' is true, then '$(\iota x)(\varphi x)$' must exist.

The concept of 'existence' in this sense (i.e. E!) also has to be interpreted in connection with Russell's differentiation of two concepts of cognition. In [131] he pointed out the fundamental difference between 'acquaintance' and 'knowledge'. We may speak of acquaintance if we are in contact with objects that are being described by means of denoting phrases. But objects that we reconstruct solely on the basis of a denoting phrase need not necessarily be connected with 'acquaintance'. 'Acquaintance' need not of course be interpreted in a narrowly empirical spirit, although that kind of interpretation is not far from Russell's conception. Russell himself, unlike certain of his adherents and followers, did not interpret 'acquaintance' quite so narrowly and did admit the possibility of some mediation. For example, he stressed that there was no reason for excluding the possibility that other people's thinking is inaccessible in principle for our acquaintance. The only important thing is that all our thinking springs from the foundation of that which is accessible on the basis of 'acquaintance'. Thus the concept of existence (i.e. E!), which corresponds to that of 'acquaintance', might be better conveyed by the term 'accessible existence'.

If then we say of anything that it 'exists' (E!), this attribute can only apply to something which may be denoted by a denoting phrase that is a definite description. In other words, the existence (E!) of an object $(\iota x)(\varphi x)$ can be inferred from a true statement containing $(\iota x)(\varphi x)$, provided $(\iota x)(\varphi x)$ is a definite description.

The connection between the conceptions of 'existence' and individual descriptions, whereby the existence of $(\iota x)(\varphi x)$ follows from the truth of $\psi(\iota x)(\varphi x)$, might appear to justify Descartes' well-known inference 'cogito, ergo sum'. However, a deeper analysis of this apparent inference shows that the concept of 'existence' cannot follow from the truth of a statement containing a definite description. First of all it is the case that 'sum' does not carry the sense of 'accessible existence'. So even if the antecedent contained a true statement with a definite description (which is equally controversial), it is impossible to infer from it 'esse', i.e. existence in the metaphysical or ontological sense. Another reason why this is only an apparent inference resides in the fact that 'cogito' is not a statement containing a denoting phrase which is a definite description. In other words, the fact or experience of one's own psyché does not automatically imply its 'accessibility', or the accessibility of the subject of the psychic activity to other subjects. It is quite a different situation, however, if the psychic activity is objectivised and is objectively accessible. If we say 'person XY thought about this and that and wrote, spoke etc. about it', then if the sentence is true, it follows that XY existed.[27]

This handful of positive features of the theory of descriptions as evolved by Russell in no way alter the fact that from the semantic point of view (and unquestionably also from the gnoseological and methodological points of view) the theory of descriptions leads on to a number of serious difficulties. We have already shown that this theory does in the end assume a class of privileged unique objects which are accessible in terms of what is expressed by the concept 'acquaintance'. Although this assumption is not stated explicitly and can only be inferred in Russell's case, it is actually hardly tenable anyway from the point of view of the present level of science. The concept of an 'accessible object' is relative and it is also governed by the interaction of an object (independent of the observer) and the 'observer channel', the latter being deemed to embrace the empirical, experimental and theoretical equipment of anyone who happens in a given context to be the subject of cognition.

Another reservation concerns what have been called non-existent objects, the names of which, as implied by Russell's conception, have no denotation. Russell has shown that the object called 'the present king of France' cannot have assigned to it meaningfully either the property 'bald' or its negation. In other words, if we decompose all objects into the 'bald' and the 'non-bald', an object designated as the 'present king of France' will not be included in either of the two classes. Nevertheless we may still ask: Is it genuinely beyond the realms of possibility meaningfully to assign properties to non-existent (in Russell's sense) objects? The question cannot be answered in the negative in all circumstances.

To survey the question we may take some of the examples of non-existent objects (in Russell's sense) that are known from mythology. One such object is the 'centaur'. We know that this non-existent creature was credited with many properties in Greek mythology. We cannot therefore say that it has no properties, but that it has some of the properties of man, some of the horse, etc. Not that Russell's theory of descriptions felt helpless in the face of such examples. Of the two possible solutions the first at least is at one with Russell's conception.

(1) The first solution is based on the fact that expressions of the form $\psi(\iota x)(\varphi x)$ (e.g. 'the present king of France is bald', 'the centaur has some properties of man and some of the horse' etc.) or of the form E! $(\iota x)(\varphi x)$,

are not taken in isolation, but as components of statements of the form $f[\psi(\iota x)(\varphi x)]$ (e.g. 'I have heard that ...', 'the Greek legends say that ...'). In cases of this kind Russell speaks not of 'existence', but of 'secondary occurrence'. If, for example, φx is 'x is the present king of France', Russell insists that

$$\sim[\mathrm{E}!(\iota x)(\varphi x)]$$

or

$$\sim[(\exists a)\cdot a=(\iota x)(\varphi x)],$$

in otherwords, he insists on the non-existence of anything that might be denoted 'the present king of France'. Thus in no circumstances can we admit the truth of a statement that includes as a component the denoting phrase $(\iota x)(\varphi x)$ if $\sim[\mathrm{E}!(\iota x)(\varphi x)]$. On the other hand we can, from the truth of the statement $f[\psi(\iota x)(\varphi x)]$, say something about the secondary occurrence of $(\iota x)(\varphi x)$, though not about the existence of $(\iota x)(\varphi x)$.

While this differentiation of 'existence' (tied in Russell's conception to the concept of 'acquaintance') and 'secondary occurrence' may appear to be quite in order formally, there are some serious objections to it. If we applied the differentiation consistently, we should have to conclude that we would be unjustified in considering as existing objects those of which we know only from the testimony of others. This would make the historian's propositions about the existence of most historical events practically inadmissible. But even in the experimental sciences we might find ourselves in situations that would be hard to explain. For example, if an experimenter makes his calculations using the findings of his colleagues as well, is he exposing himself a priori to a greater risk than if he were to operate solely with his own results? In other words, do one's own experimental results necessarily enjoy some privileged status? Experience tells us that it is wise to be at least as wary of our own results as of those of others. This means that we may consider existence in the objective sense (not in the sense of Russell's 'E!') both on the basis of 'acquaintance' and on the basis of secondary occurrence, but in both cases the considerations are relative, being conditioned by many other factors, such as the equipment of the observer or experimenter, the reliability of the 'observer channel' etc.

(2) The second solution is based on the fact that we are dealing with indirect speech, which, from the logical point of view, is not an extensional context. This means that the truth value of the expressions 'I think that S', 'it is possible that S', 'XY believes that S' and the like does not depend

on the truth value of *S*. This second solution opens up a number of partial alternative solutions, but these will be left until the more detailed account of the method of extension and intension.

Russell's theory of descriptions can in no way be treated as being independent of the gnoseological point of view. In a nut-shell this viewpoint may be expressed in the principle that all knowledge is based on acquaintance – the principle of the reducibility of knowledge to acquaintance. This may also be expressed by saying that it is assumed as a principle that all sentences can be translated into sentences (Russell has 'propositions') containing individual descriptions. The ultimate phase of this kind of translation are sentences that contain purely demonstrative signs. L.S. Stebbing, who has given a critical survey of Russell's theory of descriptions [143, pp. 144–158, 502–505], describes these as 'pictorial sentences'. This means that Russell assumes that any sentence may in the end be translated into this kind of pictorial sentence. Corresponding to this is what Russell described as the 'basic principle in the analysis of propositions containing descriptions': 'Any proposition which we are capable of understanding must be entirely composed of constituents with which we are acquainted' [132, p. 59].[28] While Russell in his later works variously altered the formulation of his gnoseological viewpoint, he basically remained faithful to it.

This principle, which is only a very one-sided conception of empiricism and something of a linguistic and logical analogy of mechanism in the natural sciences, is hardly compatible with the contemporary state of science. Just as it is not possible to recognise any constituent elements or unchanging building blocks in the hall of knowledge of the natural sciences, it is impossible to concede that we should operate with such basic building blocks in the realm of language and cognition. These are the main reasons why the theory of descriptions, especially when associated with this kind of gnoseological conception, cannot be treated as a particularly useful instrument for the analysis of the name relation.[29]

3. THE METHOD OF EXTENSION AND INTENSION

Frege's differentiation between sense and denotation is a semantic conception that enables us to resolve some of the difficulties connected with the identity of names. The differentiation is then applied in the semantic analysis of sentences with the difference that abstract objects have to be constructed to correspond to the sense and denotation of a name. The method of intension and extension, which was evolved primarily by Carnap [23] but is also suggested, though in a rather different form, in certain of the works of C.I. Lewis, is a kind of expansion of Frege's conception to take in all expressions of language for which relatively independent semantic analysis is justified. The method of extension and intension is fundamentally related to Frege's approach: 'extension' corresponds to 'denotation', and 'intension' to 'sense'. As with Frege's, this too is basically a logical method which disregards any psychological connections.

Unlike Frege's approach, the method of extension and intension assumes that for all expressions of language where semantic analysis is justified it is possible to construct abstract objects corresponding to the concept of 'proposition'. Carnap uses in this connection the term 'concept', which is a constructed or postulated abstract object having certain properties. Therefore the term 'concept' is a common denominator not only for what we call in everyday speech a 'property' or 'relation', but also for that which corresponds to them in the case of individuals (Carnap here uses the term 'individual concept'.)

While Frege's conception sets out from certain intuitive ideas on the role of a *name* and on the possible generalisation of the name relation, the method of extension and intension is based, intuitively speaking, on the role of *adjectives*. For many adjectives of natural language are simultaneously the name of what we call a 'property' and the name of a class. If we say that 'Socrates is mortal', we understand it to mean both that to the person called 'Socrates' we are attributing a particular property, and that the individual called 'Socrates' is being assigned to the class of individuals that are mortal.[30]

[*A*] *The Concepts of 'Class' and 'Property'*

The analysis of adjectives does not obviously coincide semantically with everything covered by the vague term *property*. If I say 'Socrates is mortal', I am ascribing to the individual called 'Socrates' a particular property. If I say 'Socrates is a man', then while it might seem at first sight that the attention is focussed on the assignment of the individual called 'Socrates' to the class of 'men', in fact the difference of form between the two statements may be semantically quite insignificant. In both cases it is possible that the main concern is the assignment of some property or other to some person.

From the point of view of the formalised language that we have constructed, this assignment of a property to an individual is made by the concatenation of a predicate and the individual, or, in the case of multiplace predicates, the relevant number of ordered n-tuples of individuals. If Socrates has the property 'be a man', this also means that he is an element of the class that is determined by the property 'be a man'. This in turn means that we can transfer from the assignment of a property to an individual or ordered n-tuples of individuals to the relation 'be an element of a class'. All the individuals that have the property P may be brought together in the class that is made up of all the individuals that have the property P. Thus for each property there is a corresponding class, which may even be an empty class.[31] Since an n-place predicate can be understood as a property of ordered n-tuples, this property also constitutes a class, of which the elements are all the ordered n-tuples that have the given property.

A class made up of all the individuals that have the property $P^{(1)}$ will be written down in the notation as $(\hat{x})Px$; the class that consists of all the ordered pairs that have the property $P^{(2)}$ will be written as $(\hat{x})(\hat{y})Pxy$, and so forth.

This procedure also demonstrates that predicates may be interpreted in principle in two ways: as expressions of properties (in the general sense) and as expressions for classes.

The same or fundamentally similar results are reached if we use the procedure suggested by Church [35]. His procedure is based on the operator of abstraction, the starting point being a well formed formula containing a free variable. If '... x ...' is one such formula with a free variable, then by binding that free variable by the operator of abstraction (Church called the procedure λ-conversion) we obtain the abstractum '$(\lambda x)(\ldots x \ldots)$'. In a sense this is a more general procedure in that it starts from well formed formulae of a given $\mathscr{L}$, and not from predicates as in the previously outlined method. Here too it can be said that expressions of the kind $(\lambda x)(\ldots x \ldots)$ are in principle interpretable in two ways – as expressions for classes and as expressions for properties.[32]

In natural language the interpretation of these and similar expressions usually depends on the context, or also on the grammatical form, and from these it can be understood whether the interpretation is in the sense of property or in the sense of class. In formalised languages there are three possible procedures: (a) in translating into the metalanguage expressions of the object language that are to be interpreted either as properties or as classes, some specific indication may be given as to which interpretation is meant; (b) expressions for properties and classes may be actually distinguished in the object language (this is effectively the method employed in the *Principia Mathematica* system by the differentiation of $\varphi\hat{x}$ and $\hat{x}(\varphi x)$); (c) a so-called neutral metalanguage may be introduced which eliminates the differentiation of classes and properties. Carnap analyses methods (a) and (c), preferring method (c), which is a kind of semantic analogy to Occam's razor ('entia non sunt multiplicanda', here: 'It is better not to duplicate interpretations').

If we proceed by method (a), some indication must be given of the connection in which the difference between the two interpretations figures as being important, i.e. the difference between interpreting the expression $(\lambda x)(\dots x \dots)$ as a property and interpreting it as a class. The main difference lies in the conditions of identity. This means that the motivation for differentiating the two interpretations is similar to that for the differentiation of sense and denotation in Frege's conception.

It is intuitively obvious that two classes may be treated as identicial if they contain the same elements.[33] The problem of the identity of properties is rather more complex and not without certain difficulties. First of all it must be appreciated that the term 'property' is ambiguous, which calls for a clear statement of the sense in which it is being used. What we have in mind in the present account is the concept of '*property in the semantic sense*', i.e. a particular circumstance, quality or complex of qualities that may be expressed by a particular name, irrespective of whether the circumstance or quality have been ascertained empirically or theoretically, or whether they have only been postulated (in the light of which there are no substantial differences between the properties expressed by the terms 'be red', 'have the value *n*', 'be almighty' etc.), irrespective of whether they are expressed by a single sign or a sequence of signs, and irrespective of whether they are assignable to real objects or some other constructive objects.

Therefore the concept of a property in the semantic sense cannot be identified with other explications of the term. Some of the other explications are:

(a) the traditional view, whereby 'property' is distinguished as something that can be expressed by a one-place predicate from 'relation' as something that is usually expressed by a multi-place predicate.[34] From the semantic point of view there are no grounds for raising a barrier between the two types of expressions and not treating both as properties.

Moreover, we may have properties of individuals (e.g. the property 'be green'), ordered pairs (e.g. 'be more northerly than'), or ordered n-tuples generally.

(b) the empirical view which usually treats as a property that which can be said of an object investigated in the light of given empirical or experimental equipment. We naturally take a different view of what a property is when we simply apply the senses as opposed to when we make observations under an electron microscope or using a Wilson chamber. The main thing here is what we might call *delicacy of discrimination*. This situation can be illustrated by the so-called communication model, which assumes that observation (or experimentation) takes place through a particular channel (the 'observer channel') which itself can be characterised by certain properties (capacities, memories, delays). Under this view it is important to have a precise idea of the concept of identical properties, which depends on the facilities for guessing the state of the input on the basis of data at the output of the channel.

(c) the purely abstract view of the term 'property'. We can, for instance, speak of properties of the concept of information (e.g. additivity) without it being possible to assume that properties of this kind can be investigated empirically or experimentally. Here too it is of course possible to assume at least some theoretical equipment in the person operating with the term 'property' in this sense and being capable of meaningfully assigning this kind of property to certain constructive objects.

If we take any of the above cases (or possibly other properties, which might be the creations of what we call imagination or fantasy) and express them by a predicate (whether one- or multi-place), we know them as properties in the semantic sense if we understand the predicate used to denote the particular property. This intuitive characterisation does of course call for some refinement. For the purposes of giving some refinement to it we may consider the question of the identity of properties.

In order to tackle this question we must first start from the problem of the identity of classes. We have said that we treat classes as identical if they contain the same elements. But since a class is constituted by a particular predicate, it holds that:

$$(\forall x)\{[x \in (\hat{y})Py \equiv x \in (\hat{y})Qy] \longrightarrow (\hat{y})Py = (\hat{y})Qy\}$$

This principle[35] determines the conditions for the identity of classes and is based on the equivalence of what constitutes a class. Both classes, i.e. $(\hat{y})Py$ and $(\hat{y})Qy$, may be determined by different properties, but if both classes have the same elements they may be considered identical. In other words, it is not necessarily identity of properties that is a condition for the identity of classes (e.g. not the identity of the properties determined by the expressions 'man' and 'rational being capable of manufacturing tools and using language for communication'), but the fact that the properties determine classes with the same elements.

We may say then that classes are identical if and only if the expressions[36] that determine the classes are equivalent. We may also speak of the equivalence of expressions which *denote* identical classes.

The conditions for the identity of properties are more strict. Properties (in the semantic sense) are identical if and only if the expressions which *signify*[37] the properties are L-equivalent. This in turn corresponds to the intuitive requirement that the relation between properties should be demonstrable in the language itself, without recourse to extralinguistic devices.

These conditions for the identity of classes and the identity of properties (in the semantic sense) now make it possible to introduce the concepts of extension and intension. If two (or more) predicates relate to the same individuals, or, to put it another way, if they are equivalent, we may say that they have the same extension. This means that two (or more) predicates denote the same classes.

As for the concept of intension as corresponding to the concept of property (in the semantic sense), it is expedient to assume that two properties are identical if their identity can be demonstrated by logical means, disregarding non-logical factors. Hence two predicates signify the same property, i.e. have the same intension, if they are L-equivalent.

In interpreting the concepts of 'extension' and 'intension' with respect to predicates there are two possible lines of approach. Either we are content with determining what it means for two predicates to have the same extension and intension, or on top of that we state what we understand by extension and intension with respect to the predicates. In the former case we can disregard the association of 'extension' and 'intension' with 'class' and 'property', while in the latter we can give a more concrete conception of 'extension' and 'intension' with respect to the predicates, or abstracta, and so arrive at the concepts of 'class' and 'property'; in this way these are of course constructive objects that are the fruit of the semantic analysis of the expressions.

If we stay with the first procedure, two conventions are sufficient for introducing the concepts 'have the same extension' and 'have the same intension':

(C_1 – V 3 A) Two predicates have the same extension if and only if they are equivalent.
(C_2 – V 3 A) Two predicates have the same intension if and only if they are L-equivalent.

If we adhere to the second method, we can add to the two conventions that, where predicates are concerned, we understand by the extension of a predicate the class which the particular predicate denotes, and by the intension of a predicate the property which a given predicate signifies.

[*B*] *The Concepts of 'Extension' And 'Intension'*

The procedure here suggested for predicates and abstracta may be broadened to take in other expressions of language. Carnap calls these expressions designators and includes among them all expressions of language for which semantic analysis in the sense suggested can be applied. What are meant are expressions for individuals and sentences.

The basic assumption is that, from the semantic point of view, all these expressions can be characterised by what corresponds to property on the one hand (in logic it has been customary to use the term 'concept' to denote the abstract objects which by their nature correspond to the concept of property in the semantic sense), and class on the other. This assumption is analogous to that of Frege, in whose conception the intuitive starting point is the concept of name, with the further assumption that even sentences are names sui generis. It has to be said that neither Frege nor Carnap prove or emphasise any such assumption. Thus we ought to pause over the assumption and decide whether it is justified.

Language forms a hierarchically ordered structure, the different kinds of language expression being somehow – depending on the nature of the language – categorisable. The idea of a hierarchically ordered structure the elements of which can be incorporated into the different levels does not of course mean that if, syntactically, the elements of one level are made up of elements of another (lower) level, the semantic properties of the elements of one level are also made up out of the semantic properties of elements from a lower level.

Thus it appears distinctly unwise to take a pattern of semantic analysis that is adequate for elements of one level and use it unaltered in any way for the analysis of elements of some other level. In this sense there is justification for the scepticism voiced against a procedure by which the semantic analysis of one type of expression (in Frege 'name', and in Carnap an expression corresponding to one- or multi-place predicates) is considered the sole possible starting point. Such scepticism is fully warranted if the semantic analysis is pursued in the manner of naive realism – roughly along the lines of 'what corresponds to…'– where that which corresponds to all kinds of words and also to sentences and any other kinds of expressions is always understood to be an empirically apprehensible entity.

On the other hand, and here we must see a certain justification for Frege's and Carnap's assumptions, we are capable of *understanding independently* the language expressions of a certain category given certain habits of language use or certain rules of language. This means that in the case of such language expressions we are capable of assigning to them a *relatively independent meaning.* In harmony with most modern logicians, Carnap stresses that only sentences, and one kind – declarative, indicative – sentences at that, have an independent meaning in the strict sense of the word. Other expressions, e.g. predicates, abstracta, individual expressions etc., derive their meaning from the manner in which they share in determining the meanings of the sentences in which they occur. This, and any other similar, viewpoint cannot be taken as absolute, since situations occur frequently in the use of natural language sentences where a sentence's meaning cannot be defined from the sentence alone, some recognition having to be paid to the context, however delimited. Hence even with sentences (at least in natural language) the meaning has to be assumed to be only relatively independent. Moreover the criterion of its independence is basically pragmatic, i.e. it defies semantic analysis as such.

However, since the degree of independence may vary, the boundaries of the expressions to which we are justified – with respect to the aims of the particular use of language – in assigning a meaning (i.e. both denotation and sense), may also vary. This also means that language has a characteristic dual hierarchy: (a) on the one hand the hierarchy of the different kinds of language expressions that can be categorised into the different hierarchically ordered levels, (b) the hierarchy of the points of view from which analysis of different expressions is made. It can be said on the basis of hierarchy (b) that the degree of independence of semantic analysis depends basically on pragmatic considerations and that the semantic points of view are accordingly subordinate to pragmatic points of view.

Therefore insofar as the following account considers the relatively independent semantic analysis of individual expressions, predicates, abstracta and sentences, the decision to do so will depend on the nature of the language we have chosen, but also on the chosen aim, i.e. the task of justifying the possibility of analysing the expressions in a given language.

In the next section we shall also show that methods used in the semantic analysis of these kinds of expressions can also be used in the analysis of certain kinds of sentence contexts.

Thus far we have considered the concepts of 'same extension' and 'same intension' in predicates and abstracta. The procedure followed can be generalised to cover all expressions where we consider relatively independent semantic analysis to be justified. We can then tender the following definitions:

(D_1 – V 3 B) Δ_i and Δ_j have the same extension (in $\mathscr{L}$) = df Δ_i and Δ_j are equivalent.

(D_2 – V 3 B) Δ_i and Δ_j have the same intension (in $\mathscr{L}$) = df Δ_i and Δ_j are L-equivalent.

As in the case of predicates, it is expedient to go beyond being merely content with defining 'have the same extension' and 'have the same intension', but to specify the concepts of 'extension' and 'intension' for the different individual categories of language expressions that may have relatively independent meaning.

In the case of *sentences* the specification of extension and intension is analogous to the definition of sense and denotation. Two sentences are equivalent if they have the same truth value. It is therefore useful to treat what two equivalent sentences have in common, i.e. their *truth value*, as their *extension*.

In deciding on the intension of a sentence a similar line may be followed, i.e. we may consider what they have in common if they are L-equivalent. Intuitively speaking, two sentences are L-equivalent if they signify the same thing, and that which they signify is usually characterised as a proposition or judgement. Thus we may take as the *intension of a sentence* the (corresponding) *proposition*.

As with 'property in the semantic sense', 'proposition', as we hinted in the analysis of Frege's conception, is a special kind of concept. This means that it is not a psychological category or an expression for a particular subjective experience, but a kind of abstract entity.

This conception of 'proposition' is not universally recognised in the literature of logic and logical semantics. There are many controversial questions here that are very like those of psychologism and logicism in traditional logic. One of the most difficult questions is the understanding of propositions in the case of untrue sentences. The problem is roughly this: If propositions are entities of an objectively logical nature, or, as it has also sometimes been put, if propositions represent certain 'facts', is it possible to admit of the 'existence of negative facts'? That is how the question was put by Russell in particular, and he finds the recognition of the existence of negative facts inadmissible. For these reasons Russell interprets the sense of a sentence in terms of subjective psychology and therefore also treats what is here expressed by the term 'proposition' as a subjective category (Russell's terminology itself does not have 'proposition' in this sense). In Russell sentences have the same sense if they express the same 'belief', belief being a state of the language user's awareness as applied in a certain activity or in relation to certain objects or acts outside the language user. This also means that the belief must be objectively manifested not only in language behaviour, but also in other forms of objectively apprehensible behaviour. In addition, Russell finds it expedient to link truth and untruth not with sentence, but with belief.

So Russell sees no alternative for resolving the problem of untrue sentences than to link their sense with belief, i.e. to give basically a subjective psychological interpretation. This is of course the sole possibility insofar as we see as the alternative the identification of propositions with 'facts' (the case of, for example, C.J. Ducasse) or with a way of 'making sure about the existence of the facts'.

One way in which to satisfy ourselves that these are not the only two possibilities is to compare a number of questions which (variously) bring sentences into relations that can by no means be generally identified.

(1) What corresponds to an untrue sentence if it is certain facts that corresponds to a true sentence?

(2) What does it mean to understand a sentence? (a) To know what facts correspond to it? (b) To know whether it is true? (c) To know in what circumstances or under what conditions it can be true?

(3) What does it mean to believe an untrue sentence? (This need not be the same thing as believing that a particular sentence is untrue.)

The first question, which operates with the ambiguous term 'corresponds', is unclear until it has been defined in 'what world' this or that corresponds to the sentence. In principle there are three distinct areas potentially involved: (α) the domain of objects that are accessible to the language user through objective means, (β) the domain of objects of an abstract, conceptual, nature which can be constructed or postulated with respect to the elements or expressions of the given language to which relatively independent meanings can be assigned, (γ) the domain of the language user's psychological processes or experiences. Thus we can say that question (1) is itself semantically unclear and that if facts or events in sense (α) can be assigned to true sentences, this does not imply at all that facts or events in senses (β) and (γ) cannot be assigned to untrue sentences.

For the second question, none of the suggested answers can be treated as fully or universally satisfactory. We are capable of understanding the sentence 'the universe is infinite' without knowing what corresponds to it, whether or not it is true or under what circumstances it can be true.[38] The term 'understand' in natural language has many aspects, but only some of them assume the application of methods of semantic analysis. Interpretation of the term 'understand' must usually require a 'someone' who is capable of understanding, i.e. a certain language user, and certain tasks in which the language is used. Semantically speaking, what stands at the focus of attention is undoubtedly that which corresponds to the concept of the 'sense' or 'intension' of the sentence. We understand a sentence if we recognise that property of the sentence that remains intact under any correct translation of the sentence. That property, which is similar in nature to the analogous property of predicates, is of course an empirical property. It is an abstract property that is the outcome of the application of a particular theoretical viewpoint, in the present case a semantic viewpoint. These are the reasons why it is better to treat the proposition which we consider the sentence's intension not as a subjective psychological category (like Russell's 'belief'), but as a postulated or constructed object conceptual in nature.

The third question concerns the so-called belief sentences, which are a complex subject that still awaits a uniform solution. Whatever solution may be found, it should be appreciated that it can have no major effect on the conception of propositions as supported here.

The extension and intension of *expressions for individuals* can be established in two ways. Either all such expressions can be treated as names, which matches Frege's conception, or a distinction can be made between proper names in the narrow sense and individual descriptions in Russell's sense. If Q_1 is a one-place predicate for the property 'be the largest city in Bohemia' and p a constant for 'Prague', it is obvious that

$$(\iota x)(Q_1 x) = p$$

By Frege's conception both $(\iota x)(Q_1 x)$ and p are names. With a description, however, it is possible to separate the description from what is described (the descriptum). By contrast, it is not usual to make this separation in the case of a proper name. In natural language appeal can be made to the fact that when a proper name is used, ostensive definition or other like procedures offer themselves, making the establishment of the descriptum by means of language redundant.

This also means that the relation of an expression for an individual to a particular denotate need not be of the same nature in every instance; sometimes it may be a direct relation, and at other times it may be more or less mediated. Also the reasons for an individual name being born may differ considerably: it is one thing to name a new-born child, another to name a newly discovered elementary particle. Thus in natural language it is hardly possible to draw a clear boundary between individual descriptions and names, and there is here the additional reason that some individual names arise on the basis of descriptions. A totally unequivocal differentiation between individual names and individual descriptions is therefore only possible in formalised languages, where the distinction is actually set up at the time when the vocabulary of a formalised language is being defined.

The individual descriptions 'the largest city in Bohemia', 'the capital of Czechoslovakia', 'the largest town on the Moldau' and the (proper) name 'Prague' are equivalent. All these expressions, whether individual descriptions or individual names, denote the same individual. Hence as the extension of all these expressions we may take the corresponding individual denoted by a given expression. Therefore with respect to individual descriptions or individual names an individual has the same standing as a class with respect to predicates. Here too we actually have a class, but one containing exactly one element.

For the intension of expressions for individuals the procedure to follow may be as in the preceding cases: the postulation of abstract objects of a conceptual nature.[39]

The possibility of assigning extension and intension to predicates and expressions for individuals means that any sentence containing at least one predicate and at least one expression for an individual may be translated into a metalanguage in at least four ways. If we have the sentence $P_1 p$, we can translate it into a metalanguage (which is a part of the English language) as:

(1) 'the individual Prague is an element of the class of all the places that are in Bohemia';

(2) 'the individual concept Prague has the property that it is in Bohemia';

(3) 'the individual Prague has the property that it is in Bohemia';

(4) 'the individual concept Prague is an element of the class of all the places that are in Bohemia'.

[*C*] *Extensional and Intensional Context*

Even in Frege's conception a part is played, as we have suggested, by context, especially in connection with so-called indirect speech and partially in connection with certain kinds of sentences in which a particular kind of context can be assumed; although not stated explicitly, the context may be reconstructed.

The method of extension and intension provides some advantageous facilities for the analysis of context. It cannot afford an exhaustive analysis of context in natural language, for besides the syntactic and grammatical aspects of context, there are here many other important considerations, primarily of the kind that come within the pragmatic level of language analysis and the analysis of communication. It is important to realise that the concepts of 'extension' and 'intension' were given their precise meanings on the basis of Leibniz's criterion of identity, which operates with the concept of substitutability, while the concepts 'have the same extension' and 'have the same intension' have been defined by means of the concepts of equivalence and L-equivalence.

In the semantic analysis of context by means of the method of extension and intension it should also be realised that we are considering the meaning of a language expression to which a relatively independent meaning can be assigned not just in isolation, but with respect to its concatenation with another language expression to which a relatively independent meaning can likewise be assigned. Thus the concept of (*contextual*) *concatenation* corresponds to that of the concatenation of signs, with the difference that components in (contextual) concatenation always necessarily have to be assumed to be language expressions to which a relatively independent meaning can be assigned. The concept of contextual concatenation will be denoted by the symbol $\overset{C}{\frown}$, e.g. $\Delta_i \overset{C}{\frown} \Delta_j$.

If we replace the expression Δ_j in the expression $\Delta_i \overset{C}{\frown} \Delta_j$ by the expression Δ'_j, which is equivalent to Δ_j, we speak of its *substitutability* with respect to Δ_i; if we replace Δ_j by Δ'_j which is L-equivalent to Δ_j, we speak of its L-*substitutability* with respect to Δ_i.

This characterisation of the concepts of substitutability and L-substitutability relativised to a particular expression can be generalised so that instead of a particular expression (i.e. Δ_i) we take any sentence of the object language for the relativisation. Then it becomes possible to define substitutability and L-substitutability with respect to contextual concatenation with any sentence (in $\mathscr{L}$):

(D_1 – V 3 C) Δ_j is *substitutable* by Δ'_j (in $\mathscr{L}$) = df Δ_j is, in contextual concatenation with any sentence (in $\mathscr{L}$), substitutable by Δ'_j (i.e. Δ_j and Δ'_j are equivalent).

(D_2 – V 3 C) Δ_j is *L-substitutable* by Δ'_j (in $\mathscr{L}$) = df Δ_j is, in contextual concatenation with any sentence (in $\mathscr{L}$), L-substitutable by Δ'_j (i.e. Δ_j and Δ'_j are L-equivalent).

On the basis of these definitions we can clarify the concept of extensional context: If the extension of Δ_i in the expression $\Delta_i \overset{C}{\frown} \Delta_j$ does not change unless the extension of Δ_j changes, we may say that Δ_i in context with Δ_j is extensional. This characterisation of extensional context is relativised to a certain Δ_j. It can be generalised if the same characterisation is possible with respect to any Δ (in $\mathscr{L}$). This can be expressed in the definitions:

(D_3 – V 3 C) The expression Δ_i is *extensional with respect to the context* $\Delta_i \overset{C}{\frown} \Delta_j$ (in $\mathscr{L}$) = df in the context $\Delta_i \overset{C}{\frown} \Delta_j$ Δ_j is substitutable by any expression that is equivalent to Δ_j.

(D_4 – V 3 C) An expression Δ_i is *extensional* = df Δ_i is extensional with respect to the context of any Δ (in $\mathscr{L}$) with Δ_i.

There is an advantage in further introducing the concept of an extensional language, i.e. a language that admits only of the kind of (contextual) concatenation of sentences that is extensional:

(D_5 – V 3 C) *A language* $\mathscr{L}$ is *extensional* = df any sentence in $\mathscr{L}$ is extensional.

The concept of an extensional language is not, as is obvious, identical with that of a formalised language, although many formalised systems (such as those in *Principia Mathematica*) are extensional in character. There are plenty of well known formalised systems, such as those with strict implication, causal implication, model systems and others, which are not extensional.

The concept of extensional context in relation to the logical functions was introduced by Whitehead and Russell in the system of their *Principia Mathematica*, although the term 'context' is not used there.[40] The difficulties arise with the analysis of a context which is not extensional. In the *Principia Mathematica* system the concept of intensional context is identified with the kind of context that is not extensional. Hence this sphere also includes indirect speech, belief sentences, statements of conviction or opinion, propositional attitudes and so forth, as can be deduced from the examples quoted in *Principia Mathematica*.

It is to Carnap's credit that he put forward a much narrower definition of intensional context, based on L-equivalence [23]. We may formulate the following definitions, which are analogous to definitions 3, 4 and 5 of this section:

(D_6 – V 3 C) An expression Δ_i is intensional with respect to the context $\Delta_i \overset{C}{\frown} \Delta_j$ (in $\mathscr{L}$) = df Δ_i is, in the context $\Delta_i \overset{C}{\frown} \Delta_j$, not extensional, and in that context Δ_j is L-substitutable by any expression that is L-equivalent to Δ_j.

It should be stressed that (D_6 – V 3 C) actually contains two conditions for intensionality (unlike the system of *Principia Mathematica*, where the first alone suffices): 1. non-extensionality and 2. the L-substitutability of the relevant component.

In (D_6 – V 3 C) the characterisation of intensionality is relativised to a particular Δ. This may be generalised in the following definitions:

(D_7 – V 3 C) An expression Δ_i is intensional (in $\mathscr{L}$) = df Δ_i is, with respect to any context of which Δ_i is a component, either extensional or intensional, and it is intensional with respect to one context.

(D_8 – V 3 C) A language $\mathscr{L}$ is intensional = df any sentence in $\mathscr{L}$ is either extensional or intensional, at least one being intensional

An extensional context is used by formalised languages which are based on classical logic. Examples of intensional context are languages which are based on modality in logic. If N is a sign for modal necessity, the expression $N(\Delta)$ is non-extensional with respect to concatenation with Δ. If instead of Δ we take $\Delta \vee \sim \Delta$, i.e. an L-true sentence, and replace it by an L-equivalent sentence, we will obtain, by (D_6 – V 3 C), the expression $N(\Delta \vee \sim \Delta)$, which is intensional with respect to $\Delta \vee \sim \Delta$.

A somewhat special situation which denies interpretation from the point of view of either extensional or intensional context is provided by what have been called *belief sentences*. Examples of these are:

XY believes that Σ.

John Brown is convinced that Prague is in Africa.

The inhabitants of the Himalayan village are sure that this is the highest mountain in the world.

In analysing these sentences we are faced with two possibilities:

(1) The first possibility is to treat this kind of sentence as statements by the language users about certain attitudes to the facts expressed by the relevant sentences. So it is a case of relations between language users and language expressions, or, more exactly, between language users and a particular semantic characteristic of the sentences, relations of a pragmatic character. This means that concepts such as 'belief', 'conviction', 'certainty' and the like are constructed as expressions for certain pragmatic relations.

(2) The second possibility is to treat this kind of sentence as statements about the states of certain persons, the persons, or certain of their states which can be observed or reconstructed on the basis of the persons' behaviour, being treated as an integral part of the universe to which the given object language relates. This is essentially how Carnap analyses belief sentences in [23].

Both courses are admissible, but they are not and cannot be seen simply as alternative solutions; they represent solutions to two rather different situations. In the case of the first procedure we have to assume a class of language users which includes those who are expressing their relations to various kinds of sentences. There is still no systematic analysis and categorisation of these relations, which are usually expressed by terms such as 'pronounces', 'accepts', 'claims', 'denies', 'believes' etc., nor does there exist any system of theoretical pragmatics capable of providing a consistent analysis of all these relations.[41]

If we take the second course and treat belief sentences as sentences about certain persons' states, we are justified in investigating the nature of that context. If we take the sentence. 'John Brown thinks that Prague is in Czechoslovakia', we can replace the sentence component 'Prague is in Czechoslovakia' by the equivalent sentence 'the largest city on the Moldau is in Czechoslovakia'. But John Brown's geographical knowledge might be wanting,[42] so the truth value of the whole sentence may change even by substituting the equivalent sentence for the sentence component. To put it generally: Since the sentence 'XY believes that Σ' is in fact a sentence about a particular *state* of XY, the truth value of the whole sentence need not depend on a sentence component, but on whether that state occurs or not. This means that the sentence 'XY believes that Σ' *is not extensional.*

We will arrive at a similarly negative result if we replace the sentence component Σ by the sentence Σ′ which is L-equivalent to Σ. But since we cannot assume XY to be versed in logic, we cannot be sure whether the state of XY might not change in consequence of the substitution.

We may conclude then that in general terms the context in belief sentences is neither extensional nor intensional. We are not deprived, however, of possible subsidiary instances in which a context of a particular type is fixed in sentences of this kind by the introduction of extra conditions, e.g. conditions defining circumstances or necessary attributes of the state of the persons concerned and so forth.[43]

[*D*] *A Possible Amplification of the Method of Extension and Intension*

There can be no avoiding the question whether a semantic conception of the type represented by the method of extension and intension, or possibly also the type that assumes a dual semantic characterisation of names, i.e. sense and denotation, provides adequate facilities for semantic analysis. Some objections may be brought against both conceptions, which are after all fairly closely related, from the standpoint of a kind of psychologism. For both the conceptions analysed here deliberately disregard any psychological aspects. It is of course true that a semantic relation as a relation between expressions of language and extralinguistic entities (whether they be entities that can be defined in an 'actual world' or in 'possible worlds') need not require us to take the involvement of psychological aspects into account. But the fact that these aspects are genuinely involved is unquestionably of great importance in the semantic analysis of texts of natural language.

We are therefore faced with the possibility of somehow refining our analysis in such a way as to have it respect this involvement at least as far as is necessary, while not having to bring in psychological criteria as such. One attempt to amplify the method of extension and intension (though conceptually not directly dependent on Frege's theory or Russell's and Carnap's conceptions) is the semantic theory of C.I. Lewis. The theory is briefly recorded in Lewis's study on modes of meaning [83] and is expounded in greater detail his ample book [84].

It may be said that one motive for Frege's conception was the analysis of name and the name relation, and that the motive for the method of extension and intension was the analysis of predicates and the abstracta that correspond to predicates. In a sense Lewis's conception also starts out from the analysis of an abstraction, notably the analysis of so-called abstract terms and their role in natural language. It should be noted that the term 'abstraction' is ambiguous and that λ-conversion only captures one aspect of the process of abstraction in natural language. Compare the following sentences:

(1) Prague is in Bohemia.

(2) Prague has a hundred towers.

(3) Prague is hundred-towered.

In cases (1) and (2) an abstractum can be created by means of λ-conversion, by inserting in place of the individual name 'Prague' a variable that we then bind by the operator of abstraction. In this way we obtain expressions that would seem to be most conveniently expressed by means of the infinitive, i.e. the expressions 'be in Bohemia' and 'have a hundred towers' respectively. In case (3), however, natural language offers another possibility, which might be expressed by the term 'hundred-toweredness'.[44] This branching of the ways of abstraction, and the possible existence of yet other ways, may be taken as a sign of the redundancy or formal imperfection of natural language. This standpoint does not resolve the semantic questions that arise from a comparison of sentences (1) and (2) with sentence (3).

Some of the questions that arise here are competently solved by Lewis's conception of semantics. For this reason his conception is much closer to the problems of natural languages than are the conceptions of Frege, Russell, Carnap and others. Lewis tends to base his conception on the semantics of terms, although he does also deal with the semantics of sentences.

In handling the semantics of terms[45] Lewis distinguishes four basic 'modes of meaning':

(1) the denotation of a term, which is the class of all actual or actually existing things to which the term is applied;

(2) the comprehension of a term, which is the classification of all

possible and consistently thinkable things to which the term could be correctly applied;

(3) the signification of a term or its meaning in the narrower sense of the word, which is that property of a thing the presence of which shows that the term is being used correctly, its absence showing that the term has not been used correctly;

(4) the intension of a term, which, in formal terms, is identical with the conjunction of all other terms of which each is applicable to that to which the given term may be correctly applied.

Denotation therefore corresponds to extension. If the concept of term as understood by Lewis is a concept that embraces both individual expressions and predicates, it is easy on the whole to admit the possibility of classes that have no elements, classes that have one or more elements, or even classes that contain all the elements of the given universe. Thus it is possible to distinguish:

(a) zero denotation (e.g. 'unicorn', 'Apollo'),

(b) singular denotation (e.g. 'Prague', 'this evening'),

(c) denotation of a class that is not empty.

Membership of a class, and so non-zero denotation, is limited, according to Lewis, to things that *exist*. It goes without saying that tying denotation to the ambiguous term existence is distinctly problematic. Lewis, like Meinong before him, assumes three kinds of 'things' or objects: (α) actual objects, i.e. objects denoted by terms having non-zero denotation, (β) non-actual but possible objects, i.e. objects denoted by terms with zero denotation (for example, the 'unicorn' does not exist in modern zoology, but there is no *logical* impediment to finding or imagining such an object), (γ) non-actual and also non-possible objects, i.e. objects denoted by terms with zero denotation which in principle cannot be exchanged for non-zero denotation (e.g. 'round square').

In order to be able to distinguish case (β) and case (γ) also semantically, we have to resort to the second mode of meaning, comprehension. From the point of view of comprehension it is again possible to consider zero and non-zero values: the term 'unicorn' has indeed zero denotation, but its comprehension is non-zero. By contrast a 'round square' has not only zero denotation, but also zero comprehension. Since comprehension is the classification of all possible and consistently thinkable things, the comprehension of the term square embraces all actual and possible or thinkable squares, but not a single 'round square'.

The kind of differentiation of denotation and comprehension of a term as required by Lewis's conception must invariably be relativised – to the given level of knowledge, i.e. the given knowledge of the universe to which the individual terms can be related, and to the logical base of the given language. Moreover, it is also worth noting that case (β), i.e. the problem of terms with zero denotation and non-zero comprehension, can be resolved in another different way, namely by means of the conception of indirect denotation suggested by Frege or by respecting the main context. Carnap too, it may be added, rejected any suggestion that his approach is incapable of solving this situation: if we treat a class as the extension of a predicate, we can distinguish a class which (α) is not empty (i.e. for terms with non-zero denotation), (β) a class which is F-empty, but not L-empty (i.e. for terms with zero denotation but non-zero comprehension), and (γ) a class which is L-empty (i.e. for terms with zero denotation and zero comprehension).

The concept of intension or connotation in Lewis's conception of semantics does not differ greatly from intension as we have described it earlier, although Lewis uses other forms of explication. In Lewis's conception the intension of a term is laid down by a kind of proper definition of the term, or by what was described in scholastic logic as the 'essentia' (as opposed to 'existentia'). We can use logical equivalence to specify these terms: the definiendum and definiens have to be required to be not just equivalent, but L-equivalent. Since two terms have the same intension if and only if they are L-equivalent, Lewis's explication, based on the notion of definition, is consonant with the explication of intension in the sense described earlier.

The concept of signification, connected with a significant property, was brought in by Lewis as a device to separate abstract terms from concrete terms. For with abstract terms denotation and signification coincide, abstract terms being those which name that which they themselves mean, i.e. they name their own significant property. For example, 'roundness' is the name of the property that is significant for all round things. On the other hand, terms whose denotation and signification differ, are called by Lewis concrete. This distinction of what he calls abstract and concrete terms, which can only be sensibly applied for the terms of natural language, is of course highly verbal. Moreover it would not be hard to prove by a few examples that the boundary between abstract and concrete terms can hardly be established unequivocally (not to mention the very ambiguity of the expressions, which in the actual use of natural language is also conditioned by pragmatic considerations, such as the level of the language user's knowledge).

It cannot be denied that this extension of the modes of meaning may – at least in the semantic analysis of natural language – be of some use in the appraisal of many individual situations. But it is doubtful whether it can be made sufficiently general. If the concepts of extension and intension could be defined with adequate accuracy, the same level of accuracy could not be achieved with the other modes of meaning.

Lewis applies this same differentiation of modes of meaning to statements and propositions.[46] Here again the strength of his conception lies in extension and intension, which in the main agrees with the above analysis of the extension and intension of sentences.

The fact that certain individual situations permit a more detailed differentiation of modes of meaning confirms the possibility of a double interpretation of the concept of 'intension' or what has been called intensional meaning. Intension can after all be constituted in two ways: (1) on the basis of analytical relations (e.g. definitions) of a given term or sentence with other elements or expressions in the language, (2) on the basis of a criterion that enables us to decide that a given expression has a particular extension.

Taking Lewis's terminology, we might call the first interpretation of 'intension' *linguistic intension*.[47] 'Intension' as defined on the basis of the L-equivalence of expressions is therefore linguistic intension. This means that the primary concept here is that of 'have the same intension', which can be specified by means of certain relations between different expressions of language. If we disregard the nature of the context, i.e. if we take (relatively) isolated expressions, this characterisation of the concept of linguistic intension is enough. But in characterising linguistic intension we cannot really ignore the nature of the context. In the case of a language that is extensional, all expressions that have the same intension are L-substitutable. In the assimilation of the meaning of individual expressions of natural language where intensional meaning is concerned, linguistic intension corresponds to the so-called direct method. But in the use of the direct method two conditions must be respected: (a) the intension of the expressions used as the basis for the explication of the intension of a new or unknown expression must be known; (b) the relation between the expression whose intension is to be specified and the expressions used to that end must satisfy the relevant conditions.[48] In a case where an expression of natural language is being used in a specific context, the nature of that context must be respected.[49]

The concept of linguistic intension has also been characterised sometimes as 'syntactic meaning' (see for example [88]). This is justified only insofar as we recognise that the pillars of this interpretation of intension are relations, however characterised (e.g. 'analytical relations', relations of 'L-equivalence'), between different language expressions. If, in determining linguistic intension, we can operate with means available in the language itself (i.e. means of syntax and semantics provided they are expressions whose intension is taken as known), the concept of *sense intension* assumes, by contrast, some criterion which the language user is capable of applying. It needs to be stressed that the concepts of linguistic intension and sense intension are not two alternative solutions to the problem of the sense of language expressions; they are not only not mutually exclusive, but may even be mutually complementary.

The formulation of a criterion for the sense intension of a language expression (or, simply, a conception of a criterion of sense) is not an easy matter. Most of the attempts to find a uniform criterion of sense have run into difficulties, which has made the literature quite rich in various conceptions. They include the verifiability theory and other empiricist theories of a criterion of sense, the operationalist conception, various psychological conceptions, a pragmatic conception and others. A more detailed analysis of the problem of sense intension and a criterion of sense will follow in the next chapter.

The distinction of linguistic and sense intension shows that the method of extension and intension does not itself solve the problem of a criterion of sense, and it does not after all set out to do so. It would be inappropriate to accuse this method of failing to solve the problem of sense intension, the problem of a criterion of sense, when it is only concerned with intension in the sense of linguistic intension.

4. THE PROBLEM OF NAMING

[*A*] *The Name Relation*

Both Frege's conception of sense and denotation and the method of extension and intension amount virtually to a starting point for the semantic analysis of names (in the broadest sense of the word). The concept of name always assumes a semantic relation between some language expression to which a relatively independent meaning, i.e. extension *and* intension, can be ascribed and a certain object (of whatever nature). This object as one of the arguments in a two-place name relation (the other argument being the *name*) is sometimes also called the *nominatum.* A nominatum is then an object to which a name relates, the relation between them being, as we have already noted, expressible by a number of different terms.[50]

Carnap [23] points out that three fundamental principles are usually assumed when the name relation is applied, although they are not always formulated explicitly. In Carnap's view these principles are an indispensable prerequisite for the name relation to be used reliably. That is not to say that the principles cannot have exceptions, at least in natural language. In addition, the validity of some of the principles may be narrowed to cover only a particular category of names, and so they may be used as a means of differentiating kinds of names. The principles of the name relation may be described as (1) the principle of univocality, (2) the principle of subject matter, (3) the principle of interchangeability.

The *principle of univocality* can in principle be formulated in two ways: (a) the first, stronger formulation resides in the fact that each language expression used in a communication process is the name of only one object, i.e. nominatum. This sole object may be an individual, a class or a property that determines a class etc. This strong formulation of the univocality principle essentially requires us to apply to all language expressions that may have the function of a name the same demands that we make of proper names. (b) The second, weaker formulation of the principle of univocality really amounts to just the negative requirement that the language expressions that feature as names in a given process of communication should be rid of ambiguity.

The principle of univocality and its role in language cannot be adjudged from the fact that it is not fully observed, that there is in language ambiguity which does affect unambiguity throughout the range of a name's use, or vagueness which effects unambiguity in the so-called fringe cases. As a principle, and disregarding the exceptions, disregarding ambiguity and vagueness, it does actually guarantee the comprehensibility of language expressions; it guarantees that these expressions may be used within a certain community of language users. This means that despite the possible reservations there are here pragmatic reasons for respecting the principle of univocality, at least in its weaker formulation.

In formulating the principle of univocality we must also take into account the fact that unambiguity may be relativised in varying degrees. A common objection to the principle is that some expressions, for example predicates, cannot be treated as the names of objects, since the meanings of these expressions are tied to a particular context. Outside this context it is impossible, as is regularly asserted, to consider the expressions as having relatively independent meaning and therefore it is impossible to consider what the nominatum represents. (For example, the expression 'Soviet' may enter into concatenation with the expressions 'Union', 'product', 'citizen' etc.) If we do disregard context, there are two possible stands: either expressions of this kind represent names, but the principle of univocality is thereby broken giving rise to ambiguity, or we can treat them as expressions that differ from names in principle, thereby of course reducing the competence of the name relation to the realm of proper names in the narrow sense of the word. Neither of these possibilities seems particularly acceptable. Although Carnap tried to draft his method of extension and intension in such a way as to obviate the necessity of considering context – as in assessing the extension of predicates – it is obvious that this viewpoint is only feasible in a particular conception of formalised language. On the other hand, many of Carnap's critics, especially the more energetic adherents of the nominalist viewpoint (notably Quine and Goodman), point out that anything that corresponds to denotation or extension, in our case nominatum, has to be relativised to context. This line is clearly better suited to the idea of the semantics of natural languages. Goodman, for example, describes the extension of a predicate itself as primary extension, the extension of a complex expression containing the predicate secondary extension, and so on.

We may conclude by saying that the principle of univocality, for all the possible reservations and objections, is a mainstay of the name relation. It is expedient, especially in the analysis of natural languages, to relativise the principle to context.

The *principle of subject matter* may be formulated thus: any sentence in which names occur treats of or relates to that which constitutes the nominata of those names. For example, the sentence 'The capital of Czechoslovakia lies on the Moldau' contains two names. It deals with the nature of the relations between what is denoted by the name 'the capital of Czechoslovakia' and what is denoted by the name 'the Moldau'. This characterisation of the principle of subject matter is distinctly imprecise, as will follow from the rest of what will be said about the ambiguity of the name relation. In addition, this characterisation assumes not just a particular conception of language as a system relating to extra-linguistic entities, but also a conception of the structure of those entities.[51]

The *principle of interchangeability* is actually a kind of (not very precise) formulation of Leibniz's criterion of identity: if two language expressions are names of the same entity, or, to put it another way, if they have the same nominatum, either can replace the other, the sentence in which this occurs not undergoing any change of truth value. As pointed out above in the account of extension and intension, this principle has to be relativised in two directions: (a) with respect to the extension-intension distinction of language expressions; (b) with respect to the nature of the context in which the substitution can be made. In the simplest instance it holds that two expressions which have the same extension (i.e. they are equivalent) are interchangeable in an extensional context. This we may call the *reduced principle of interchangeability*.

This look at the basic principles of the name relation reveals a number of difficulties. Some of these were pointed out by Quine [115], who distinguishes two ways in which names are used. (1) The first will be called the *purely denotational use* and includes cases where the name relates to the nominatum, the name's object, directly. (2) The second use is that which is not purely denotational. The following examples may serve to underline the difference:

(1) Hus = Master John. (The reference is to John Hus, the Bohemian Protestant reformer of the fifteenth century, who was also a master of the University of Prague and is well known in Czech under this name.)

(2) Hus was so named after his native parish of Husinec.

Both statements may be treated as true. But if we build (1) into (2) we obtain something that cannot be treated as true. This shows that substituting a purely denotationally used name in (2), where a name with the same denotation, an equivalent name, is used but not purely denotationally, is not admissible. In order to distinguish the two kinds of name-use, we shall leave the purely denotational use (that which corresponds to the original function of names) without inverted commas, reserving inverted commas for the second kind of use. Sentence (2) can then be reformulated as:

(3) Hus was named 'Hus' after his native parish of Husinec.

Sentence (3) contains the same name twice, but used differently in each occurrence. But natural language of course cannot always avail itself of formal devices to discriminate unequivocally between the two kinds of use. It is usually the practice simply to conclude from the context which use is meant, but there is the genuine possibility of cases where the difference between the purely and not purely denotational use of a name is a matter of degree or transition.

Other difficulties arising out of the application of these principles of the name relation stem from the fact that there are two possible interpretations of the name relation.[52] Already in talking about the analysis of predicates we have seen the alternative possibilities of interpreting predicates as the names of either classes or properties. These might be described as the extensional and intensional interpretation respectively, but the possible double interpretation applies as much to other names as to predicates.

In natural language the suitability of one or other interpretation usually transpires from the overall context. In many instances both interpretations are possible. But the two possible interpretations can in formalised language be kept separate by distinguishing the formation of expressions relating to extension (e.g. to classes) from that of expressions relating to intension (e.g. to properties).[53] For example, if we take the sentence which we obtain from the previously introduced interpretation of the predicates P_1 and P'_1, the dependence between them can be recorded in two forms of notation:

(1) $(\hat{x})P_1x \subset (\hat{x})P'_1x$

(2) $(\forall x)(P_1x \longrightarrow P'_1x)$

In case (1) we have the statement that the class '$(\hat{x})P_1x$' is contained in the class '$(\hat{x})P'_1x$'. In case (2) we have the statement that anything which has the property 'be in Bohemia' also has the property 'be in Europe'. This of course leads to the 'superfluous duplication of names', giving rise at the same time to the question of whether the principle '*nomina non sunt multiplicanda*' ought not also to apply, by analogy with Occam's razor.

Whitehead and Russell clearly did not recognise this principle. By contrast Carnap noted that the usefulness of the principle becomes apparent especially in higher types. If the zero type is represented by individuals (points on the surface of the Earth in our example), type one is represented by predicates that can be interpreted as classes or properties. Names occurring at this level are classified as type one. If we pass to the next higher type, i.e. type 2, the situation is complicated in consequence of the double possible interpretation, which calls for the differentiation of four categories of names:

names of classes of classes,
names of properties of classes,
names of classes of properties,
names of properties of properties.

At the n-th level, i.e. with names of the n-th type, it is necessary to assume 2^n kinds of entities corresponding to a single name of the n-th type.

It is obvious that in using the simple name relation without stating what kinds of uses of names are involved, without stating which of the possible interpretations are being taken into account and also without stating the nature of the context, there can be no avoiding the difficulties of the kind just described. But if we add rules for the interpretation of intension and extension to our semantic rules we can achieve two results: (a) the principle of 'nomina non sunt multiplicanda' can be respected; (b) the ambiguity of the name relation can be eliminated.

Assuming that the semantic rules do include both rules for extension and rules for intension for all language expressions that can function as names, there is the unavoidable question of which of them have priority. The question can also be formulated by saying what comes first in the interpretation of a name. In principle there are three possible viewpoints here:

(A) The question may be simply rejected on the grounds that the relation between the extension and intension of a name cannot be determined accurately and unequivocally enough for priority to be given either way. Grounds for this 'neutral' attitude can be found in the semantic analysis of natural languages, but the view itself actually evades the principal issue.

(B) The second viewpoint, which comes from some of Wittgenstein's ideas in the *Tractatus* as well as from some of the logical positivist ideas of the 'Vienna' period, apparently prefers extension. By these conceptions 'understanding a sentence' is the same thing as 'knowing the truth conditions of sentences'. (This same standpoint lies behind the so-called verifiability criterion of sense, the analysis of which will be presented at a later stage.) This can also be formulated by saying that 'understanding a sentence' assumes knowing the conditions for deciding about the *extension* of a sentence. The assimilation of the meaning of a language expression is here linked primarily to extension. It must be added that, wide-spread as it was in its time, this viewpoint has been essentially quite left behind.

(C) The third viewpoint is actually a product of the second. If 'understanding a sentence' (and generally 'understanding any language expression to which a relatively independent sense can be assigned') is not and cannot be the same thing as knowing the conditions for deciding about the extension of a sentence, it is possibly better to link 'understanding' a sentence, and understanding generally any expression with a relatively independent sense, with the intension of the expressions instead.[54]

It might seem that viewpoint (B) and the ideas of Wittgenstein's *Tractatus* and of the Vienna Circle's conception of language that are connected with it, come somehow closer to the materialist conception of language whereby the basis and starting point of language is the objective world; therefore it might also seem that it is the possibility of assigning events or phenomena of that world to language expressions that is decisive to understanding the expressions. In fact, faced with understanding an expression of language, we take the very expression itself as the first thing to go on. In semantic analysis we also find what Karl Marx described in his mathematical manuscripts as the 'Umschlag in der Methode'.[55] This means that the question of the priority of one or other step cannot be asked as an abstraction, but only in connection with the nature of the task within the framework of which it is asked. This is also the main reason why the third viewpoint (C) is best suited to the semantic analysis of names and the name relation.

[*B*] *Antinomies of the Name Relation*

The most serious problems of the name relation are connected with the principle of interchangeability. Some of these were mentioned when we distinguished the purely from the non-purely denotational use of names. The principle of interchangeability can only be respected if the purely denotational use of names is kept unimpaired. But it is also possible to use the principle of interchangeability to construct antinomies of the name relation, i.e. situations where contradictory results can be reached. These antinomies are usually called the antinomies of synonymous names.[56] They are most easily constructed by means of modal sentences containing the expression 'necessary' in the sense of logical necessity.

Quine [115] gives a number of examples of analytical sentences introduced by the expression 'is necessary' and containing names which can be exchanged for names relating to the same nominata. For example the sentence:

(1) $9 > 7$ is an analytical sentence which can also be formulated using the term 'necessary' (in the sense of logically necessary):

(2) 9 is necessarily greater than 7.

This sentence is true, its truth depending solely on the semantic rules of the given language; it is then an analytically true sentence.

On the basis of the identity 'the number of planets = 9' we can, by an insertion into (2), obtain:

(3) The number of planets is necessarily greater than 7.

But because this fact has been demonstrated empirically, its truth does not depend solely and exclusively on the semantic rules of the given language. In other words, while this sentence is true, it is not true analytically. Therefore:

(4) The number of planets is not necessarily greater than 7.

Another example of how an antinomy can be reached: It is a known fact that the River Moldau flows only through Bohemian territory. It therefore holds that:

(5) The largest town on the Moldau is necessarily in Bohemia.

But it is a known fact that

(6) The largest town on the Moldau = the capital of Czechoslovakia.

By inserting (6) into (5) we obtain:

(7) The capital of Czechoslovakia is necessarily in Bohemia.

However, where a capital city lies depends on historical traditions and is enshrined in a constitution. Changes may arise and it can therefore be said that it is not necessary for the capital of Czechoslovakia to be in just one particular part of the country. Therefore:

(8) The capital of Czechoslovakia need not necessarily be in Bohemia.

Sentences (3) and (4) and sentences (7) and (8) are obviously contradictory although in both cases the contradictory conclusions are arrived at from the same name relation and the principle of interchangeability.

The solution to antinomies of this kind has basically been suggested already in the preceding account. It could be said that Frege's conception of sense and denotation, Russell's analysis of the denoting phrase, the method of extension and intension and all other similar conceptions can also be seen as contributions to the solution of antinomies in the name relation. The solution can be broken up into several stages:

(1) The first step is without question the distinction of sense and denotation, or extension and intension. This distinction assumes some categorisation of names, such as the above-mentioned differentiation of names of individuals, predicates of various degrees etc.

(2) The second step is the extension of the differentiation to the sphere of sentences. The basic prerequisite for this is an acceptance that the semantic analysis of a sentence can be made in a manner analogous to that of expressions to which can be assigned a relatively independent meaning.

(3) The next extension of the analysis is the semantic analysis of context. Although one cannot overlook here the attempts at the so-called reduction of context,[57] such experiments cannot yet be treated as having been entirely successful. Hence in applying the principle of interchangeability we always have to respect the nature of the context, as was demonstrated in the account of the method of extension and intension.

(4) Finally in the application of the principle of interchangeability to the name relation, especially when it is natural language that is being used, it is necessary to take into account certain factors of a pragmatic nature. This applies particularly to situations where the same language expression is used in connection with different motives or aims of communication in language.

5. SYNONYMITY

[A] *The Concept of 'The Same Meaning'*

In an analysis of meaning and the modes of meaning it is impossible to overlook the question of what is meant by 'having the same meaning'. The concept of 'the same meaning' appears intuitively to provide the key to the analysis of synonymity. But of course the problem of synonymity is not an exclusively semantic problem; it is one of the most widely discussed problems in linguistics and cannot be solved merely by the means afforded by semantics. There are certain pragmatic aspects that cannot be overlooked along with all that can be included in them. Hence the general conception of synonymity in communication and in the use of natural languages is a broader topic which in the context of the analysis of meaning and the differentiation of the modes of meaning can be solved only in part.

From the layman's point of view the whole problem can be dismissed with the bald statement that two (or more) language expressions are synonymous if and only if they have the same meaning. (For example, 'spinster' and 'elderly unmarried woman', 'brother' and 'male sibling', 'Prague' and 'the largest town on the Moldau' and 'the capital of Czechoslovakia' and so on seem to be synonyms because they have the same meaning.) But this statement cannot be taken as a satisfactory explication of synonymity for many reasons. First and foremost it can easily be proved that the concept of 'the same meaning' is semantically indefinite, so an explicans which operates with the term is not and cannot be any clearer, more comprehensible or more definite than the explicandum containing the concept of synonymity. Other objections arising out of the indefiniteness of the above statement may come about because there is no reference to or explicit assumption of any *relativisation of the concept of 'the same meaning'*. This relativisation can be of many different degrees and may include:

(a) relativisation to a particular language or languages (here the distinction has to be drawn between expressions having the same meaning within one language and those having the same meaning but in different languages);

(b) relativisation to a particular differentiation of language expressions (here it is worth keeping separate the synonymity of individual names, predicates, sentences, sentence contexts etc.);

(c) relativisation to the user or users of the language (here we need to distinguish situations where there is a single language user, a class of language users sharing the same equipment, e.g. having a command of the same language and therefore sharing the associated socially and culturally governed ways of using the language, or a class of users with different equipment);

(d) relativisation to some particular conception of meaning and modes of meaning, to a particular method of ascertaining meaning from a criterion of meaning (criterion of sense) etc.

This relativisation, which could be extended further, is not always sufficiently respected in systems of semantics that are conceived in purely linguistic terms. Greater attention tends to be paid to the discrimination of greater and lesser degrees of synonymity and similar problems. For example, S. Ullmann [157] distinguishes what he calls pure synonyms and pseudosynonyms or homoionyms. Pseudosynonyms are expressions that are interchangeable only in certain contexts, pure synonyms being interchangeable in all contexts. He divides pseudosynonyms into those which are interchangeable in certain contexts (e.g. 'help – aid – assistance') and those that are interchangeable wherever it is a case of cognitive meaning, but not where emotive meaning is involved (e.g. 'liberty – freedom', 'dustman – refuse disposal officer'). This is not an entirely satisfactory view and can really only apply in a descriptive approach to natural language. It is actually unsatisfactory as a conception of synonymity for any more exact approach. It is therefore clear that the concepts of 'the same meaning' and 'interchangeability' are not enough, unless the nature of the context in which a substitution is to be made is stated exactly. This is why we need to define more exactly the assumptions that lie behind synonymity and the nature of the relativisation which limits the use of the concept 'the same meaning'.

It must be realised above all that there is only any point in considering synonymity if we are operating with expressions to which a relatively independent meaning can be assigned. This means that we dismiss at the outset those expressions which may well have something in common with other expressions of the language while not having any such relatively independent meaning assignable to them, either because they are expressions that have no sense, or because they have a sense only in a particular concatenation with other expressions.

Another assumption is that which might be expressed by the dispositional term 'translatability'.[58] The connection between the concepts of synonymity and translation was pointed out by B. Mates [91], who was right in emphasising that this means translation in the general sense of the word. It is not essential to be talking about expressions from two different languages; expressions from the same language are also involved as soon as certain of them are less comprehensible than others for particular users. Seen in this light even the role of a popularisation of a scientific text is a kind of translation, since it calls for the selection of synonymous (or at least relatively synonymous) expressions which are more readily comprehended by the intended circle of language users.[59]

Translatability of an expression also means that under certain circumstances, or if certain conditions are satisfied (and these are conditions of both a semantic and a pragmatic nature), there is a user or users capable of giving an interpretation of the expression. Hence any interpretation of an expression that is the object of the interpretation (an interpretandum) is actually the translation of that expression into others (the interpretans). At the same time it has to be assumed that the interpretans fulfils certain conditions which are usually stated beforehand, such as that it is given in the required language, that it is more accessible or more comprehensible to a particular circle of language users and so forth. In general terms an interpretation is a kind of transmission of information taken in the semantic, and possible also the pragmatic, sense, and it fulfils certain pre-stated conditions. This is also to say that in the course of the transmission through the 'channel of interpretation' no new information may emerge, and that it is also accepted that there may be some loss of information due to 'noise' in the 'channel of interpretation'.[60]

After these introductory remarks we can pass to the question of what is meant when we say that two (or more) expressions are synonymous. We are still assuming that we can assign relatively independent meaning to the two expressions. For this question to be semantically precise we have to relativise it to a particular object language $\mathscr{L}$. Then a more exact formulation of the original question is: What does it mean to say that two expressions in $\mathscr{L}$ are synonymous? What is the criterion of their synonymity? If we take synonymity to be a semantic concept, and it does assume the name relation,[61] we have to assume that we are dealing with a language which is not a semantically closed language in Tarski's sense. Hence the concept of 'synonymous in $\mathscr{L}$' is also not a concept of the object language but of the metalanguage. A much more complicated situation arises if we have relativisation not to a single object language but to a number of them. There is every justification for the question: What does it mean to say that Δ in $\mathscr{L}_1$ is synonymous with Δ' in $\mathscr{L}_2$? If in the first case it is usual to operate with interchangeability within one and the same language, the second case assumes the semantic concept of translation from $\mathscr{L}_1$ into $\mathscr{L}_2$, for which more exact criteria can be formulated provided we can avail ourselves of a metalanguage that is common to $\mathscr{L}_1$ and $\mathscr{L}_2$.

Let us now look at the simplest version of the question, i.e. what does 'synonymity in $\mathscr{L}$' mean, or what are the criteria for 'synonymity in $\mathscr{L}$'. In addition to this relativisation (i.e. to a single object language) let us further assume relativisation to a single language user or, which may lead to the same outcome, to a totally uniform class of language users. Then the interchangeability of synonymous expressions means that the language form changes but something remains preserved. But here we are faced by two questions:

(1) In what conditions or, more precisely, in *what context* is it possible to consider the interchangeability of synonymous expressions?

(2) If two synonymous language expressions are substitutable for each other, then *salvo quo*?

For any conception of synonymity these questions are of paramount importance, despite the fact that in many discussions of synonymity they have tended to be accepted tacitly rather than formulated explicitly.[62]

The answers commonly assumed are:

(1) a *sentence context*, (2) *salva veritate*. Accepting these answers we obtain a fairly simple criterion of synonymity:

Two expressions are synonymous in $\mathscr{L}$ if and only if they are mutually substitutable in an occurrence in any sentence of $\mathscr{L}$ without a change in the truth value of the sentence.

It can of course be shown easily that this is not a reliable criterion. For the first question it is sufficient to take languages which contain modal operators or other instances of non-extensional contexts and to show that one cannot take an occurrence in 'any sentence' in $\mathscr{L}$ without taking these facts into consideration. As for the question *salvo quo*, the answer *salva veritate* assumes only an extensional approach to the sentence. The possible antinomies which are sometimes inappropriately designated as the 'antinomies of synonymous names' and which may occur under this approach were discussed above in the section on the name relation. The problems of synonymity, which is closely connected with the name relation, may therefore lead to exactly the same kind of problems as those encountered with the name relation. In fact synonymity is really a special case of the name problem. These are also the main reasons why, in trying to bring some precision to the concept of synonymity, we should take the view that the intensional approach will take us further and bring us closer to the heart of the matter than the extensional approach.

[*B*] *Synonymity, Intension and Intensional Isomorphism*

The above view that the extensional approach (which in the case of the occurrence of expressions in sentences is expressed by the criterion of *salva veritate*) is not sufficient for specifying the concept of synonymity is really only a negative outcome. This same negative outcome can also be expressed by saying that the equivalence of expressions is not capable of guaranteeing the mutual substitutability of expressions that may be treated as synonyms.

If the equivalence of the two expressions Δ_1 and Δ_2 cannot be sufficient guarantee for the synonymity of Δ_1 and Δ_2, the stronger guarantee based on L-equivalence might present itself as more acceptable.[63] But this solution is not without its drawbacks. Quine pointed out that this solution to the problem of synonymity is only any good for what he described as cognitive synonymity. This means that no account is taken of relativisation to the users of the language.

The intensional conception of synonymity in the form suggested by Quine therefore concerns only one (although still a fairly broad) concept of synonymity. From this point of view there are three different concepts to be distinguished:

(a) cognitive synonymity based on the L-equivalence of expressions but ignoring pragmatic aspects;

(b) linguistic synonymity, which has to do with relations within a language and does respect pragmatic aspects (e.g. different attitudes of language users to certain expressions, expecially those expressed in belief sentences; this is a reason why the criterion of cognitive synonymity cannot cope with contexts that are in belief sentences or other sentences like them);

(c) interlingual synonymity, which also takes into account the relations between expressions of different languages.

The concept of cognitive synonymity, by which Δ_1 is synonymous with Δ_2 if and only if $\Delta_1 \equiv \Delta_2$ is L-true, i.e. if Δ_1 and Δ_2 have the same intension, is actually too broad. This is why we often find attempts to narrow it down to a form more suited to the explication of (cognitive) synonymity.

C.I. Lewis put forward a conception of synonymity which, while also being based on intension, is narrower than the one which says that synonyms are expressions having the same intension. Two expressions can be designated as synonymous (in the case of sentences it is usual to use the term *equipollent* in the same sense) if and only if the two expressions have the same intension and that intension is neither zero nor universal [83, p. 59].[64] Thus Lewis removes from consideration the synonymity of L-determined expressions (in the sense of [23]). In the case of expressions with zero or universal intension Lewis concedes that it is possible to speak of synonymity only if the expressions are 'analytically comparable'. He explains this notion by saying that two expressions are analytically comparable if (1) at least one of them is an elementary (i.e. not further divisible, or in a formalised language a primitive) expression and both expressions have the same intension which is neither zero nor universal; and (2) if they are complex expressions and they can be analysed into individual expression components in such a way that each component can be replaced by another with the same intension, the components themselves having an intension that is neither zero nor universal, and if both complex expressions have the same syntactic structure [84, p. 85]. For example, the expression $\Sigma \vee \sim \Sigma$ has zero intension in Lewis's sense. Thus it is not possible to consider the synonymity of the expressions

$\Delta_1 \vee \sim \Delta_1$ and Δ_2 if Δ_2 is an elementary expression. But in the sense of condition (2) it is possible to consider the synonymity of the expressions $\Delta_1 \vee \sim \Delta_1$ and $\Delta'_1 \vee \sim \Delta'_1$, even if both have zero intension, if Δ_1 and Δ'_1 have the same intension and that intension is neither zero nor universal. It is simultaneously obvious that both expressions have the same 'syntactic structure', at least in the sense that the concatenation of the individual expressions is alike.

Carnap's conception of synonymity comes very close to that of Lewis. Carnap [23] bases his conception on the idea of '*intensional isomorphism*', which is an explication of the less clear concept of the synonymity of two expressions. This concept of intensional isomorphism is also based on L-equivalence. Intuitively it can be described in the following terms: two expressions[65] are intensionally isomorphous or, to put it another way, have the same intensional structure, if the condition of the L-equivalence of the two expressions as wholes is satisfied and if the components that correspond to each other in the two expressions are also L-equivalent. We are still of course talking about components to which a relatively independent meaning can be assigned. As an example Carnap has the expressions '$2 + 5 > 3$' and 'Gr [sum (II, V), III]' as intensionally isomorphous, while '$7 > 3$' and the second of the previous two expressions are not intensionally isomorphous since while the components '7' and 'sum (II, V)' are equivalent, they are not L-equivalent. The concept of L-equivalence can also be applied in relation to the logical constants. Intensionally isomorphous are, for example, the expressions '$\sim (\Sigma_i \vee \Sigma_j)$' and '**NApq**' in the notation commonly used by the Polish logicians. So the concept of intensional isomorphism assumes an analytical approach to language expressions, presupposing a corresponding categorisation of language expressions, constants, operators, variables etc. Hence an exact definition of intensional isomorphism must be a recursive definition and respect syntactic rules and type differences. Only in the case of elementary expressions that cannot be further reduced to components or, to be more exact, whose components cannot be treated as language expressions to which a relatively independent meaning can be assigned, can L-equivalence suffice. In cases like that the concepts of L-equivalence and intensional isomorphism coincide.

Lewis's and Carnap's solution does, however, have a serious weakness. It does not always ensure interchangeability in those contexts which Quine describes as *contexts of propositional attitudes* [122, p. 202]. One special case of that type of context are the so-called belief sentences.

These are contexts concerning certain states of the language user and include an attitude to the sentences or their sense, i.e. to the propositions. The attitudes involved here are those commonly expressed by such terms as 'believes that', 'thinks that', 'denies that', 'is convinced that'[66] and others. Carnap himself analysed this type of context as non-extensional and also non-intensional. But in situations of this kind it would seem much better to apply solutions which do not adhere to the purely semantic approach, consistently disregarding all other aspects, especially the pragmatic. Another fact to be taken into account here is the differentiation of intension in the sense suggested by Lewis. If we were to cling to the purely semantic approach it would be possible to operate only with linguistic intension. But taking pragmatic aspects into consideration, such as the kind of criteria of sense used by different users of a language, we cannot avoid respecting the various criteria of sense as we try to give precision to the concept of synonymity.[67]

[C] *Synonymity, Pragmatic Intension and Pragmatic Criteria*

Pragmatic considerations can be brought into the specification of synonymity either by taking account of relativisation to different language users or by a relativisation to different conceptions of meaning and different criteria and ways of ascertaining meaning. Or both might be combined. In the first case we can operate with various pragmatic concepts concerning the states or properties of language users, usually understood as kinds of behaviour or dispositions for certain behaviour on the basis of which conclusions can be drawn about the language users' attitudes to the language expressions or the sense of those expressions. In the second case, it is usual to assume concepts of meaning or modes of meaning (e.g. 'pragmatic extension', 'pragmatic intension') which are considered as the pragmatic counterparts of the respective semantic concepts. In most semantic literature these two ways of dealing with synonymity have been little more than hinted at.

One attempt to resolve synonymity and a criterion for it along lines corresponding to the first of the ways suggested above is Quine's conception of 'stimulus' synonymity [122]. Two expressions are 'stimulus-synonymous' if in one or more language users they prompt the same 'stimulus meaning'. Quine defines 'stimulus meaning' separately for the affirmative and negative versions of it. Affirmative stimulus meaning is characterised with respect to a given language expression and a given language user as the class of all stimulations that would prompt a user's assent. Negative stimulus meaning is described analogously, except that it is dissent which is prompted. Stimulus meaning is therefore an ordered pair of affirmative and negative stimulus meanings. This conception presupposes the behaviourist psychology of stimuli and responses and a schema of conditionality, i.e. the same response schema (or just a disposition to a particular kind of response) to a given spectrum of stimuli. The idea of stimulus meaning further presupposes a measure of stability of that spectrum with respect to the various language users. At the same time, however, stimulus meaning (in relation to one and the same expression of language) may differ quite considerably from one language user to another. Quine quotes an example [122, p. 79]: An investigator doing a survey of the Himalayas discovered that the mountain seen from Tibet and called 'Everest' is the same as the mountain seen from Nepal and called 'Gaurisangar'. In sentences which express empirically accessible actions (Quine's 'occasion sentences') the expressions 'Gaurisangar' and 'Everest' are not and cannot be synonymous because each is associated with a different stimulus meaning. Only when the two expressions are used from the point of view of introduced geographical knowledge (which is also empirical in nature, though of a higher empirical level) can the expressions be used as synonyms. Therefore stimulus synonymity must be relativised to different levels of the empirical and to the empirical and conceptual equipment of the language users.

In addition to 'stimulus' synonymity Quine introduces certain other concepts which are of particular importance for the application of context. The main one is 'structural synonymity', which differs from Carnap's 'intensional isomorphism' primarily in that it allows components in a context to be replaced by other expressions that are 'stimulus-synonymous' with the original components.[68] Hence 'structural synonymity' in Quine's conception can be considered a concept that presupposes the application of pragmatic criteria for verifying synonymity. It was from this point of view that Quine and certain other authors criticised any definition of synonymity which is based solely on semantic concepts, notably 'intension' and 'intensional isomorphism'.

Carnap, who, from the purely semantic point of view, put forward a fairly exact conception of synonymity based on 'intension', found it necessary to adapt his original idea somewhat. In doing so he was effectively responding to those of his critics whose arguments were based mainly on pragmatic considerations. He let 'intension' retain its decisive status but broadened it to take in certain pragmatic aspects. By the conclusion which he reached the semantic concepts of intension and extension had corresponding pragmatic concepts, i.e. pragmatic intension and pragmatic extension [28].[69]

It is debatable whether pragmatic viewpoints can be applied equally to extension and its affines, i.e. the concepts included by Quine in the theory of reference, and to intension and the other concepts included by Quine in the theory of meaning. While Quine admits the application of pragmatic considerations only in the case of concepts from the theory of reference, Carnap is convinced that there is equal justification for operating also with the concept of pragmatic intension. The heart of the controversy is to be found in certain matters of methodology, especially the question of the admissibility of the mentalist approach in psychology, linguistics and the behavioural sciences generally, alongside the purely behaviourist approach. Carnap's 'pragmatic extension' and Quine's corresponding 'stimulus meaning' assume a class of stimuli or, more precisely, a class of actual stimuli. On the other hand the determination of pragmatic intension entails consideration of all logically possible (admissible, 'thinkable' etc.) stimuli or situations. It can be objected that with this characterisation of pragmatic intension it is impossible to avoid certain mentalist terms, such as 'thinkable' with respect to the equipment and knowledge of the language user, which have to be suppressed in a strictly behaviourist approach. We may quote as an example of the use of mentalist terms Carnap's definition of pragmatic intension for predicates [28, p. 242]: The intension of a predicate Q for a speaker XY is the general condition which an object y must fulfil in order for XY to be willing to ascribe the predicate Q to y.[70]

This characterisation of pragmatic intension is not very definite. The same lack of definition then passes to other concepts, including that of synonymity based on pragmatic intension and relativised to a single speaker. Two expressions Δ_1 and Δ_2 in $\mathscr{L}$ are synonymous for a speaker XY if for XY they have the same intension in $\mathscr{L}$.

The concept of 'intension' presupposes, as will have been appreciated from the preceding account, that all logically possible instances are taken into consideration. By contrast, 'pragmatic intension' requires that account be taken of only those logically possible instances that the language user is able to take into his decision processes. This solution cannot avoid using certain mentalist terms if it particularly has to respect the subjective abilities and facilities of the language user. In endeavouring to avoid these consequences and the frank admission of a mentalist methodology Carnap offers two possible solutions [28]: above all he assumes that language can be characterised as a system of dispositions for the use of particular expressions. Then one can either (1) bring in all the possible conditions by which to ascertain all possible responses (thereby establishing what is meant by the 'general condition' in the above definition of pragmatic intension), or (2) investigate all the possible states of the language user. Carnap describes solution (1) as a behaviourist method and solution (2) as a method of structural analysis. He sees the method of structural analysis as particularly well suited to a definition of the pragmatic intension of a language expression used by a robot or by some technical apparatus equipped to use language.

These conceptions, which are in any case little more than suggested and are not worked out in any consistent manner, can be met by a number of objections. The pragmatic conception of intension, synonymity, analyticity and other allied concepts (i.e. concepts of the theory of meaning in the Quine sense), which presupposes consideration of all logically possible situations, all logically possible instances, amounts to a confusion of two kinds of criteria: purely logical criteria that totally disregard anything non-logical on the one hand, and empirical or behavioural criteria on the other. There are therefore two possibilities open to us: either to respect the pragmatic aspects, which leaves no alternative but to surrender those aspects which are typical of intension and the other concepts from the theory of meaning, or to respect the viewpoint of 'same intension' and then disregard the pragmatic aspects.

This critique of the concept of pragmatic intension does not imply that in applying pragmatic aspects to the solution of synonymity the intensional approach has simply to be cast away. It should be recalled in this connection that intension can be understood in two ways – as linguistic intension and as what has been called sense intension. The above critique of pragmatic intension applied only to linguistic intension (for only linguistic intension belongs to the theory of meaning in the Quine sense), and not to sense intension.

So if we link the concept of synonymity to that of sense intension, we can define synonymity as follows:

Two language expressions are synonymous if and only if their synonymity can be verified by the same means. In other words, two expressions are synonymous if it is possible to verify the identity of their sense.[71]

Our analysis of synonymity so far enables us to draw the following conclusion: To determine the synonymity of two or more expressions, the expressions cannot be assessed in isolation; various kinds of relativisation have to be respected. The actual range of types of relativisation may be broadened or narrowed down as the case requires. The further we broaden the range, i.e. the greater the exactitude or completeness with which we define the *conditions for synonymity*, the trickier and more demanding it becomes to respect the conditions and to verify that they have been respected. We have here therefore a relation not unlike the complementary relation between the theoretical growth of accuracy in one respect and the practical possibilities of maintaining or checking that accuracy in another.

Conditions for synonymity are not of course an arbitrary matter. They depend on the demands placed on synonymity. Therefore synonymity must be conceived so as to contain not only data on the object language expressions which we consider synonymous, but also data on the range of relativisation considered. This means that synonymity is not a relation between expressions taken just as they are, but is governed by the conditions in which the given expressions occur and in which their synonymity is being considered. These conditions may be called 'marginal references'.[72] Then synonymity may be treated as the relation of two or more object language expressions taken in certain defined conditions. Moreover the determination of these conditions, i.e. the statement of the marginal references, may depend on all manner of semantic, but especially pragmatic, circumstances. The following examples show the 'marginal references' in *italics* and the relevant object language expressions in inverted commas:

(1) *In Tarski's conception* 'equality' and 'identity' mean the same thing *in arithmetic.*

(2) *In journalese it is usually possible to consider* 'the USSR' and 'Soviet Russia' as synonyms.

(3) *In classical physics* what is meant by 'mass' is the 'measure of the inertial and gravitational effects of matter'.

The synonymity in these examples is not a relation between two expressions of $\mathscr{L}$ taken in isolation, i.e. the relation between Δ_i and Δ_j, but the relation between these expressions taken in their marginal conditions. If we denote these conditions M_1 and M_2, synonymity becomes a metalanguage predicate whose arguments are the two expressions taken in their marginal conditions, i.e. $(\Delta_i M_1)$ and $(\Delta_j M_2)$, which may be written as

$$\mathbf{Syn}\,[(\Delta_i M_1), (\Delta_j M_2)]$$

Naturally the most common kind of situation is that where $M_1 = M_2$.[73] Marginal conditions may include, for example, the occurrence of language expressions in a particular text, their use by a certain person or body of persons, use in a particular situation and so forth. Synonymity in the strict sense of the word means that any occurrence of a particular language expression in certain circumstances is synonymous with any occurrence of another expression in the same or other defined conditions. This conception of synonymity in the strict sense can of course be weakened. For example, it may be merely assumed that in certain marginal conditions there is at least one occurrence of the relevant expression which is synonymous with any occurrence of another expression in defined conditions. This kind of weakening is of course a disadvantage, since it leads to a reduction of accuracy, and it is therefore expedient also to limit the definition of the marginal conditions in such a way as to make it possible to consider any occurrence of the given expression.

In the above notation the concept of synonymity is taken as a two-place metalanguage predicate. In order to define synonymity, we need to investigate what is necessary for a decision about synonymity of Δ_i and Δ_j, taken in their respective marginal conditions, to be taken. To arrive at a decision we need to be able to give an interpretation to the expressions. And in making an interpretation of (ΔM_1), we have to assume that we are capable of assigning to the expression in the given conditions (we shall call it the interpretandum) another expression (the interpretans). The interpretation too, however specified whether just semantically or, which is the more common case, pragmatically as well, ought also to respect certain marginal conditions. The interpretation can also be treated as a relation, which can be recorded as:

$$\mathbf{Inter}\,[(\Delta M_1), \Gamma],$$

where (ΔM_1) is the interpretans and Γ the interpretandum.

On the basis of the concept of interpretation we can define synonymity and characterise some of its properties:

$$(D_1 - V\,5\,C)\quad \mathbf{Syn}\,[(\Delta_j M_1), (\Delta_j M_2)] = \mathrm{df}\,\mathbf{Inter}\,[(\Delta_j M_1), \Gamma_1] \cdot \mathbf{Inter}\,[(\Delta_j M_2), \Gamma_2] \cdot (\Gamma_1 \underset{\mathrm{id}}{=\!=} \Gamma_2)$$

Synonymity defined in this way is a relation that is reflexive, symmetrical and transitive. If for Δ_j we insert Δ_i, we obtain:

$$\textbf{Syn}\left[(\Delta_i M_1), (\Delta_i M_2)\right] \equiv \textbf{Inter}\left[(\Delta_i M_1), \Gamma_1\right] \cdot \textbf{Inter}\left[(\Delta_i M_2), \Gamma_2\right] \cdot (\Gamma_1 \underset{\text{id}}{=\!=} \Gamma_2).$$

It holds further that

$$\textbf{Syn}\left[(\Delta_i M_1), (\Delta_j M_2)\right] \equiv \textbf{Syn}\left[(\Delta_j M_2), (\Delta_i M_1)\right].$$

Finally it holds that

$$\textbf{Syn}\left[(\Delta_i M_1), (\Delta_j M_2)\right] \cdot \textbf{Syn}\left[(\Delta_j M_2), (\Delta_k M_3)\right] \longrightarrow \textbf{Syn}\left[(\Delta_i M_1), (\Delta_k M_3)\right]$$

The concept of synonymity is, in linguistic writings, usually given in opposition to 'heteronymity'. This concept may also be defined as a two-place metalanguage predicate, and again the definition can be based on 'interpretation':

$$(D_2 - \text{V 5 C}) \quad \textbf{Heter}\left[(\Delta_i M_1), (\Delta_j M_2)\right] = \text{df}\, \textbf{Inter}\left[(\Delta_i M_1), \Gamma_1\right] \cdot \textbf{Inter}\left[(\Delta_j M_2), \Gamma_2\right] \cdot \sim (\Gamma_1 \underset{\text{id}}{=\!=} \Gamma_2).$$

Unlike synonymity, heteronymity cannot be reflexive. If we were to insist on **Heter** $\left[(\Delta_i M_1), (\Delta_i M_2)\right]$, we would break the principle of univocality, which is a major presupposition for communicability. This applies particularly if $M_1 = M_2$. But at the same time it is useful to limit the validity of the principle of univocality to just the exactly defined marginal conditions. This also makes it possible to limit the ban on **Heter** $\left[(\Delta_i M_1), (\Delta_i M_2)\right]$. If the language does not contain an absolute ban on **Heter** $\left[(\Delta_i M_1), (\Delta_i M_2)\right]$, this means that there occur in it expressions that can be described as ambiguous.

Unlike the concepts of synonymity and heteronymity, that of 'ambiguity' can be characterised as a one-place metalanguage predicate.[74] An expression is ambiguous if there are at least two different interpretations of it:

$$(D_3 - \text{V 5 C}) \quad \textbf{Amb}(\Delta) = \text{df}\, \textbf{Inter}(\Delta, \Gamma_1) \cdot \textbf{Inter}(\Delta, \Gamma_2) \cdot \sim (\Gamma_1 \underset{\text{id}}{=\!=} \Gamma_2)$$

Our analysis of the concept of 'synonymity' may be concluded in these terms:

(1) 'Synonymity' must be linked to expressions to which a relatively

independent meaning can be assigned, that is expressions for which an interpretation can be made.

(2) In defining synonymity we have to respect marginal conditions.

(3) The determination of those marginal conditions is governed by not only semantic, but also pragmatic considerations.

(4) If the marginal conditions can be drawn up in various ways it must be assumed that there are more than one concepts of 'synonymity', each tied to a different statement of marginal conditions.

VI. THE CRITERION OF SENSE

1. THE FORMULATION OF THE PROBLEM

[A] *Conceptions of a Criterion of Sense*

In our account of the sense or intension of a sentence we said that we understand a sentence if we assimilate that property of the sentence which is preserved in any correct translation of the sentence. To say that we are dealing with the assignment of an abstract object of a conceptual nature does nothing in itself to solve the problem of a criterion of sense. Instead it merely hints at the problem, albeit only implicitly. For it assumes that the language user is capable of performing, actually or potentially, certain operations on the basis of which the assignment or non-assignment of a sense can be decided.

(1) Is there one single type of operation that the language user is capable of performing with the aim of ascertaining, verifying or otherwise testing sense? In the evolution of the problem of a criterion of sense there have been many attempts to give a positive answer to this question. Various conceptions that put forward a positive answer to the question may also be described as *unitary theories of sense*. They include notably the verifiability conception of sense, the operationist conception, the translation conception and others. These conceptions specify an operation, such as verification, measurement (or other related operations), translation etc., on the basis of which it is possible to decide on the assignment or non-assignment of a sense. The literature dealing with the problems of sense tends to be dominated by unitary theories of a criterion of sense, although, as we shall see in due course, serious objections can be raised against some of them. Many of the attempts at a unitary theory of sense and a conception of a criterion of sense may be said to have been motivated by subsidiary concerns instead of by some positive endeavour to resolve the central issue. (Examples of this are the efforts of adherents of the verifiability conception of a criterion of sense to eliminate 'metaphysical statements', efforts to reduce all data in science to 'observable data', or efforts to apply the notion of 'observability in principle' and so forth.)

(2) If we answer the first question in the negative it becomes possible to ask further questions: If there are more than one types of operations which the language user is capable of performing in order to decide on the assignment or non-assignment of sense, is the selection of this or that operation an arbitrary choice on the part of the language user, a matter of (linguistic) convention, or something that depends on other (objective) circumstances that are extralinguistic in nature and outside the subjective equipment of the language user? The latter part of this question may also be formulated as follows: Are criteria of sense only subjective, or can they also be objective?

In what follows we shall try to justify the negative answer to the original question, i.e. including a critique of the attempts to put forward a unitary theory of sense, and we shall also try to suggest that a criterion of sense must also be taken objectively.

[*B*] *The Criterion of Sense and the Problem of Linguistic and Sense Intension*

The problem of the criterion of sense concerns all expressions to which a relatively independent meaning can be assigned. In formalised languages it is possible to restrict matters just to sentences or to define exactly the kinds of expressions assumed to be open to relatively independent semantic analysis.[1] In natural languages it is expedient to link the problem of a criterion of sense above all to those expressions which (a) may not only be thought of as relatively independent objects of semantic analysis, but (b) may also fill a relatively independent role in the process of communication.

Requirement (b) narrows the applicability of the criterion of sense to clauses, or sentences, to use the common categorisation of expressions of language. It follows from this requirement, which in the copious literature on the topic tends to be assumed implicitly, that any criterion of sense cannot ignore the syntactic structure of the language.

In natural language we have to respect not only that which corresponds to formation rules, but also other rules specific to the language. Therefore, wherever we mention sentences in what follows, we shall mean well formed sentences whose formation and, where relevant, external form respect the relevant rules.[2]

Thus if we exclude all sentences that break the relevant rules, we can, so it seems, divide the remaining sentences into two subsets: sentences that have a sense and those that do not. We are then offered the possibility of formulating a criterion of sense on the basis of the common property of all the sentences that can be included in the subset of sentences with sense.[3] If all these sentences have some property in common we need to ask after the nature of that property and after the nature of the means of language by which to express it. It has been a common practice of adherents of the unitary theory of sense to start out from the assumption that it is an empirical property or one that is empirically verifiable. This standpoint has however, as we shall show in the more detailed analysis of the verifiability criterion of sense, led to insoluble difficulties.

The scope of the problems connected with the question of the nature of the property possessed by all sentences that can be treated as having a sense is of course much broader. We might mention the following questions in particular:

(a) By what means can the property be ascertained or verified?

(b) Are these means uniform for the entire set of sentences that have sense, or is it essential to use different means in different situations and tasks with which the use of language is connected?[4]

(c) If having sense is a property of certain sentences, is the language expression for the property, i.e. a predicate, vague or clear-cut? In other words: Is there a sharp divide between the class of sentences that have sense and its complement?

(d) Is the predicate 'have sense' a one- or many-placed predicate? We take it for granted this is of course not an object language predicate, but a predicate in the metalanguage. This means that its arguments are the names of sentences of the object language. If it is a many-place predicate, we have to ask what its other arguments are.

For reasons that will be stated briefly, it must be seen as expedient to narrow further the competence of these questions. First and foremost we can leave side analytical sentences in the true sense of the word, as well as all contradictions. It should be noted in this connection that the definition of analyticity depends on the nature of a given expression and its interpretation or the interpretation of its components. There is a substantial difference between the analyticity of the sentence '$S \vee \sim S$' and the sentence 'all spinsters are unmarried women'. Only the former is analytical in the true sense of the word, whatever the interpretation put on its components. This is the reason for leaving aside analytical sentences in the true sense of the word and for concentrating only on those sentences that are usually described as 'having cognitive sense'.

The term 'having cognitive sense' has been defined variously and not always with adequate precision. It is usually held to apply to synthetic sentences, 'cognitive sense' being identified with 'empirical sense'. This kind of view need not in principle be narrow and wrong if the concept of an empirical starting point or empirical basis of all human knowledge is treated sufficiently generally and if the empirical basis is described objectively and not in subjectivist terms.[5]

It is possible to point to certain assumptions that taken en bloc enable us to ascertain and verify that sentences do have a cognitive sense. Whether we can decide on the sense of a sentence usually assumes that we can decide on the sense of those expressions of language of which the sentence consists.[6] But this is a necessary, though not adequate, condition, for the decidability of the sense of all a sentence's components does not of itself imply the decidability of the sentence as a whole.

If we assume that we are considering the sense of sentences used in any communication, there is no overlooking the fact that the concept of 'sense of a sentence' has some important pragmatic aspects. Writings on semantics usually point to one major dispositional property of a sentence that has a sense, namely translatability. 'Having a sense' is therefore relativised to a given set of language users who are capable in translation of preserving that property which remains constant in any correct translation.[7]

In describing the method of extension and intension we pointed out certain possibilities for refining and expanding the method (suggested in particular in the work of C.I. Lewis) by distinguishing between sense intension and linguistic intension. Thus there is no avoiding the question of whether 'having a cognitive sense' concerns only sense intension, only linguistic intension, or both. If sense is relativised to a given set of language users, we are assuming that they are capable of deciding whether the sentences make sense on the basis of certain specified operations or the possibility of those operations. There are two attitudes that may be taken in this context:

(A) Any such operation, or the possibility of it, must be somehow qualified. It is usually emphasised that it must be an operation only of a certain kind, e.g. verification or verifiability, translation or translatability, measurability etc. This viewpoint is taken by most adherents of the unitary theory of sense. It is, however, a narrow view in that it connects the problem of making sense and a criterion of sense only with one kind of activity on the part of the language user. This is what may lead to certain subjectivist and essentially idealistic conclusions. Moreover this viewpoint accentuates only some aspect of intension: sense intension where the verifiability theory of sense is applied, linguistic intension in the case of the translatability concept and so forth.

(B) As opposed to the above viewpoint, we find more appropriate and more justified the second view, whereby the operation on the basis of which a decision on the sense of a sentence is reached need not be somehow generally qualified of necessity. The operation settled upon is relative, i.e. it depends on the nature of the sentence in question as well as on both objective circumstances connected with the nature of the particular universe, and subjective circumstances pertaining to the possibilities and capabilities of the given group of language users. What is fundamental is however the possibility of assessing the *quality* or *reliability* of the operation with respect to the particular aims, or findings, associated with each type of decision. For these reasons we are not justified in binding a criterion of sense to any one type of operation. Instead we shall think in abstract terms of a particular decision process which may be qualified in some way, e.g. on the basis of its results. This viewpoint not only does not lead to subjectivist conclusions (although it obviously cannot eliminate the role of the user or interpreter of the language itself, or, to put it abstractly, the role of a subject in the activity), but it brings to the fore *objective criteria*. And it need not always mean an utter denial *a limine* of all conceptions of sense and a criterion of sense associated with the first viewpoint. In individual special circumstances it is useful not only to respect, but also practically to apply partial conceptions. But difficulties usually arise the moment these partial conceptions claim to offer a more universal solution. In the next section we shall look at some of the more important conceptions that are attempts to provide a unitary theory of sense.

2. THE OPERATIONIST CRITERION OF SENSE

The characteristic feature of most attempts to create a unitary theory of sense is the endeavour to link an operation, whose purpose is to ascertain or verify the sense of an expression, with the principles of empiricism. One of the most significant principles of empiricism is the conviction that all knowledge expressed in sentences that are not analytical must be based on an empirical starting point. In itself this principle is unobjectionable, but what are dubious are the various conceptions of that starting point. This is also why what is sometimes described as the 'empirical criterion of sense' is not a single united conception, but a body of very different viewpoints, each with their own distinctive idea of the empirical starting point.

In trying to bring some precision into the empirical criterion of sense there is a danger of running to one of two extremes: either the criterion is defined so narrowly that it has to be supplemented by other means (e.g. schemas of 'translatability', 'reduction' etc.)[8] or it is defined so broadly that what is called the empirical starting point is capable of embracing anything at all, including, for example, 'experience' as subjectivistically conceived by certain adherents of pragmatism who are even prepared to go so far as to consider 'religious experience'.

It is quite natural that those inclined more towards the natural sciences, who were resistant to such a broadening of the empirical criterion and the empirical starting point of human knowledge, found the narrower, and hence more strict, definition of a criterion of sense more acceptable. This is particularly the case of the operationist criterion of sense whose author was the physicist P.W. Bridgman. His conceptions were met with considerable sympathy, especially in the empirical sciences.[9]

Of course Bridgman was not originally concerned with the solution of semantic problems: his main concern was to elucidate the nature the scientific concepts and the problems of defining them. Nevertheless, Bridgman's schema of 'operational definition', although it was only put forward in conceptual form, may be seen as one version of an empirical criterion of sense.[10]

[*A*] *Einstein's Definition of Simultaneity and the Operationist Criterion of Sense*

The basic layout of an operational definition, and hence of the operational criterion of sense, was Einstein's well-known generalisation of the concept of 'simultaneity'. Until Einstein *simultaneity* had been thought of as a property of two or more events. The property had been definable in terms of Newton's conception of absolute time. Einstein subjected this conception of simultaneity, as associated with Newton's conception of absolute time, to a critical analysis and proved that the simultaneity of two events, the spatial distance between which is not negligible, may be defined by means of measuring the time-space coincidence of signals of the same kind emitted from the places where the two events occur and meeting midway between them. Einstein was chiefly concerned to prove that simultaneity is not an absolute property of two events, but that it is a concept that should be relativised with respect to measuring procedures performed by an 'observer'.[11]

Bridgman considered Einstein's generalisation of the concept of simultaneity a model for his schema of operational definition. His basic idea was this: A concept is synonymous with the corresponding class of operations [16].

From the semantic point of view it is important to notice that the original version of the operationist criterion of sense does not primarily concern sentences, but terms of a descriptive nature, i.e. terms that serve in the language of science to described actions, events, processes etc. Bridgman considered that the only sentences that make sense are those which contain terms that are definable in the spirit of the schema of operational definition. Thus, for example, a sentence that contains the concept of absolute time is neither true nor untrue, but is meaningless. So the operationist criterion of sense replaced a sentence about the non-existence of absolute time by one that said that the concept of 'absolute time' is meaningless.

In the spirit of this suggested schema of operational definition Bridgman analysed the basic types of physical parameters. Later supporters of operationism subsequently extended the schema to take in other categories of scientific thought, i.e. laws, hypotheses etc.

[*B*] *Critical Comments on the Operationist Criterion of Sense*

The original version of the operationist criterion of sense was fairly simple. This conception brought about a very radical limitation of the ordinary nomenclature of the natural sciences and eliminated from it all terms that could not be linked to some empirical or experimental activity. Thus it took a negative stance to all concepts that were more connected with the theoretical aspect, with ideas arising out of hypothesis and so forth. The history of physics, as well as some other sciences, tells us that many important concepts (including those of atom, electron, elementary particle, spin and others) passed through a stage where they were linked only to certain theoretical notions, rather than to the results of measuring, observation or experiment.

The operationist attitude to scientific nomenclature therefore usually means a great narrowing or restriction of the nomenclature which may prove harmful and unproductive. This is not to deny that the operationist stand may have certain positive features. It may reinforce the requirement that all components of the nomenclature in empirical sciences be so constituted as to make plain their connection with empirically verifiable components, to give an exact formulation of that connection and so forth. But it is clear that the operationist criterion of sense is not sufficient in itself, in that it has to be supplemented or expanded by other means to guarantee the invariant transfer of sense under translation. For operationism to claim to be treated as the sole universal conception of the empirical starting point of science and of the cognitive meaningfulness of scientific nomenclature, is a claim that must be described as exaggerated and unwarranted.

It is sometimes argued against the operationist conception of scientific nomenclature and the operationist criterion of sense that such conceptions are an assertion of subjectivism or subjective empiricism. The objections are fully justified if what we are thinking of is something that might be described as the homocentric version of operationism or *homocentric operationism.* By this version the properties of *objects* are effectively reduced to operations, qualified in some way, on the part of a subject, i.e. to some kind of subjective activity of man.[12]

The homocentric version of operationism may appear as the sole acceptable solution to the interpretation of terms in the empirical sciences only insofar as we insist on the dilemma: naive realism – operationism. Naive realism treats descriptive concepts or empirical predicates as expressions of 'absolute properties' of objects. According to this conception, properties such as 'have a given temperature', 'have certain spatio-temporal properties', 'have a certain trajectory' may be assigned to any object in the same sense. But in terms of modern physics it would be controversial to consider, for instance, the 'temperature' of a single elementary particle. The situation is further complicated by the degree of metrical refinement of measuring apparatus, the interaction between the object measured and the measuring apparatus, the inevitability of applying quantum principles of measuring, etc. Thus it is understandable that within the natural sciences themselves there could not fail to be resistance to naive realism and its absolutisation of all qualities and reification of all elements of scientific nomenclature. The abandonment of the naive realist conception, with its system of unchanging things and absolute qualities, is not, however, grounds for justifying the purely subjectivist conception of the language of science and the homocentric conception of operationism.

Against the homocentric version of operationism we might erect a conception that treats an empirical or experimental operation itself as an objective process the basic characteristics of which can be determined objectively. This applies, for instance, to measuring procedures, various types of observation made with the aid of apparatus, and so on. It is useful in this connection to employ, in the objective analysis of these and other empirical procedures, the communication model in accordance with which an object under observation is a source of messages, these being accessible through a given channel (e.g. by means of material apparatus) with certain properties, e.g. certain limits of discrimination, a certain 'capacity', 'memory', etc. And there is no doubt that both the message source and the operation of the channel, in which there may in addition be some noise, can be treated as objective. Subjectivism is not even justified in situations brought about by the limitations of transmissibility or deformations caused by noise or the retroaction of the channel on the message source.[13] Finally, it is still possible to assume on this cascade an 'observer channel' which is also, in principle, capable of being characterised objectively and the decisions of which are relativised to certain restricted 'capacities', 'memories' and 'delays'.

There are still more objections that can be raised against the operationist criterion of sense, especially if connected with the homocentric version of operationism. If the sense of a term is synonymous with the corresponding class of operations, then this may lead to unwarranted extension of scientific nomenclature. It is well known that today we have quite a rich and differentiated range of ways of measuring heat. If the sense of the term 'temperature' is synonymous with the corresponding class of metrical operations, then 'temperature' as measured by an ordinary thermometer is one thing, and 'temperature' as ascertained by other means, such as in the thermodynamics of gases, is another. It then follows from the operationist criterion of sense that we are not entitled to operate with the concept of 'temperature', but only with as many different concepts as there are basically distinct ways of measuring heat. This trend is at variance with two requirements that apply (usually elementally) in the formation of scientific nomenclature: (1) the requirement of interpreting objectively properties that are ascertainable empirically (and with it the requirement of identifying the elements of the message source on the basis of the different channels used and the various kinds of output data); (2) the requirement of a measure of economy in the naming relation, i.e. the requirement of the non-proliferation – where avoidable – of names for the same nominata.[14]

In any critical analysis of the operationist criterion of sense it is impossible to overlook Bridgman's requirement that the sense of a term be *synonymous* with the corresponding class of operations. Firstly, it may not be entirely clear what is to be understood by the term 'corresponding'. For example, in Bridgman's conception it was taken as a matter of course that a measuring of length by means of certain metrical apparatus 'corresponded' to the term 'length'. But given that each act of measuring (just as any other empirical procedure) takes place in time and space, it need not be unequivocally obvious just where the correspondence is. The original versions of operationism made light of this problem. Other problems arise in connection with the requirement of synonymity: Bridgman's formulation required that expressions for properties (e.g. 'simultaneity', 'temperature', 'length' etc.) and expressions for the relevant metrical or other empirical procedures be synonymous. Yet it is easily demonstrated that this requirement may lead to confusion, if we were to treat both types of expression as intensionally isomorphous or L-substitutable for one another. Clearly, this formulation operates with synonymity as a hazy and ill-defined term. Hence it is essential to bring more precision to the formulation so that the sense of the demands placed on the operationist criterion of sense may be clearer.

Here too it is advantageous to employ the communication model of empirical procedures [110]. Any empirical procedure assumes traffic along a channel between the source of the message (the input of the channel) and the channel's output. The channel works by making decisions about the input possible on the basis of data at the output. For example, experimental apparatus for carrying out spectrum analysis must work in such a way as to enable us to make the most accurate possible estimation of the structure of the source on the basis of the relevant bands in the spectrum.[15] At this point we should clarify the demands that have to be placed on the channel's operation. Two requirements may be formulated:

(1) Data on the state at the input are *deducible* on the basis of data ascertainable at the output of the channel;

(2) *Meaning* remains *invariant* under transmission.[16]

These requirements correspond more to what ought to be done by the actions of an empirically of experimentally working scientist than to a description of the actual practice of the scientist.

Thus we assume the availability of some procedure whereby to deduce logically the state at the input on the basis of the state ascertained at the output. This is the syntactic aspect of these requirements. We assume further that we are capable of deciding as to the universe that is accessible by means of the channel, again on the basis of the state at the output. This is the semantic aspect of the requirements.

This conception of empiricism and empirical procedures in science is a fairly broad one and does not usually use the communication model, but it does have some fundamental shortcomings:

(a) Above all it is restricted to just the syntactic analysis of an empirical procedure and does not take sufficient account of the semantic and, most importantly, the pragmatic aspects;

(b) The conception ignores the fact that the channel which is interpolated between input and output is generally affected by noise, that both this channel and the observer channel have certain by no means unlimited properties ('capacities', 'memories', 'delays' etc.), and that the dependence of states at the output on states at the input is generally a stochastic type of dependence.

The communication model of an empirical procedure presents some profitable means not only of overcoming these shortcomings, but also of making more precise the above requirements. Any empirical procedure must facilitate the kind of *decision process* (or conjecture) about states at the input on the basis of states at the output that will ensure – with respect to the goals or tasks set – a quality of decision-making that is compatible with the goal or task. In other words, this model is also capable of taking into account the pragmatic aspects of an empirical procedure. Moreover, this model does respect the basic fact that the facilities and degree of accuracy, capacity and other features of the metrical or experimental apparatus are not unlimited, but limited. These are the main reasons why it is expedient to alter (and in one respect generalise) the above requirements that have to be made of a channel that models empirical procedures:

(1a) Data on the state at the input are *decidable* on the basis of data ascertainable at the output of the channel;

(2a) The decision space is a subset of the space of all possible states at the input, or, in the extreme case, it coincides with the latter (invariance of meaning);

(3a) The risk connected with the non-satisfaction of (2a) – we call it the *semantic* risk – does not on average exceed the level compatible with the aim of the given class of empirical procedures or tasks.

This conception of the empirical starting point of human knowledge and of the basic nature of empirical procedures in science can surmount the subjectivist and homocentric consequences of the original version of operationism. At the same time it may also be the starting point for a quite *objective* analysis of any empirical procedure (e.g. measuring or observation), an analysis, moreover, that far from ignoring the relative a priori limitations of the 'observer' or 'experimenter', actually has an inbuilt allowance for them.

3. THE VERIFIABILITY CRITERION OF SENSE

[*A*] *Motives for the Original Version of the Verifiability Criterion*

Verification of a sentence as a possible criterion of sense is usually given in the literature in the context of the beginnings of the Vienna Circle. In fact the idea of verifiability is not an original idea of the members of the Vienna Circle, having been formulated – although not in all clarity – earlier by L. Wittgenstein (166, Theorem 4,024): To understand a proposition means to know what is the case if it is true. This means we are not required to know whether a sentence is true or untrue, but to know the circumstances that would enable us to ascertain its truth value.

This principle of verification was adopted without any major changes by the adherents of the Vienna Circle. And yet the actual impulse behind the adoption of Wittgenstein's view was neither the desire to resolve the criterion of sense, nor the desire to define the property that is common to all the well formed sentences that can be treated as having a sense. What was more at work here was the desire to adduce compelling arguments against the kind of 'school' philosophy, usually demonstrated on texts from the German idealists, and to demonstrate that that philosophy 'had no sense'. Then because the concept of 'having a cognitive sense' was identified with that of 'having an empirical sense', and because the possibility of verification was originally understood as the possibility of direct verification based on straight observation and on sentences about the facts observed, it was fairly easy to show that the texts selected did not 'have a sense'.

We do not wish to appear to be claiming that any criticism of the metaphysically and idealistically oriented philosophers from this angle is unjustified, but it is apparent that the verifiability criterion in this form amounts to an extremely narrow and simplified criterion of sense. Although the question of the translatability of such texts may be a matter of controversy,[17] there is no question that such translations exist. There are both interpretations of the passages in more straightforward language, and literal translations into other languages. But it is questionable how far these passages are translatable into sentences that can be subjected to direct empirical verification in the form demanded by supporters of the empirical criterion of sense.

There is no doubt, however, that many of the sentences from such texts may have some emotional effect, the play of terms and the conglomeration of words generally may have certain psychologically and esthetically important consequences, and so they do 'have sense', at least in this respect. But if we were to demand that the sentence 'das Absolute ist das absolute Absolute' be verifiable by the same schema of verification as the sentence 'on May 20 the bus leaves London for Southampton at 8:05 a.m.', then the former sentence is clearly not verifiable.

Thus the verifiability criterion of sense was conceived as an empirical criterion, the empirical starting point usually having been conceived very narrowly – it was tied to 'sentences about observation' (Beobachtungssätze). It should be added that the concept of Beobachtungssatz was never – at least during the early period of the Vienna Circle – defined exactly, let alone uniformly, which means it was only a vague concept. This in turn means that the boundary between the set of sentences that can be treated as 'Beobachtungssätze' and the set of those that cannot is indefinite. For it is necessary to distinguish the normal observer with normal sensory equipment from the observer with a defect in some area (e.g. the blind, deaf-mute or colour-blind), and those from the observer whose facilities for observation and discrimination etc. are extended or expanded by metrical or experimental apparatus.[18]

If we disregard the difficulties connected with defining the concept of 'Beobachtungssatz' and treat it as an exactly defined concept (which is not the case in reality), we might attempt to give a closer specification of the verifiability criterion of sense. The usual examples given of the use of this criterion are not sentences that assume *actual observation*, but sentences about observation that assume a non-actual observer (sentences such as: 'There are mountains on the far side of the Moon'. 'The man who lived in this house at Pompeii centuries ago was dark-haired'. 'In the year 2000 rocket-planes will be used for human transportation'. etc.). This means that the first step in specifying the verifiability criterion of sense has to be the differentiation of 'actual observer' and 'language user'. Using the ordinary terms of communication theory, this could also be expressed by saying that the empirically relevant properties of the channel of an actual observer and the channel of a language user (capacities, memories, delays, order of accuracy etc.) need not be identical. Hence the adherents of the verifiability criterion of sense never demanded any actual verification to be made (which would follow from identity of 'language user' and 'observer'), but merely that verification be *possible*. It was customary to use in this context the term 'verifiable'. In other words, a sentence has sense not only if verified, but also if verifiable.

The interpretation of the concept of 'verifiable' was not originally uniform. 'Verifiable' might be taken to mean, for example, that verification is *conceivable*. It is conceivable – and this is something that has come about – that some kind of satellite may photograph the far side of the Moon. It is conceivable that there were other people alive at the time in Pompeii who could check for themselves whether the man in question was dark-haired or not. Equally it is easy to conceive of the man in the year 2000 who will fly in a rocket-plane. This conception has then no problems over the sense of sentences about the past or future, provided we can conceive of an observer or the 'expectation' of an event. For example, M. Schlick [140] stressed that 'expectation' is a perfectly legitimate method of verification. This will lead inevitably to the question: What about the situation where we cannot imagine any human observer, for example in the description of an event in the Palaeozoic period, when Man did not exist? There would seem to be an analogy here with the conception of 'prinzipielle Koordination' of the early version of positivism.[19]

The concept 'conceivable' has certain other drawbacks. One can conceive of a supernatural being with a human head and an animal's body – and man has proved capable of giving such conceptions shape in art – and we can conceive of an observer with infinitely greater facilities for observation than our contemporary scientist with his electron microscope, spectrum analyser, telescope and so on. It is probably impossible to set any exact limit on the human imagination. This is why the original intention of the supporters of the verifiability criterion of sense misfired: on the one hand there was a desire to segregate cognitively or empirically meaningful sentences, while on the other the gate was thrown wide open to practically unimpeded arbitrary imagination, fantasy and illusion.

For these reasons later attempts to give a more exact formulation of the verifiability criterion of sense usually abandoned such terms as might lead to this kind of indefinite psychologising interpretation. The commonest version of the criterion is that which is based on the concept of the 'logically possible': a sentence has sense if and only if its *verification is logically possible* (occasionally less precise expressions were used such as possible 'in principle' and the like).

The attribute 'logically' plays an important part here. In his critical analysis of the verifiability criterion of sense [82] C.I. Lewis had already demonstrated that the concept of 'verifiability', or 'possibility of verification', was ambiguous in itself. In particular it could mean (a) the possibility of an empirically based decision about the semantic characteristic of a given sentence, where the decision process might not be actually feasible (because, for example, of limited empirical or experimental facilities), although it might be feasible ultimately, (b) a logically based decision that rests on the exclusion of anything that is at variance with the logical structure of the given language. Interpretation (a) obviously need not differ very much form the earlier interpretation that was tied to the concept of 'conceivable'.

Schlick, who was obliged to accept Lewis's criticisms, also distinguished between the 'empirical possibility' and the 'logical possibility' of verification. According to his conception anything is 'empirically possible' that is not at variance with the laws of nature [140, p. 152]. This conception of the empirical is therefore relativised to a given or attained level of knowledge, fixed first and foremost in the laws revealed by the natural sciences. One obvious immediate objection is that this conception of the empirical and the 'empirically possible', which implicitly brings in certain a priori criteria, denies the possibility of new experience which disagrees with the given level of theoretical knowledge – it denies the 'possibility' of 'emergent facts'. Yet it is of course these circumstances that have often played a very important part in the history of scientific cognition and have also been a major influence on theoretical thinking. But it was not Schlick's wish to tie the concept of sense to verifiability in the sense of empirically possible verification; what he meant was the logical possibility. His basic standpoint, which is typical for the conception of a criterion of sense in the Vienna Circle period, says:

> "Verifiability, which is the sufficient and necessary condition of meaning, is a possibility of the logical order; it is created by constructing the sentence in accordance with the rules by which the terms are defined"
>
> [140, p. 155].

Schlick was also convinced that this dependence of the criterion of sense on the logical possibility of verification created a very sharp watershed between the properties 'have sense' and 'not have sense'. There can be, as he says, no graduated transitions between the two, which constitute a principle of excluded middle. We cannot let this pass without a few critical comments.

(a) Linking the verifiability criterion of sense to the concept of 'logical possibility', which is not really exactly defined either by Schlick or other adherents of this conception, brings about a connection between the logical structure of language (set primarily by the syntactic rules of each particular language) and the criterion of sense. This means that the criterion, which is expected to be an empirical criterion of sense, is dependent on what is logically admissible in terms of the syntactic rules of a given language. The trickiness of the situation can be partially overcome by introducing and respecting meaning postulates. If we take Schlick's example of a sentence that has no sense: 'Show me the country where the sky is three times as blue as in England', it is obvious that we can only achieve 'logical impossibility' after the introduction of certain postulates, such as one to establish the logical incompatibility of the terms 'blue' and 'three times'. Such postulates may, however, be a product of experience or of an experience-based dependence among the extralogical constants of the given language. This can be demonstrated with greater precision on the example of the formalised language described earlier. If that language has the valid meaning postulate

$$(\forall x)(P_1 x \longrightarrow P'_1 x).$$

then the sentence '$P_1 p \cdot \sim P'_1 p$' (i.e. Prague is in Bohemia and Prague is not in Europe) obviously has no sense. Similarly, it holds that in consequence of the postulate about the non-reflexive nature of the predicate P_2, i.e.

$$(\forall x) \sim P_2 x, x$$

the sentence '$P_2 p, p$' (i.e. Prague is more northerly than Prague) also has no sense.

So if the concept of 'logical impossibility' has to respect a semantically relevant dependence between extralogical constants of the given language, which is to all intents and purposes the case of every natural language that adherents of the verifiability criterion of sense have used to demonstrate the property 'not have sense', then the concept depends on how the meaning postulates are introduced. We have already stated that the various ways may be tied to the results of some experience, but they may be connected with other conditions as well.

(b) The properties 'have sense' and 'not have sense' may be treated as predicates of the semantic metalanguage. Also the sentence of the metalanguage 'it is logically impossible to verify Σ' can only be decided upon if we know the semantic nature of the object-language sentence that is being expressed by Σ in the metalanguage. This places us in a vicious circle: a sentence has a sense if it is logically possible to verify it. A sentence can be logically verified if and only if it has a sense.

(c) According to Schlick's original conception sentences that are contradictory obviously have no sense. The contradiction may be on the basis of purely syntactic rules of the given language, or on the basis of the meaning postulates respected.[20] If we leave aside analytical sentences (in both senses of the concept 'analytical'), we must assume that the sentences of the semantic metalanguage 'Σ has a sense' or 'Σ has no sense' are synthetic sentences.

In this situation there is no avoiding the question of the nature of the metalinguistic predicate 'have a sense' or 'have no sense' as the case may be.[21] There are two possibilities here: Either it is an empirical predicate, in which case the synthetic sentence 'Σ has a sense' has to be verified, or it is not an empirical predicate and the synthetic nature of the metalanguage sentence becomes questionable. This simultaneously makes the very verifiability criterion of sense questionable.[22]

If 'have a sense' were an empirical property, we could not avoid treating the metalanguage predicate 'have a sense' as *vague*. In other words we cannot exclude the possibility that one section of the language users may decide that the given object-language sentence makes sense, while another section may decide the other way. But this clearly brings with it the downfall of the ambition of adherents of the verifiability criterion to see in that criterion a means for making a clear-cut division between the complementary sets of sentences that do and do not have a sense.

Intuitively of course this property is not an empirical one. However, this does not make it impossible to use the criterion of sense on itself. (At the same time we may be dealing, understandably, with a criterion of sense of a 'higher level', which is conceived in principle along the same lines as the criterion of sense of a 'lower level'.)

Thus the verifiability criterion, based on the concept of the 'logical possibility of verification', is fraught with insoluble problems. The concept of the 'logical possibility of verification', which was supposed to replace that of the 'conceivability of verification', has thus not proved any more suitable and has been incapable of accurately defining the property of 'have a sense'.

[B] *Different Degrees of the Verifiability Criterion*

The original version of the verifiability criterion of sense arose out of the desire to see a clear-cut distinction between syntactically correct sentences with a sense and sentences to which no sense could be assigned. At the same time verification was understood to mean *complete verification.* The requirement of the logical possibility of complete verification was also connected with the problematical radicalism favoured in the main by Schlick. In the discussion of 'report sentences' (Protokollsätze), in which he himself denied these sentences the role of a starting point in cognition, Schlick said that the basis of human empirical knowledge are assertions (Konstatierungen), as he called sentences about what is 'simultaneously perceived'. Such sentences, according to Schlick, are definitely determinable with the same unambiguity as sentences of an analytical nature. It was on this basis that he then raised the requirement of complete verification, which might be formulated as follows:

A sentence Σ has sense if and only if it is neither an analytical sentence, nor a contradiction, and if it follows logically from a consistent finite class of sentences Φ, the elements of that class being observation sentences.[23]

Against this version of the criterion of sense, based on the requirement of the possibility of complete verification, a number of serious objections may be raised.[24] We know that the verifiability criterion of sense was formulated as a device by which radically and unambiguously to separate meaningful sentences from 'metaphysical' sentences that do not have a sense. But if we were to use this version of the criterion, with its requirement of complete verification, in relation to the discoveries of the natural sciences, especially their laws, we would reach the paradoxical conclusion that even the sentences that express those laws are a 'metaphysics' lacking sense. For this version of the verifiability criterion of sense denies the possibility of verifying certain sentences that could be recorded in a form using the general quantifier. Let us take the simple example of Newton's law of gravitation: For any two bodies it holds that they will be drawn together by a force that is directly proportional to the product of their masses and inversely proportional to the square of the distance between them. If we wish to apply the requirement of complete verification here, we would have to conclude that it cannot be applied. A like conclusion would be reached with all sentences that can be formalised by means of the general quantifier wherever the quantifier relates to an infinite domain.

If the foregoing objection has led to the conclusion that the laws of the natural sciences have no sense (in the spirit of the verifiability criterion of sense), we can equally easily show that the converse is true, namely that even the so-called metaphysical sentences, or more precisely sentences whose components are metaphysical sentences, do have a sense. Let us assume two sentences: Σ_1 (which is an abbreviation for the sentence 'The absolute is the absolute absolute') and Σ_2 (which is an abbreviation of the sentence 'On May 20 the bus leaves London for Southampton at 8:05 a.m.'). At first sight Σ_1 does not satisfy the verifiability criterion of sense, while Σ_2 does. But the molecular sentence that is created by the disjunctive connection of the two sentence, i.e. $\Sigma_1 \vee \Sigma_2$, does satisfy this version of the verifiability criterion fully.

Finally it can be shown that this version of the verifiability criterion of sense conflicts with the classical conception of negation.[25] This may be demonstrated on a sentence with the existential quantifier which contains an empirical predicate. If, for example, P is a one-place empirical predicate, then the sentence $(\exists x) Px$ does – in the spirit of the given verifiability criterion – have a sense. But if we wish to negate the sentence, at the same time using the general quantifier, we will find ourselves in the situation described previously. If this quantifier relates to an infinite domain, we may be led to the conclusion that some sentences have a sense while their negations do not.

The difficulties surrounding the verifiability criterion of sense when it is tied to the requirement of complete verification lead to attempts at certain reforms of the original version whereby the difficulties might be overcome. Such attempts generally toned down, or even eliminated completely, the requirement of complete verification. It has to be said, of course, that such reforms conflicted with the original ambitions of the verifiability criterion of sense, namely the desire to find a sharp dividing line between 'metaphysics' and the world of meaningful sentences.

One revised interpretation in particular was that attempted by A.J. Ayer:[26] A sentence has a sense if and only if it is possible, by means of further hypotheses, to deduce from that sentence observation sentences that of themselves do not follow from the hypotheses. To give it a more exact formulation, a sentence 'Σ' is meaningful if and only if there exists a finite number of other sentences $\Phi_1, \Phi_2, \ldots \Phi_n$ such that from the conjunction $\Sigma \cdot \Phi_1, \Phi_2, \ldots \Phi_n$ one observation sentence 'Σ_1' can be deduced, the sentence 'Σ_1' itself not being deducible from $\Phi_1, \Phi_2, \ldots \Phi_n$.

This version of the verifiability criterion can also be shown fairly simply to be capable of arguing for any sentence that it has a sense. For example, if Σ is the metalanguage name for the sentence 'the absolute is the absolute absolute', it is enough to select a hypothesis Φ, by which Σ_1, which is an observation sentence, is deducible on the basis of Σ and Φ. Let us say that Φ is the metalanguage name for the sentence 'If the absolute is the absolute absolute, then on May 20 the bus leaves London for Southampton at 8:05 a.m.'. It is therefore possible to deduce, by the modus ponens, the sentence 'On May 20 the bus leaves London for Southampton at 8:05 a.m., which is clearly an observation sentence.[27]

An even more far-reaching modification of the verifiability criterion was put forward by Carnap [21]. It has two fundamental features:

(1) The requirement of complete verification is abandoned as a matter of principle and in its place we have a more 'moderate', but not of course clear-cutting, set of means for verifying the semantic characteristics of a sentence: By *verification* what had been understood was the finite and unequivocal ascertainment of the truth or untruth of a synthetic sentence. In the case of questions that could be recorded using the general quantifier complete verification in this sense is often not feasible. Hence Carnap brought in to counter verification and the requirement of complete verification the concepts of '*confirmation*' (Bewährung) and '*testing*'. By confirmation was meant a graduated process that increases the accuracy of our knowledge of a sentence's semantic characteristic step by step. Testing meant using some quite definite method for ascertaining that characteristic.

(2) None of these means can be legitimately considered until some language has been exactly constructed syntactically and semantically. An example of a language which has these properties and in which these means can be employed is the empiricist Ding-Sprache outlined by Carnap. Its vocabulary contains as its primitive extralogical expressions the names of spatio-temporal points and what he calls observation predicates (Beobachtungspredikate). The constitution of other extralogical expressions is carried out by so-called reduction sentences. Basically these sentences specify the semantically significant relations among extralogical expressions and therefore play the part of meaning postulates.[28]

This modification throws a completely new light on the search for a universal conception of the verifiability criterion of sense. In the system of language recommended by Carnap the concept of a 'meaningful sentence' is actually assumed provided semantic rules and meaning postulates are set out. Effectively this means that the role of the verifiability criterion in its original form has been suspended.

Carnap's modification of the verifiability criterion leads to the legitimate conclusion that the 'metaphysical sentences' that ought to be eliminated by the verifiability criterion and be pronounced as sentences that 'have no sense' are unlikely to be found in language systems that are exactly constructed in syntactic and semantic terms. This is the obvious reason why the original aims were not, and could not be, attained. Bearing in mind that the verifiability criterion of sense was originally conceived as a universal criterion, connected with an attempt to find what we have described as a unitary theory of sense, we can see that Carnap's modification in *Testability and Meaning* [21] was effectively the first step in the direction of the abandonment of the universal criterion.

[C] *The Logical Nature of the Verifiability Criterion*

According to the conception of the verifiability criterion, the metalanguage predicate 'have a sense' is connected with the predicate 'verifiable'.[29] This gives rise to the question of the nature of the revelation that 'all sentences that have sense are verifiable'. If 'verifiability' is a common property of the set of sentences that are well formed and in addition have a sense, then this discovery is a product of inductive generalisation of the same kind as say 'all objects with a lower specific gravity than water are capable of floating on water'. But this conflicts with the principle that follows from the consistent application of the requirements of the verifiability criterion and has it that 'verifiable' and 'having a sense' are synonyms. In the same way, the expressions 'have a lower specific gravity than water' and 'be capable of floating on water' are not synonyms.

If then the sentence that formulates the verifiability criterion of sense is treated as an inductive generalisation, i.e. a synthetic sentence, it has to be conceded that 'have a sense' and 'verifiable' are not synonymous predicates.

An even tricker situation arises if we treat the sentence that formulates the verifiability criterion of sense as an analytical sentence. Then there is nothing for it but to recognise that the sentence itself is unverifiable and that it therefore has no sense. But the formulation of the verifiability criterion of sense is clearly not an analytical sentence. If it is a synthetic sentence, we must face the question of whether the sentence itself has a sense. In other words, can the criterion of sense be applied to itself? If it can, then it is easy to prove[30] that it follows from the truth of the criterion of sense that the criterion itself has no sense. An obvious possible objection is that this means a mixing of sentences from the object language and sentences from the metalanguage (or from the metalanguage and meta-metalanguage as the case may be). If this is to be avoided, there is no alternative but to assume an infinite series of verifiability criteria of various degrees and hence infinite regress. This also confirms of course what we have already stated: setting up a universal criterion of sense cannot be successful and leads to controversy. Or we might put it a different way: If there *is* a universal criterion of sense, it cannot be applied to all languages; and if a criterion of sense cannot be applied to all languages, it is not a universal criterion.

4. THE TRANSLATABILITY CRITERION OF SENSE

[*A*] *The Concept of 'Having a Sense' as a Primitive Concept of the Semantic Metalanguage*

Attempts to find a unitary theory of sense on the basis of the verifiability criterion have demonstrably failed. This does not of course remove the need for distinguishing between sentences that have a sense from those that do not. The situation may be expressed by a preliminary division of well formed sentences of a given language into three groups:

(1) sentences that obviously have a sense;

(2) sentences where it has to be decided – by some operation – whether or not they have a sense;

(3) sentences that obviously have no sense.

Clearly, this kind of classification may be treated as provisional and indefinite, since after the relevant operations have been performed the second group further divides into two sub-groups. However, this division does typify the situations where the problem of the criterion of sense may arise.

Let us for the time being leave to one side the question of those sentences that obviously have a sense. This question is by no means as simple as might at first sight appear. Most supporters of empiricism have usually assumed that these are sentences about the immediately observable, although there have been, and are, considerable differences of conception here. In principle, however, it is not necessary to treat only observation sentences as belonging to this first group. This depends on the nature of the language itself and on its relations to a given universe, on the aims of communication processes, and on the language users' facilities for making reliable decisions about the relations of language expressions to elements of the given universe. Let us merely accept that there are sentences of which we know for certain that they have a sense. Under these circumstances it is expedient to treat the concept of 'having a sense' as a primitive concept of the semantic metalanguage.

By accepting the above assumptions we have done nothing to eradicate the problem of a criterion of sense. Its function and competence are, however, somewhat narrowed down: it is now restricted to sentences of the second group, of which we have to decide whether they have a sense, or, to put it more exactly, we have to decide about them on the basis of sentences of which we know that they do have a sense.

Under these assumptions we may pronounce the so-called translatability criterion of sense: The translatability criterion of sense, while not resolving the problem of a criterion of sense generally, can work under the assumption that some sentences already have assigned to them the property 'have a sense'.

P. Marhenke, who hinted at (without fully elaborating) the possibility of a translatability criterion of sense [87], gave three alternative formulations of this criterion:

(1) A sentence is significant if and only if it is translatable into a significant sentence;

(2) A sentence is significant if and only if it is a member of a class of synonymous sentences;

(3) A sentence is significant if and only if it is well-formed or transformable into a well-formed sentence.

These formulations, which Marhenke neither analyses nor comments on, do, of course, have different values. The first is actually the least precise and at the same time closest to a quite intuitive conception: it assumes that translation means a transition from one expression of language to another with the same sense. The concept of synonymity, on which the second formulation is based, is more precise provided that the concept is duly specified. The third formulation may be interpreted as meaning that we are to assume some formalised language defined by both syntactic and semantic rules (and possibly also meaning postulates, if relations among extralogical constants have to be respected). The term 'transform' is to be understood here in the sense of respecting the relevant inference rules.

[*B*] *Pragmatic Limitations of the Translatability Criterion*

The translatability criterion, too, rests on a dispositional predicate, the concept of 'translatable', i.e. the 'possibility of translation'. But what does the 'possibility of translation' mean? The question may be expounded in a variety of ways. In the literature of logical semantics we find – without their always being directly implied – two different viewpoints:

(1) By 'possibility of translation' is meant the translatability of sentences about which it is to be decided whether they have a sense, into sentences of an empiricist language. This viewpoint is actually the starting point of Carnap's *Testability and Meaning* [21]. For this case the translatability criterion may be formulated as follows: A sentence has a sense if and only if it is translatable into empiricist language.

(2) By 'possibility of translation' is simply meant the translatability of a sentence, whose sense or otherwise has to be decided, into a sentence that is recognised as having a sense, either because it obviously has a sense, or because the concept of a 'sentence with a sense' has been accurately defined in the given language, and so forth.

The first of these viewpoints, although amounting to something of a retreat from the original ambitions of the verifiability criterion of sense, does nevertheless seek to preserve some of its characteristic features. The second viewpoint[31] is obviously more general and hence more acceptable. While the former assumes the necessary existence of some basic or initial level of language expressions (which is usually taken to mean expressions that can be linked to immediate observation), and with it the possibility of reduction to that level, the second viewpoint is not tied by any such constraining assumptions.

This is not of course to say that the second viewpoint precludes any kind of link between the problems of sense and those of the empirical starting point of human knowledge. Such a link may of course be mediated in various degrees.[32] Moreover, this empirical starting point must be understood broadly enough as corresponding to the contemporary level of cognition in the natural sciences, furnished as they are with ever improving techniques for measurement and experimentation. Here we can no longer make do with the traditional concept of 'observable' as tied to the limited sensory capabilities of man.

But the second viewpoint needs further complementation and specification. The concept of 'translatable' assumes that we take into consideration certain pragmatic factors. For instance, we allow for the fact that there exists a certain non-empty set of language users capable of performing the translation. And here too we may return to the question of the interpretability of some of the texts that used to be quoted as examples of metaphysical sentences that have no sense. Then it becomes desirable to relativise 'having sense' in pragmatic terms as well, i.e. with respect to a somehow defined set of language users. This in turn means there are certain pragmatic limits to the translatability criterion.

It would be legitimate to object that this kind of relativisation of the concept of 'having sense' carries with it a certain risk of subjectivist arbitrariness or of a purely conventionalist conception and so on. But if there are such pragmatic limits, it still does not mean that any pragmatic limits are acceptable from the scientific point of view and from the point of view of the use of certain expressions in science. As a rule each science creates its own specific requirements as connected with the subject and fundamental tasks of that science. At the same time the empirical sciences require the sense of certain expressions to be tied to certain data of measurement, observation or experimentation, and the relations between this kind of expression and other expressions to be exactly specified. Most important among the other expressions are the so-called theoretical concepts.[33] On the other hand, in fields that work with constructive elements (e.g. automata theory) the concept of 'having sense' can be determined in other ways, notably by formal rules. For the set of language users it is desirable that each should have a command of the relevant terminology and the relevant operations; operations can of course hardly be characterised generally since they have their own peculiarities in each domain.

Emphasising the pragmatic limits of the translatability criterion (and ultimately any criterion of sense however else conceived) may give rise to the further objection that seeks to suggest that this viewpoint might amount to the same thing as the so-called psychological conception of the sense of sentences. This latter conception was evolved by Russell in his *Inquiry into Meaning and Truth* [130]. Its basis is once more the concept of 'belief' to which we have already referred. A sentence Σ has a sense if and only if there can exist a psychological and somatic state that can be described by the words 'XY believes Σ'. Russell summarises his psychological theory of the sense of sentences as follows [130, p. 193]:

"There are states which may be called states of 'believing'; these states do not essentially involve words. Two states of believing may be so related that we call them instances of the same belief. In a man with suitable language-habits, one of the states which is an instance of a given belief is that in which he utters a certain sentence. When the utterance of a certain sentence is an instance of a certain belief, the sentence is said to 'express' the belief. A spoken sentence is 'significant' when there is a possible belief that it 'expresses'." A sentence which we perceive (hear, read etc.) is significant if it prompts one of the three states described by Russell as '*S*', 'not-*S*' and 'perhaps *S*'. This entire theory is clearly based on the attitude of the language users to the content or sense of sentences. Russell sees the virtues of the theory as lying, amongst other things, in the fact it is independent of truth or untruth.

From this point of view Russell's conception of the sense of sentences represents the second extreme – the exact opposite of the verifiability conception of Schlick, and also Wittgenstein. The verifiability conception tried to fix the dependence of the concepts of 'having sense' and 'truth': a sentence has a sense if we can state the conditions under which it can be true (Wittgenstein), or if the sentence is logically verifiable (Schlick). On the other hand Russell's psychological conception is a total dismissal of a connection between the properties of sentences that we label as 'truth' and 'having a sense'.

The drawback of both conceptions is that they put forward a universal solution, with on the one hand unequivocal dependence between the concepts of 'truth' and 'having sense', and on the other their total independence. In the analysis of language systems (whether natural or formalised) under various conditions and in relation to various kinds of tasks (e.g. in ordinary conversation, in the language of certain scientific systems or in literature) we encounter a variety of situations that suggest varying degrees of dependence whichever approach to the analysis of sentences is adopted. Any universal solution that aspires to fix one type of dependence and so resolve all situations must, in some situations, prove inadequate.

These considerations may be summarised in a few points:

(a) The concepts 'truth' (or 'untruth') and 'having sense' (or its opposite, the absence of sense etc.) do, if related to the sentences of a given language, represent metalanguage characterisations of the sentences, but they are characterisations from different points of view. This does not mean that in certain formalised languages they may not sometimes coincide.

(b) The dependence of the two characteristics on each other may vary in degree: from complete independence, which can be asserted in certain extreme cases, to semantically relevant dependence in other cases. This may depend, for instance, on whether we take into account language and its use, e.g. in fairy tales or scientific disciplines. Thus we have to respect the aims or tasks associated with the different types of language use. At the same time it is natural that we should not dismiss this dependence altogether in the language of the natural sciences; certain kinds of dependence have to be respected. If such dependences are relatively stable, they can be fixed by, for example, meaning postulates.[34]

[C] *The Concept of 'Having a Sense' as a Many-Place Predicate*

Having analysed the pragmatic limits of the criterion of sense we may return to some of the questions raised at the beginning of the present chapter. Let us first look at the question of whether the predicate 'have a sense' is a one- or many-place predicate.

In the original version of the verifiability conception of sense this question was not raised. It was seen as a matter of course that this is a 'property' of a sentence and is therefore a one-place predicate. It is to the credit of Carnap's modification of the verifiability criterion (see [21] in particular) that it pointed out the necessity of relativising the concept of 'having a sense' to a language conceived in a certain way (in Carnap this was the so-called empiricist language). But this forces us to abandon any idea that 'having a sense' is a one-place predicate.

In the account of the pragmatic limitations of a criterion of sense we additionally pointed to the example of certain other types of relativisation, especially relativisation to a certain class of language users and their apparatus (in the broadest sense of the word, including their knowledge, customs, the empirical and experimental components of their equipment, etc.), as well as relativisation to certain specific tasks or aims involved in the use of a given language.

It follows from all this that the question of whether a given sentence has a sense has been formulated narrowly and hence inexactly. The question needs to be completed by a statement of the language system in which the sentence was uttered (or written), one on the body of users of that system, and one on the specific tasks and aims of the relevant language communication.

The predicate 'have a sense' is therefore better treated as a many-place predicate. That is not to say that it would be possible, in certain circumstances, to abstract from some of its possible arguments. Such would be the case in, for example, the semantic analysis of formalised languages where we usually disregard the pragmatic aspects. But it is always necessary to respect relativisation to the given language, which makes 'have a sense' at least a two-place predicate.

In the next question concerning the predicate 'have a sense' the point is to find whether there is or is not a sharp dividing line between sentences that have a sense and those that do not. In other words the question is whether the predicate 'have a sense' is clear-cut or vague. There is no unequivocal answer to cover all situations.

As we have seen, some adherents of the verifiability conception (e.g. Schlick) aspired to make the divide so sharp as to eliminate all uncertainty. This aspiration, especially when applied to sentences of natural language selected at random, was thwarted. Generally speaking, it cannot be proved that the predicate 'have a sense' is other than vague. This is particularly so in situations where it is necessary to respect relativisation to a certain set of language users whose equipment is not and need not be identical, and/or relativisation to certain tasks and aims attaching to the use of the particular language, where the tasks and aims need not be understood identically by all users.

However, this does not preclude the possibility that under certain exactly defined circumstances the predicate 'have a sense' might operate as clear-cut. This applies primarily to the semantic analysis of formalised languages, where the concept of a sentence that makes sense is precisely defined. It is in circumstances where the predicate's other arguments (i.e. except for the sentence and language system) can be ignored that there is the best guarantee that 'have a sense' will not be vague.

It goes without saying that the language of the empirical sciences also requires the most exact possible differentiation between sentences that do and do not have a sense. It is therefore expedient to eliminate, or reduce to a minimum, any differences of conception of what 'having a sense' means caused by, for instance, different equipment available to the language users. If in the language of science we use expressions taken over from everyday speech, we will naturally be faced by certain difficulties. In the thirties it was these difficulties that contributed to the movement to unify science, especially its language, in the conception of 'physicalism' and the later attempts to construct a uniform scientific language on an empiricist basis. These efforts were unsuccessful. Carnap himself later (see [31]) renounced his conception of a unified scientific language on a purely empiricist basis.

A rather different programme of a 'non-pragmatic language of science' was declared by V.A. Uspenskij [179]. Part of the inspiration for this programme came from practical needs, such as those connected with automatic retrieval of information, information languages and information machines. This programme, which has since been gradually put into effect, did not initially count on the creation of a unified non-pragmatic informational language for all disciplines, but a whole range of distinct languages constructed in accordance with the different characters of the individual scientific disciplines. There is no doubt that this is a more productive programme, since it is allied to exactly constructed languages whose use is tied to quite definite practical objectives.

5. SENSE AND THE EMPIRICAL

[*A*] *The Concept of 'Empirical Predicates'*

We have shown that the translatability criterion of sense can be conceived generally and that it does not have to be tied to translatability into empiricist language. This is not to say that a knowledge of the forms such translatability might take is not useful – especially in the empirical sciences. Hence the problem of the relation between sense and the empirical in the use of language in the empirical sciences and other domains connected with the empirical is an important problem for semantics and simultaneously a problem of the theory of cognition. It must of course be stressed beforehand that the problems of sense and those of the empirical basis of human knowledge cannot be identified, which was what some of the supporters of the verifiability criterion of sense tried to do.

If we are to set up the concept of a sentence about observation with something like accuracy, we must, in our analysis of these problems, start out from the concept of empirical predicate (observation predicate, observable). Adherents of empiricism have usually assumed that the extralogical elements in the vocabulary of the language of science can be divided into a minimum of two groups: empirical terms and theoretical terms.[35] Leaving aside for the time being the question of how far such a division is at all feasible, we must face the problem of how to study the interrelations of empirical and theoretical terms as well as those of sentences that operate with empirical and theoretical terms. Customarily only sentences that contain empirical terms alone in addition to logical terms have been treated as observation sentences.

The definition of empirical terms in such a way as to make possible the formation of sentences on the basis of them, i.e. the definition of empirical predicates, is by no means an easy problem, as it had seemed to be in the early days of the Vienna Circle. In those days it was treated as a matter of course that one could manage with what was essentially a subjective psychological conception of empiricism. The basis of this version of empiricism, which usually made appeal to 'elementary experiences', 'sensory data' and the like, was the conviction that those properties of things that are bound to these 'atoms' of the human psyche (and they were conceived differently at different times), could be considered 'observable.' Hence the names of those properties were taken as the basis of the 'logical structure of the world', or, more precisely, the logico-linguistic structure of the world.[36]

Carnap later modified this subjectivist conception of the empirical foundations of human knowledge in two respects: (a) he made a relativisation of the predicates that could be described as emprical to certain conditions, i.e. to a given language and given language users; (b) he switched from the subjective psychological conception of the empirical to the behavioral conception [21].

The basis of this conception of the empirical starting point, and so too of the concept of 'observational predicate', are two concepts that are not defined but merely explained, namely '*observable*' and '*realisable*'. As Carnap points out, it is behavioral psychology, or, more exactly, the behavioral theory of language, that is competent to provide more exact definitions. This means that the language user is considered a 'black box', to use the terminology of the so-called macro-approach in cybernetics, and it is his reactions to certain stimuli that are the centre of attention.

A predicate P in the language $\mathscr{L}$ can be treated as *observable* to an organism or person XY if for suitable arguments, i.e. a, and under suitable conditions XY is capable, on the basis of a few observations, of reaching a decision about the sentence 'Pa', i.e. verifying that Pa or $\sim Pa$ to such a degree that he will accept or reject Pa [21, p.63].

Carnap admits that this kind of explanation is vague and comes nowhere near to providing a sharp division between the 'observable' and the 'unobservable'. If we were to consider distinguishing colours, we would have to take into account the normal conditions, i.e. illumination by white light, the abilities of the observer, who must not be colour-blind, etc.

The concept of 'observational predicate' need not necessarily be tied to just the receptive component of the activity of the senses. This is why 'realisable' is taken as another possible basis for the concept of 'observational predicate'.

A predicate P in the language $\mathscr{L}$ is *realisable* for XY if for a suitable argument a and under suitable conditions XY is capable of ensuring that the sentence 'Pa' is true, i.e. of creating the property denoted by P to go with the given object.

For example, if $P_1 a$ means that a given individual has a temperature of 100°C, then P_1 is realisable if we can practically create the situation that the individual a has a temperature of 100°C. If $P_2 a$ means that the individual a has a temperature of -300°C, then P_2 is a non-realisable predicate since it is impossible to cool any object to a temperature lower than absolute zero.

This conception of the concept of an empirical predicate may justifiably provoke the objection that this kind of elimination of the original subjectivism is only apparent, since it is no more than a switch from the subjective psychological conception of the experiential starting point to the behavioral conception. This conception may therefore lead to the conclusion that real or only imaginary properties of an object of observation, which exist independently of the subject of the observation (or the subject of practical activity), are transferred to the behaviour of the observer. But there are objections that can be raised against this interpretation. First of all it has to be stressed that these formulations concern expressions of language, i.e. predicates, that are used as part of decision processes in tasks that relate to observation or some practical activity. Hence these formulations do not and cannot resolve the traditional controversy between phenomenalist-idealist and materialist gnoseology. If we overlook the phenomenalist interpretation[37] of these explanations as an interpretation that goes against the materialist traditions of the natural sciences, there are a number of other problems here. As early as 1940 Russell [130] was reproaching Carnap and pointing out that a decision on 'observability' could not always and under all circumstances be secured by 'a few observations'. There is no doubt that the criticism was warranted and all the more so in that the range, quality and means of the observations cannot be determined globally for all situations. It would seem better to talk in terms of an adequate number of observations made at a level that is also adequate to the given task or class of tasks. In general terms decision-making is by nature statistical, a process made more exact by trial and error or by whatever means the language user has for extending the range of his experience (and so too the quality of his decisions).

Some other objections to Carnap's conception of the 'observable' are given by H. Putnam [113]. Putnam states that in relation to a given language the concept of 'observable' and 'observation predicate' generally is taken as invariant. This is not in fact the case. We can never exclude the possible use of empirical predicates in relation to objects that are actually or in principle not observable (in the ordinary sense of the word). Putnam concludes that if an observational predicate is usable in principle only for things that are observable, then there can actually be no such things as an empirical predicate.

It must also be said that 'observability' can be related only to a certain level. We can for instance distinguish between the macroscopic and microscopic level when making decisions about things that are designated by the predicates 'have such and such a temperature', 'be red' etc. that are applicable only at the macroscopic level, and predicates such as 'have such and such a mass' or 'have such and such dimensions' that are applicable also at the microscopic level (while not being applicable at what is sometimes called the submicroscopic level).

In our view there is a possibility of using, for the interpretation of the concept of 'observation predicate', the communication model of empirical or experimental activity that we referred to in the account of the operationist criterion of sense. This model assumes in principle that the 'observer channel' has certain (not unlimited) characteristics that have to do with threshold values, degrees of accuracy, capacities, memories, delays etc. These characteristics are not absolute and unchanging, as is assumed by the subjectivist or anthropocentric conception of empiricism, and this is so for two reasons: (a) the characteristics of the observer channel can be changed by changing the conditions under which decisions are made, i.e. by using metrical, experimental or other equipment capable of higher degrees of accuracy and of raising threshold values etc.; (b) we also have to accept the possibility of modelling or imitating these procedures on technical equipment (e.g. 'observing' the far side of the Moon without ourselves flying round it, and measuring and recording metrical data in places where man cannot go and where he merely sends measuring instruments equipped with the means for storing and transmitting information). From the point of view of information theory there is no difference of principle between the two reasons. Hence the conception of 'observability in principle', on which the subjective idealist and subjective psychological conceptions of empiricism are based, is erroneous, since it makes absolute only one level of observation and petrifies the characteristics of the 'observer channel'.

Thus we may conclude by saying that the boundary between the 'observable' and the 'non-observable' is neither invariant, clear-cut, nor absolute. It is relative to the given characteristics of the 'observer channel'.

[*B*] *Dispositional Predicates*

If a language is equipped with a group of predicates $P_1, P_2, P_3 \ldots P_n$ which, with respect to given conditions and given characteristics of the 'observer channel', may be designated as empirical predicates, then the positive or negative decisions made about sentences that arise by the concatenation of these predicates and the individual constants of the given universe, i.e. for example $P_1 a_1$ or $\sim P_1 a_1$, $P_1 a_2$ or $\sim P_1 a_2$ etc., may be described as observation sentences. (Clearly these need not be just one-place predicates as is here suggested merely for the sake of simplicity.)

If in the given language we wish to introduce another term, i.e. in specific terms another predicate, which cannot, with respect to the given conditions and given characteristics of the 'observer channel', be treated as empirical, we have two possibilities:

(1) the introduction of the new predicate by definition;

(2) the introduction of the new predicate by means of the procedure known as reduction.

In terms of logic the definition method is the simpler, but it has its limitations, arising from the analytical nature of definition. Let us assume that we have some individuals that are simultaneously carriers of the properties expressed by the predicates P_1, P_2, P_3 and P_4. We can now introduce a molecular predicate on the basis of these. This is the kind of line taken in ordinary thought processes. We might say that a man has influenza if he has a high temperature, a cough, bloodshot eyes and so on. Also, if he has influenza, he will have all these symptoms.

Thus we may bring in a new predicate Q by means of definition, and that definition might have the form:[38]

$$(\forall x)(Qx \equiv P_1 x \cdot P_2 x \cdot P_3 x \cdot P_4 x)$$

or, in a simpler form, omitting the quantifier and the arguments:

$$Q \equiv P_1 \cdot P_2 \cdot P_3 \cdot P_4$$

A definition that introduces a new term as a kind of abbreviation for a complex of previously existing terms fixes the meaning of the new term once and for all and for all conditions. Very often the relations between the new term and the existing terms have to be tied to certain conditions. In such cases the definitional introduction of the new term is not entirely satisfactory. These are the occasions referred to in everyday speech as *conditional connections* and expressed by the words 'if…, then…'. The predicates that concern such conditional connections are usually called *dispositional predicates* and include 'inflammable', 'soluble in water', 'flexible', 'conductive' and the like. It should be noted that dispositional predicates are often the names of properties that are very important in the empirical and technical sciences, including such properties as resilience, strength, conductivity, corrosion resistance, permeability etc. With this kind of property it is usually important to have an exact definition of the extreme conditions for them.

On the face of it even dispositional predicates could be introduced by means of definitions. But difficulties arise if, in a definition to express a conditional connection, we resort to material implication. For example, if we say of some object that it is conductive, what we mean is that it *would* be able to carry an electric current *if* such a current were introduced into the object. If we are to attempt to define the dispositional predicate 'conductive' (Q), the following procedure presents itself: Any object is conductive if and only if, under exactly stated conditions (which we shall treat as the second argument in the formalised notation, usually interpreted as time, and denote it c), with a current (P_1) introduced through the object, the relevant measuring instruments show that the object is carrying a current (P_2). It should be noted that it is one thing for an object to have the ability to carry a current and another for it actually to carry a current. Similarly a distinction has to be drawn between the dispositional predicates 'be soluble', 'be inflammable', 'be corrosion resistant' etc., and the empirical predicates 'dissolve', 'burn', 'not rust' etc.

If we were to try to express the relation between dispositional and matching empirical predicates in a definition, we might find the following form of definition acceptable:

$$(\forall x)[Qx \equiv (P_1 x, c \longrightarrow P_2 x, c)]$$

However, this kind of definition is not satisfactory. The object has to be considered conductive without our introducing a current, water-soluble without our immersing it in water, inflammable without our igniting it. The well known properties of material implication also make it possible to deduce anything at all from a falsehood in the antecedent of an implication. Therefore this form of definition cannot be satisfactory for introducing a dispositional predicate and recording its relations to empirical predicates.

One resolution of these difficulties was put forward by Carnap [21], who suggested replacing the unsuitable definitional approach by a schema of 'reduction sentences'. The simplest form of 'bilateral reduction sentence' may be recorded by the following notation:

$$(\forall x)[P_1 x, c \longrightarrow (Qx \equiv P_2 x, c)]$$

(In words this might be expressed thus: If under certain stated conditions a current is introduced into any object, then that object is conductive if and only if the relevant measuring instruments show that under the conditions given the object carries a current.)

The bilateral reduction sentence expresses a conditional connection, i.e. a connection between the dispositional predicate and empirical predicates. But in this context it is possible to extend the idea in such a way that we can equally express the connection between the negation of the same dispositional predicate and other empirical predicates. This can be done by a pair of sentences such as Carnap described as a reduction pair. The advantage of the reduction pair is that it takes into account both the presence and absence of the dispositional predicate being introduced and so gives the conditional connection in the positive and the negative respect.

A reduction pair can be recorded in the following simplified form:

$$P_1 \longrightarrow (P_2 \longrightarrow Q)$$
$$P_3 \longrightarrow (P_4 \longrightarrow \sim Q)$$

While this procedure does remove some of the difficulties connected with the possibility of using definition to introduce dispositional predicates, it brings with it certain further problems.[39] In effect it expresses formally the principles of operational definition. Dispositional predicates[40] are characterised here as reducible to certain relationships of empirical predicates (predicates connected with empirical procedures, such as measuring or observation). But for any such procedure to be defined with anything approaching accuracy one would have to assume a relatively broad range of relevant operations. This can be expressed as follows (it will be noted that the schema uses the bilateral reduction sentence; an analogous schema could be drawn up for reduction pairs):

$$P'_1 \longrightarrow (Q \equiv P''_1)$$
$$P'_2 \longrightarrow (Q \equiv P''_2)$$
$$\dots\dots\dots\dots\dots\dots$$
$$P'_n \longrightarrow (Q \equiv P''_n)$$

where all the predicates except Q are empirical predicates. It is obvious from this that the introduction of a dispositional predicate and the closer specification of its sense on the basis of empirical predicates is usually *inductive in character*. For this reason the relevant dispositional predicate must be treated as an *open concept*.[41] This means that having taken n steps to verify the conditional relationship between elements of the class '$(\hat{x})(P'x)$' and those of the class '$(\hat{x})(P''x)$', we do not exclude, but on the contrary assume, the possibility of further steps in the course of which the same conditional connection will be verified. Thereby of course falls Carnap's original ambition to have all scientifically admissible terms completely reducible to terms of empiricist language.

The openness of the concept of 'dispositional predicate' (and, as we shall show in due course, the analogous openness of theoretical concepts) is therefore connected with the fact that a dispositional predicate can, as a rule, have assigned to it several empirical interpretations by means of different operations that can be used in connection with it.[42]

Having pointed out that the concept of 'dispositional predicate' is an open one and that the constituting procedure by which a new term is introduced is inductive in character, we have to state in addition that the above schema necessarily assumes that the class '$(\hat{x})(P'x)$' is not empty. For instance, if we consider the dispositional predicate 'soluble' in relation to a given object, we can decide about it experientially only if the object has been immersed in a liquid at least once.[43]

If $(\hat{x})(P'x)$ is an empty class, making it impossible to consider any empirically relevant sense of the predicate Q, any use of material implication is precluded. Thus dispositional predicates are connected with conditional connections that are contentual in character. This also justifies the unsuitability of using

$$Q \equiv P' \longrightarrow P''$$

for introducing the dispositional predicate Q on the basis of the empirical predicates P' and P''.

One possible way of surmounting the difficulties connected with the properties of material implication is to use causal implication.[44] Using the sign $\underset{c}{\longrightarrow}$ for causal implication, we may express the consitution of a dispositional predicate Q on the basis of the empirical predicates P' and P'', in a simplified form again, as follows:

$$Q \equiv (P'_1 \underset{c}{\longrightarrow} P''_1)\cdot(P'_2 \underset{c}{\longrightarrow} P''_2)\cdot \ldots \cdot(P'_n \underset{c}{\longrightarrow} P''_n).$$

Likewise under this solution it is advantageous to treat Q as an 'open concept' whose introduction is based on a procedure of induction.

Another solution also offers itself. Let us begin with the simple idea of the term 'inflammable'. 'Inflammable' is said of an object which, if exposed in various experiments to the different effects of higher temperatures, will, with a probability that is non-zero and may amount more or less to certainty, burst into flame. In other words, the conditional connection on which the concept of dispositional predicate rests need not of necessity be a determinist dependence (in the sense of Laplace's determinism), but a dependence of the stochastic type. It would seem that in the general case it is this kind of dependence that conditions the experiential basis of our concepts. Then by experience and the experiential basis what is meant is a probability calculation of the fact that given a certain number of experiments, a proportion of them will be successful. In other words, if I say of some object that it is 'inflammable', I will normally be expressing the expectation (based on the stochastic dependence referred to) that under the effects of higher temperatures the object will ignite.[45]

6. 'THEORETICAL CONCEPTS' AND THE RELATIVITY OF THE EMPIRICAL STARTING POINT

[A] *The Problem of Scientific Empiricism*

The concepts of 'empirical predicate', 'observation sentence' and 'empirical starting point' itself are in principle open to two kinds of interpretation.

The characteristic feature of the *subjectivist interpretation*, widespread among adherents of positivism in particular, is the conviction that empirical predicates and observation sentences relate in their semantic (content) aspect to states of the observer. What is assumed is therefore something that might be called 'subjectivist assignment'. This characteristic can be upheld even if the form of the assignment is changed or if it is conceived more objectively in some respect. In this context the evolution of the positivist conception of empiricism since the days of the Vienna Circle is typical. There has been a move from the originally purely psychological version, tied to 'experiences', 'perceptions', 'sensations' and other states of awareness however described, to a rather more behavioral characterisation of the observer's states. These concepts were assigned to certain kinds of behaviour or certain objectively discernible operations on the part of the observer and so forth. It must be stressed that these and other similar modifications do nothing to change the fundamental characteristic of this kind of interpretation.

But in science we not understand empirical predicates and observation sentences as something we assign to the states of the observer, whether these are introspectively accessible states of the awareness, or behaviour accessible to another observer. The empirical and natural sciences cling elementally to the objective, or rather the *materialist interpretation* of empiricism. This interpretation does not of course exclude the observer, nor can it overlook or dismiss his role and the importance of the sensory, experimental or theoretical equipment that he brings to bear in his empirical activity, i.e. observation, measuring etc. It was from this point of view that we have been using the concept of 'observer channel'. Nor does this interpretation stop us from describing certain elements of language, certain sentences and predicates – with respect to the given language, the observer's equipment and the objectives that he has set – as empirical predicates and observation sentences. Empiricism in sciences also assumes observer decision processes, communicable by means of language. What is most important, however, is that these decision processes relate to objects, states or processes in the objective world, i.e. the world that exists outside the awareness and language apparatus of the observer.

From the semantic point of view what is important is that observation sentences do not differ from other sentences through their content or the nature of the assignment (i.e. they do not differ from other sentences by being related to states of the observer), but through the manner in which he makes decisions about them. Such decisions, which are also always bound to the observer's equipment, being dependent on the properties of the 'observer channel', are more or less direct. This means that they are decisions about states or processes in the objective world taken directly *on the basis* of certain states at the input to the 'observer channel'.[46] It also follows from this conception of empiricism that no kind of decision can be treated as unfounded, i.e. as something in which the relatively a priori equipment of the observer is just not involved. And because this relatively a priori equipment must always be taken into account, we cannot treat this conception of empiricism and its associated conception of the empirical starting point as absolute, but invariably as relative. In other words, what may be considered empirical from the point of view of one observer and his equipment, may, from the point of view of another observer, differently equipped, be considered theoretical, and vice versa.

If we tie the conception of scientific empiricism to the concept of 'observer channel' and that of 'decision process', we are still not yet in a position to face all the objections that are commonly raised against the materialist interpretation of empiricism. Our main concern now are those objections that appeal to delusions of the senses and similar phenomena. Were we to be satisfied with just any decision process and failed to respect what might be described as the *quality of a decision* (with respect to a given task, or to a given class of decision problems), we would be in no position to draw any substantial distinction between the delusion of the senses by which we would say that a stick immersed in water is bent, holding that to be true, and empirical discovery. The conception of scientific empiricism that rests on decision making on the basis of elements at the input of the 'observer channel' also assumes, therefore, that we can evaluate decisions, having at our disposal the criteria for assessing their quality. (In the case of statistical decision, we may judge the quality of individual types of decision by being in the position, for example, to state the average loss or average risk connected with the different types of decision.)

Our analysis of the empirical starting point so far may be summarised by saying that we have found that it has to be treated as relative and conditional. Its relativity rests in the fact that the boundary between the empirical and the non-empirical is not absolute, depending on

(a) the observer's equipment, i.e. the properties of the 'observer channel',
(b) the tasks or objectives connected with a particular empirical activity, i.e. observation, measurement, experimentation etc.,
(c) the facilities for qualifying the relevant decision process, i.e. the

facilities for applying adequate criteria by which to assess individual decisions.

It remains to be added that the concept of the observer's 'equipment' embraces not only empirical and experimental assumptions, but also certain theoretical assumptions, such as that the observer is starting out from particular theoretical ideas, hypotheses and the like. In other words, the kind of activity that in science is usually described as empirical can never go without the prior execution of those steps that lead to the creation of these assumptions. Thus for this activity what is called in cybernetics the 'pre-programme' is absolutely essential.

We shall call the sum of these circumstances (i.e. 'equipment', 'objectives', the facilities for 'qualifying decisions') the *compass of observation*. Then the concept of the 'observable', if we relate it to predicates of a given language (i.e. P, P' etc.), has to be relativised to that compass. At the same time it is obvious that (for a given observer, or a given level of science and experimental technique) there is some opportunity for a limited extension, or restriction as the case may be, of the compass by, for instance, our opting for the use of an electron microscope instead of the classical microscope.

In the light of this we may return to the concepts of 'empirical' and 'dispositional' predicates. Several degrees can be distinguished (the following characterisation of those degrees will be based on the previously described communication model):

(1) The observer has access to the output data on certain phenomena, events or processes, the data being carried by *empirical predicates* (e.g. 'be green', 'be heavy', 'be painful' etc.; the examples show these may include processes of both extrareception and intrareception) $P_1, P_2 \ldots P_n, P'_1, P'_2 \ldots$ etc.

(2) The next degree is the introduction of terms that might be described as *dispositional predicates proper*. Their essential feature is that the procedures that introduce them may contain in addition to logical terms (including modal terms such as causal implication), only empirical predicates, but relativised to the given compass of observation. These procedures may take the form of reduction sentences, definitions based on causal implication, or they may be based on the means of expression of stochastic dependence.

The introduction of a dispositional predicate proper can be illustrated by the 'black box' schema, where the observer has access both to the input (i.e. states that can be expressed by means of empirical predicates P'_1,

$P'_2 \ldots P'_n$), and to the output (i.e. states that can be expressed by means of empirical predicates $P''_1, P''_2, \ldots P''_n$). If to describe the possible behaviour of a 'black box' of this kind we use the dispositional predicate Q_d and to express the relevant conditional relationship we use causal implication, then we can use definition to introduce Q_d in the following way (for brevity's sake the arguments are again omitted from the notation):

$$Q_d \equiv (P'_1 \underset{c}{\rightarrow} P''_1) \cdot (P'_1 \underset{c}{\rightarrow} P''_2) \cdot \ldots \cdot (P'_n \underset{c}{\rightarrow} P''_n)$$

It would be possible analogously to formulate the matching pairs of reduction sentences or bilateral reduction sentences in which all the non-logical terms except Q_d would be of the nature of empirical predicates.

It remains to be added that the followers of operationalism and the verifiability criterion of sense assumed that these two degrees were enough for building up the vocabulary of the empirical sciences. They were convinced that all scientific terms could be introduced as dispositional predicates proper by using the above or similar procedures. But this conception of empiricism proved to be too narrow: Science at its modern level usually operates with a number of theoretical concepts whose connection with empirical predicates is far less direct. Carnap himself, in one of his later works [31], frankly admitted that his earlier standpoint (especially in [21]), whereby all scientific concepts could be introduced as dispositional predicates on the basis of 'empiricist language', was unsatisfactory. This was one reason why he brought in a conception of relatively independent theoretical concepts.

In the language of science we may posit not only dispositional predicates proper and empirical predicates, but also what have been called theoretical dispositional predicates[47] and theoretical concepts. This in turn means that further degrees have to be distinguished from those given above:

(3) *Theoretical dispositional predicates* are also based on a kind of conditional relationship. Here the observer only has access to the output of the 'black box' (i.e. states that can be described by empirical predicates $P_1, P_2, \ldots P_n$), the states at the input being inaccessible. These states, which can be described by predicates $Q_1, Q_2, \ldots Q_n$, are assumed on the basis of a theory, hypothesis etc.

A theoretical dispositional predicate Q_{td} may be defined thus:

$$Q_{td} \equiv (Q_1 \underset{c}{\rightarrow} P_1) \cdot (Q_2 \underset{c}{\rightarrow} P_2) \cdot \ldots \cdot (Q_n \underset{c}{\rightarrow} P_n)$$

In the case of diagnosis in medicine we might interpret $P_1, P_2, \ldots P_n$ as empirically ascertainable symptoms, while $Q_1, Q_2, \ldots Q_n$ might be taken as the kind of pathological states that we assume on the basis of theoretical knowledge, hypotheses and so forth. In this case the dispositional predicate expresses the conditional relationship whereby the occurrence of certain empirically ascertainable symptoms depends on the occurrence of the relevant pathological states. This could be the case of certain illnesses or their intermediate stages. Even in this context, i.e. in connection with the introduction of theoretical dispositional predicates, it is to be noted that the strict determinist dependence expressed by causal implication may be treated as no more than the extreme case of a broader spectrum of dependences of a stochastic type.

(4) In the empirical sciences we always work with a mass of *theoretical concepts*[48] that cannot be translated without remainder into terms of empiricist language and therefore are not completely reducible to empirical predicates.[49] The introduction of theoretical concepts, and their interpretation, cannot therefore be tied to any specified empirical procedure (e.g. measurement, experiment, observation), but to the overall theoretical picture of the given universe as fixed in the laws, theories and hypotheses of science. That is not to say that there is no essential relationship between theoretical concepts on the one hand and empirical or dispositional predicates on the other. On the contrary, it is always profitable for science to be able to fix the connection between theoretical concepts and empirical predicates in the shape of *correspondence rules*. Correspondence rules may be treated as expressions of what we described in the previous paragraph as theoretical dispositions, the possible forms of the dependence between theoretical concepts and empirical predicates varying considerably, from a strict determinist dependence to the principle of indifference (the principle of statistical independence, principium rationis insufficientis).

It may also be added that even the semantically relevant relations of theoretical concepts can be expressed by the means used here for empirical and dispositional predicates. If, for example, Q_1 and Q_2 are theoretical concepts, another theoretical concept Q_3 may then be introduced by means of a bilateral reduction sentence of the type

$$Q_1 \longrightarrow (Q_2 \equiv Q_3)$$

or by means that rest on causal implication, or again means that are capable of expressing a stochastic dependence.

[B] *The Admissibility of Theoretical Concepts*

If, as has here been suggested, theoretical concepts are based on the overall theoretical picture of a given universe and cannot be linked to just empirical procedures applicable in the given universe with respect to the properties of the 'observer channel', the problem arises of whether these terms are admissible. The history of science tells us that certain theoretical concepts, such as fluidum, phlogiston, thermal matter etc., have been based on erroneous hypotheses and have had to be abandoned. Generally, the essential feature of this kind of concept is that their interpretation cannot be analysed in isolation from any context, but only in connection with some definite kind of context, i.e. some hypothetical picture of the given domain. For example, the concept of 'entelechy' features in the vitalist conception of biology. For these reasons it is better to treat the question of the admissibility or inadmissibility of theoretical concepts as a derived question, subordinate to the question of the admissibility of the relevant hypothesis or conception.

Nor can the concept of the 'admissibility' of theoretical concepts be identified with the concept of 'having a sense', for the latter is undoubtedly broader than 'admissible' or, more exactly, 'scientifically admissible'. Only if we tie the criterion of sense to translatability into empiricist language do the two concepts coincide. This is the typical standpoint adopted by followers of the operationalist or verifiability conceptions of sense.

From the logical point of view similar problems arise in the case of theoretical concepts as we have seen with dispositional concepts. In particular it is impossible to define explicitly all theoretical concepts, or introduce them by means of reduction sentences, as Carnap still assumed in [21]. Carnap was accordingly forced to abandon his original viewpoint, and in his study on the methodological character of theoretical concepts [31] he recommended that they should be treated as primitive non-logical constants of what he called theoretical language. As distinct from the empiricist language, which in addition to the ordinary terms of logic contained individuals and empirical predicates as non-logical terms, theoretical language contains as its non-logical terms theoretical concepts, moreover both terms that can be treated (in a given theoretical language) as primitive and terms that can be constituted on the basis of the primitive terms. This also means that the processes of constitution and reduction are much more far-reaching, not being admissible in one direction only, i.e. in the direction of empirical predicates, as was originally thought to be the case. In its logical aspect, empiricist language has usually been constructed as a first degree predicate calculus.[50] If the only admissible values of empiricist language are observable individuals, then this requirement has been 'liberalised' in the case of theoretical language and it can be related to any specified domain of individuals that satisfy the requirements that stem from the acceptance of a given theory or theoretical conception.[51]

Another typical feature of theoretical concepts, especially those in science, is their openness. That is to say that their complete interpretation cannot always be demanded, and for a number of reasons: (a) It cannot always be assumed that the entire domain of individuals that can be treated as variables in a given theoretical language will be accessible. In other words, the methodological competence of these concepts may relate to domains that are not, or are not yet, empirically accessible. For example, the theoretical concept 'gravitation', which is connected with the competence of the law of gravitation and is given a precise sense by means of that law, is applied by us to phenomena whose gravitational effects we can only assume theoretically, having no means of measuring them. (b) The link between theoretical concepts and an empirically accessible domain that may serve as a relative base for some partial interpretation need not be direct and may be mediated in varying degrees. At the same time the forms of scientific abstraction that lead to the creation of theoretical concepts may free themselves in varying degrees from direct dependence on empirical data, as is the case of certain theoretical concepts that are tied to an idealisation ('ideal thermal machine', 'ideal gas', 'Turing machine', etc.) or to procedures that combine abstraction on the basis of empirically ascertainable entities and of constructive objects (e.g. the concept of 'neuron network' and the like).

Thus deciding on the scientific admissibility of theoretical concepts is bound to decisions on the scientific admissibility of the relevant scientific theory. If that theory relates to a domain of which one part, which we may consider sufficiently representative, is empirically accessible, any decision as to the scientific admissibility of the theory is, of course, tied to the interpretation of the theory with respect to the empirically accessible area.

Since it is usual for a scientific theory to operate with not one, but an entire system of theoretical concepts, decisions about any one concept cannot be taken in isolation from the whole system. The concept of 'scientific admissibility' of a theoretical concept that is a metalanguage predicate of the theoretical language must therefore be relativised (a) to a scientifically accepted theory, (b) to the entire system of theoretical concepts embodied in that theory, (c) to any empirical domain over which the theory is competent, and (d) to the relevant correspondence rules. This can be demonstrated using the previous 'geographical' interpretation of $\mathscr{L}$. Let us assume that in addition to the predicates we have in $\mathscr{L}$ the individuals p (Prague), t (Tábor) and d (Děčín). The predicate constants in the language being interpreted are clearly not empirical predicates. They may be treated as theoretical concepts. These concepts must respect some 'geographical theory', or the meaning postulates capable of expressing that theory. The 'geographical theory' may be expressed by suitable meaning postulates, e.g.:

$$(\forall x)(P_1 x \longrightarrow P'_1 x)$$
$$(\forall x)(P_1 x \mid P''_1 x)$$
$$(\forall x) \sim P_2 x, x \qquad \text{etc.}$$

If we consider p, t, d to be names of individuals that can be verified empirically, the relevant correspondence rules then concern the relation between these names and theoretical predicates, or the classes that may be formed from them. Correspondence rules of this kind may take the following forms:

$$p \in (\hat{x}) P_1 x, \quad t \in (\hat{x}) P_1 x, \quad d \in (\hat{x}) P_1 x \quad \text{and so forth.}$$

Naturally the form taken by correspondence rules that record the dependences between empirical and theoretical concepts may vary in order to give the optimal expression of the nature of each dependence. The dependence may also be of the stochastic type, so the correspondence rules must be capable of expressing that too.

[C] *Ways of Interpreting Theoretical Concepts*

The interpretation of theoretical concepts can, as has already been emphasised, be considered only in the context of a theory, hypothesis or theoretical conception of the given domain. Hence the concept of 'value' may be considered in political economy, 'gravitation' in connection with the law of gravitation or the general theory of relativity, 'electromagnetic field' in connection with Maxwell's equations, 'ideal gas' in connection with the equation of state, 'ψ-function' in connection with quantum theory, and so forth. At the same time we may distinguish several ways of interpreting theoretical concepts.

(1) The *naive realist* and *naive empiricist interpretations* simply ignore the fact that there might be a difference between theoretical concepts and the other non-logical terms in language. With an empirical predicate, such as the concept 'dog', it is possible to use in its interpretation ostensive definition (Russell): 'This here is a dog'. It is possible to use explicit definition and enumerate all the observable properties of a dog that are necessary and sufficient for anything that has those properties to be called a dog. It is possible to imagine all the possible individuals to which the predicate 'be a dog' can be related. The naive realist interpretation of theoretical concepts is rooted in the simple assumption that even for a theoretical concept there is a 'corresponding' object, discernible and ascertainable in essentially the same way as in the case of objects 'corresponding' to empirical predicates. Hence the assumption that as 'something' 'corresponds' to the terms 'dog', 'be green', 'Prague' etc., so a similar 'something' 'corresponds' to the concepts 'value', 'ψ-function', 'photon', 'spin' and so on.

The naive realist conception is sometimes linked with the naive empiricist. For naive realism anything that can be denoted by a predicate 'exists'. To this it can only be added that this kind of interpretation is also naive in the sense that it assumes that a theoretical language can only be a language of the first degree – that its non-logical expressions may only be of the zero or first type. This attitude understandably renders any higher abstraction in science impossible.

(2) The *instrumentalist interpretation* reacts against some of the difficulties of naive realism. If we assume that for theoretical concepts there is a 'something' that 'corresponds' to them in the same way as in the case of empirical predicates, the impossibility of observing what 'corresponds' to theoretical concepts leads to a negative viewpoint: theoretical concepts have nothing corresponding to them. The instrumentalist viewpoint is therefore the negative product of the consistent combination of naive realism and naive empiricism in the spirit of the principle: "non percipi = non esse".

The instrumentalist interpretation of theoretical concepts does not of course exclude them from science, nor does it deny their usefulness. Theoretical concepts are merely 'instruments' of scientific thinking and serve to derive one set of observable data from another. In this context it has usually been suggested that theoretical concepts have no sense, being merely signs or sounds without an independent sense which, if connected with observation sentences and certain syntactic rules, facilitate prediction or, as is sometimes stated, the 'systematisation' of observation sentences. At the same time the problem of the reality of 'non-empirical entities', i.e. the entities that in the spirit of naive realism are assigned to theoretical concepts, is declared to be a pseudoproblem.

The instrumentalist interpretation of theoretical concepts is also closely connected with the purely syntactic conception of scientific laws, which was expressed most openly by Schlick: Scientific laws have no counterpart in reality, being merely rules of inference or transformation (the ideal rules being those that are deductive in character) which facilitate the 'calculability' of certain empirical data, certain parameters, on the basis of other parameters. Seen in this light, the laws and theories of science are 'calculation devices' and theoretical concepts are merely the means of expression of such 'calculation devices'.

The instrumentalist standpoint may be demonstrated by taking the idea introduced by Hempel [56] in support of the instrumentalist interpretation of theoretical concepts. Unlike the concepts of 'weight' and 'capacity', which can be ascertained by direct measuring (and so can be treated as empirical predicates), the concept of 'specific gravity' has to be determined as the quotient of the weight and capacity of a given body. The concept of 'specific gravity' is not an empirical predicate, but it can be constituted on the basis of empirical predicates. Then it becomes possible to formulate the sentence that a solid body will float on a liquid if its specific gravity is less than that of the liquid. On the basis of this sentence and knowledge of the relevant specific gravities, which can be established operationally, i.e. on the basis of observable properties expressed by empirical predicates, it is possible to predict another observable property, expressed by the predicate 'is capable of floating on a given liquid'. This entire reasoning can be readily formalised. Hempel deduces from the above that the assumption of 'unobservable entities' expressed by theoretical concepts is merely an instrument of 'systematisation'. Systematisation is understood to mean the scientific procedure of prediction or its converse, i.e. scientific explanation. This instrument enables us to derive 'observable entities' on the basis of other 'observable entities'. With simplifications this procedure can be expressed schematically as follows: If we have an 'observation sentence' (or a class of such sentences) Φ_i and a theory Φ_t, the simplest form of 'deductive systematisation' can be presented as follows:

$$(\Phi_t \dot{\frown} ET \dot{\frown} \Phi_i) \dot{\frown} SEQ \dot{\frown} \Phi_j \text{ [or possibly } \Phi_t \dot{\frown} SEQ \dot{\frown} (\Phi_i \dot{\frown} SEQ \dot{\frown} \Phi_j)],$$

where Φ_j is an observation that can be derived on the basis of Φ_t and Φ_i.

It is possible that Φ_t satisfies the function of systematisation, or, to use Schlick's original terminology, that it facilitates 'calculability'. This means that Φ_t fixes a certain dependence between Φ_i and Φ_j, e.g. $\Phi_i \dot{\frown} AEQ \dot{\frown} (\Phi_i \dot{\frown} SEQ \dot{\frown} \Phi_j)$. Then it is easy to imagine that:

$$[\Phi_i \dot{\frown} ET \dot{\frown} (\Phi_i \dot{\frown} SEQ \dot{\frown} \Phi_j)] \dot{\frown} SEQ \dot{\frown} \Phi_j$$

Thus instrumentalism does not exclude theoretical concepts from scientific language, nor does it deny their usefulness. It differs in this respect from the original positivist versions of empiricism which simply excluded everything that was not 'observable' (including concepts of 'atoms', 'electrons' and the like). All that instrumentalism denies is direct real interpretation (which it identifies with empirical interpretation) of theoretical concepts, recognising only 'partial interpretation', under which 'understanding' a theoretical concept, or acquiring its sense, means knowing 'how to use the concept', the word 'how' being tied to rules which enable us to operate with observable entities.

The instrumentalist standpoint, as will be apparent, denies denotation to theoretical concepts and limits their sense to the role they play in the syntactically understood context (e.g. in the form of 'deductive systematisation') of empirical predicates. But this standpoint must be left absolutely helpless in the face of situations where, with advances in science and especially with changes in some of the properties of the 'observer channel', some theoretical concepts may become 'observable' or vice versa. G. Maxwell [93], in his detailed critique of instrumentalism, makes the excellent comment that one of the fundamental features of the evolution of science is what he called the 'emergence of reference' with respect to all those new entities that usually start life as theoretical or hypothetical assumptions. The evidence is to be found in the evolution of quantum theory, the theory of elementary particles, the evolution of genetics and others. The concept of 'reference' and therefore also that of 'denotation' of course does not, in connection with theoretical concepts, mean that it is not unavoidably necessary to distinguish between the denotation of empirical and theoretical concepts from the point of view, for instance, of incorporating entities, their properties and relations into various levels, and of the relation of these elements to the 'observer channel'.

(3) Another standpoint, which might be called the *neutralist conception* of theoretical concepts, tries to find some form of reconciliation of instrumentalism and realism by suggesting that the conflict between them is not a matter of merit, but of linguistic controversy. This view was stated

openly by E. Nagel [99], who stressed that the conflict between the instrumentalist and realist conceptions was a conflict of two ways of expression. He went on to say that any assertion that this or that viewpoint was more correct or more suitable was really only a question of terminology.

The subsequent development of positivism was characterised by a retreat from the frank instrumentalism proclaimed in the thirties by Schlick in particular and connected with empiricist reductionism, and by a shift towards the neutralist viewpoint. G. Bergmann [8], for example, rejected the 'ontological' problems of theoretical concepts. He also refused to consider criteria of sense with respect to theoretical concepts, recommending instead that we should consider 'patterns' for their use. A similar view was taken by Carnap in his *Empiricism, Semantics and Ontology* [27]. In this work he declared the question of the existence or non-existence of entities assignable to theoretical concepts to be a problem of accepting a 'linguistic framework' by which to introduce such entities.

The neutralist standpoint is actually a more delicate form of instrumentalism. Clearly, the problems of selecting a proper 'linguistic framework' or 'linguistic patterns' cannot be underestimated, but decisions on the choice of a 'linguistic framework' and decisions on the semantics of theoretical concepts are not quite the same thing. Supporters of the neutralist viewpoint may use certain pragmatic considerations in arguing their case – the expediency or usefulness of choosing particular 'linguistic patterns', but other pragmatic reasons can also be adduced to argue against the neutralist conception and in favour of the realist interpretation. The realist, or more often the hypothetico-realist, interpretation of theoretical concepts that relate to (hitherto) unobservable entities may also be taken as inspiration for refining the empirical and experimental equipment in such a way as to be able to verify assumed entities empirically or experimentally. If physics at the beginning of the century had accepted without reservation Mach's negative stance on the atomist hypothesis (on the grounds that atoms are 'unobservable on principle'), the development and later use of such instruments as the Geiger counter or the Wilson chamber would have been impossible.

(4) The *realist or materialist interpretation* of theoretical concepts is obviously not to be identified with the naive realist interpretation. Naive realism, as we have indicated, could not distinguish between the questions: 'Do the entities we assign to theoretical concepts exist?' (e.g. do electrons,

value, entropy, ideal gas, mass point etc. exist) and 'Do theoretical concepts have a (cognitive) sense?'. The first question is clearly semantically imprecise. For example, if the concept of 'existence' is understood in the sense of the empirically verifiable possibility of assigning an empirical predicate, the answer to the question will differ in the case of electrons (which can be assigned properties that can be expressed by empirical predicates) from the case of the ideal gas.

The interpretation of theoretical concepts cannot then be tied to an assumption of empirically verifiable existence (e.g. in the sense of Russell's 'acquaintance'). The cognitive sense of theoretical concepts – unlike fictitious concepts, which are the product of unscientific abstraction or the unfettered fantasy or imaginings of the human intellect – rests in the assumption that the entities, of which the properties and relations assignable with more or less accuracy to theoretical concepts are *objective*, are not creations of the language user or observer, but a part of his a priori equipment.[52] It is said of the theoretical concepts in science that they are a product of abstraction, that they result from a certain conception of phenomena and processes in connection with some whole, some relatively stable conditions, or relatively stable or recurrent dependences and so forth. This is certainly correct, but with the addition that the level and direction of the abstraction may take on different forms.

For these reasons it is expedient to distinguish several categories of theoretical concepts. The most important categories include:

(a) entities of other levels than the level of approach (e.g. with a cell level approach in the analysis of biological phenomena the lower level entities include those that fall within the molecular, atomic and quantum levels);

(b) concepts that arise by idealisation (e.g. the concepts of 'ideal gas', 'mass point', 'ideal thermal machine');

(c) concepts that express abstract properties, relations and structures constituted on the basis of real systems (e.g. the concepts of 'value', 'entropy', 'information') whose elements are empirically accessible;

(d) concepts that express abstract properties, relations or structures constituted on the basis of idealised or constructive systems (e.g. 'Turing machine', 'finite automaton', 'neuron network' etc.).

In theoretical work it is obviously not only possible, but even desirable to combine the various categories of concepts. For example, it is expedient to characterise certain dependences that are demonstrable by the empirical analysis of the behaviour of gases by means of the state equation, which of course derives from the concept of ideal gas. Likewise in mapping or formally modelling properties, relations or structures of real systems it is advantageous to use concepts that relate immediately to formal systems of constructive elements.

The insistence that theoretical concepts have entities that are assignable to them and that these entities are objective does not mean that these entities have to be imagined as 'things', i.e. objects presumed to have a relatively independent 'existence' that can be ascertained in isolation. Theoretical concepts are overwhelmingly systemic in character, i.e. they have to do with the properties, relations or structures of entire systems, whether these be real or constructive systems.

This emphasis on the systemic character of theoretical concepts still needs to be supplemented by a few remarks on the syntactic aspects and the a priori element in the interpretation of these concepts. The fact that we cannot normally assign 'things', in the ordinary sense of the word, to theoretical concepts, but only the properties, relations and structures of systems, is also connected with the fact that a much more important role is played, in the case of theoretical concepts, by what is expressed in linguistic semantics as 'syntactic meaning'. In the language of science theoretical concepts always figure within a network of relations with other theoretical or empirical terms, and these relations may be expressed by not only the non-logical, but also the logical terms of the relevant language. This brings us to the question of how far theoretical terms are terms a priori. The dilemma between instrumentalism and the a priori approach is based on an absolutisation of the extreme possibilities: theoretical concepts cannot as a rule be constituted on the basis of individual empirical predicates or partial observations, but on the basis of the theoretical analysis of an entire system, yet they may also figure in science as relatively a priori means for 'organising' entities, whether these be 'observable' entities or entities expressible by other theoretical concepts. To deny the relatively a priori character of theoretical concepts which cannot of course to identified with the absolute apriorism of the Kantian or Platonising type would mean that in an analysis of a subsystem or its elements we would be giving up the fund of knowledge furnished by the system overall. Metaphorically speaking, to deny the relative a priori character of theoretical concepts would mean giving up the knowledge which only the whole ocean can provide when we want to analyse the smallest droplet of water.

7. PROBLEMS OF SENSE AND REDUCTION PROCEDURES

[*A*] *A Critique of Empiricist Reductionism*

The problems of the criterion of sense are closely connected with questions of reduction and reduction procedures. The very concept of 'reduction' is ambiguous, having its own aspects not only of semantics, but also of ontology, gnoseology and others. It was Wittgenstein who brought the problems of reduction within the broader sphere of logico-semantic problems, although in his *Tractatus* he did not use the term reduction in the semantic aspect. The problems of reduction actually lie concealed in the requirement that the language of science be extensional, i.e. that it should be possible to treat the truth value of a sentence as a function of the truth values of the basic components of the sentence. It was in this connection that two fundamental assertions, later described by Quine as the "two dogmas of empiricism" [121], were first made, albeit without absolute assurance. These were (1) the assertion that there is a difference in principle between 'analytical truth' (which in some authors, including Carnap, is identified with logical truth), which does not depend on 'facts', and 'synthetic truth', which is based on 'facts'; and (2) the assertion that any expression that has a sense is, if synthetic, equivalent to a logical construction made up of the kind of language elements that correspond to direct sensory experience. Quine, who rejects both assertions, describes the second as the "dogma of reductionism". It is really a view that could be more advantageously described as empiricist reductionism.

Empiricist reductionism (see especially [18]) is directly connected with the verifiability conception of sense. What is meant by reduction here is the process whereby language expressions of 'higher levels' are ultimately analysed as logical constructions of expressions from the 'elementary level'. At the same time it is taken for granted that this 'elementary level' is equally binding for all branches of science. In other words, an absolute reduction base is assumed for theoretical concepts from the most varied domains. As for the actual nature of these logical constructions, they were deemed to be held together by the ordinary terms of extensional logic. This view, which Quine calls "radical reductionism," is another typical characteristic of the early days of the Vienna Circle and Carnap's "logical structure of the world". Radical reductionism, to adopt Quine's term, seeks to express the assumed direct connection between language expressions of a 'higher level' and those of the 'elementary level' by means of the ordinary terms of logic. This means that any sentence that has a sense can be translated into sentences of the 'elementary level', i.e. sentences about direct sensory experience.

A measure of moderation of the original radical reductionism was brought about primarily by the logical problems which we have already discussed in connection with dispositional predicates. The impossibility of using explicit definitions and the problems with the paradoxes of material implication stimulated a number of attempts to modify the original standpoint by the introduction of reduction sentences (Carnap, especially[21]), or by the replacement of material implication in its original definitional form of a constitutional procedure by causal implication or by means to express stochastic dependence.

Neither these nor any similar attempts to tone down the original radical reductionism did anything to change the principle standpoint of empiricist reductionism, which may be summarised in the following terms:

(1) Reduction is always *unidirectional*, i.e. it can be drawn in the direction of the 'elementary level' only. (Although this 'elementary level' is conceived in different ways by different authors, they mostly agree that it is this 'elementary level' that has to be the guarantee of empiricism.) Hence constitution is also unidirectional and can be drawn in the direction of the 'higher levels'. Since the procedures of reduction and constitution are portrayed by the means afforded by modern logic and since the 'higher levels' are constituted logically, this viewpoint is also sometimes called logical empiricism.

(2) The 'reduction base', whether characterised in purely phenomenalist terms or in the spirit of physicalism, is taken as equally competent for all 'higher levels'.[53] This effectively also means that the reduction base is made absolute. This view of the essence of empiricism, taking as its basis a relatively unchanging set of sensory equipment of man and sometimes also the language expressions that come out of the elementary sensory activity of man, leads naturally to a quite unhistorical view of the essence of human cognition. By contrast, the communication model of science and scientific activity, based on the historically changing and expanding concept of the 'observer channel' which also gains in delicacy with time, is not only capable of capturing the relationship and mutual contingency of the a priori and the a posteriori in science, but also of explaining why science can make new discoveries even if its universe remains apparently unchanged.

(3) It is assumed that any sentence taken out of the context of the system within which it occurs, and out of the context of the objective being pursued, may, by means of reduction, be defined semantically in such a way as to show whether or not it is translatable into sentences of the elementary level. In other words, it is deemed possible to reduce the terms of a sentence taken in isolation irrespective of the nature of the system in which it occurs and irrespective of the nature of the task in hand.[54]

Even this summary of the characteristic features of empiricist reductionism reveals its obvious shortcomings. While it has to be acknowledged that some of the works connected with empiricist reductionism have bred some interesting logical offspring, in connection, for example, with the formalisation of operational definition or with efforts to resolve the relations between modal and extensional logical terms, it must be stated that this viewpoint was by and large not a success.

The most erroneous assumption was that of the unidirectionality of reduction. Even intuitively it is clear that progress in science is not so much based on our reducing the 'theoretical' to the 'phenomenalist', the 'essential' to the 'phenomenal', as on the reverse, our seeking the 'theoretical' behind the 'phenomenalist', working towards the 'essence' from the 'phenomenon'.[55] From the scientific point of view it is important to seek not only that which figures in an individual observation or experiment, but also that which is essential to the entire class of such observations or experiments, to a whole class of tasks. The question of 'essential parameters' is not a question of empiricist reduction, but both a theoretical and practical question.

Rejection of the unidirectionality of reduction or constitution procedures leads indirectly to rejection of an absolute or absolutely stable reduction base. It is worth stressing here that criticism of empiricist reductionism is definitely not to be understood as making light of, or completely denying the importance of, reduction in science. Reduction, whether concerning data of one particular level (and so too the expressions for this data in language) or the relations between data of different levels, is of fundamental importance in science. What is essential, however, is that any 'reduction base' is only relative, both in terms of the ontological aspects of reduction (the relations between elements of different levels), and in terms of the gnoseological, theoretical and practical aspects (reduction within a particular theoretical or practical task or decision problem).

As for the possibility of reducing the terms of a sentence taken in isolation, that is a procedure that can only be applied in the most highly simplified formalised languages. But the significance of reduction in scientific thinking usually hinges on its being one component in a complex of scientific procedures, as we shall show in due course.

In its philosophical consequences empiricist reductionism leads as a rule to subjective idealism, although of course the philosophical and semantic interpretations of unidirectional reduction towards empirical predicates need not always coincide. Identification of the philosophical and semantic interpretations, such as is characteristic of a number of positivistically inclined philosophers, logicians and linguists, shows up particularly in the failure to draw any distinction between the original version of Occam's razor ("entia non sunt multiplicanda") and what we have described as the semantic analogy to it ("nomina non sunt multiplicanda"): if the only admissible entities are taken to be the names of individuals, their properties, relations, the properties of their properties etc., i.e. if the entities of which the 'world is structured' are identified with the non-logical elements of the vocabulary of a language, then clearly the difference between the two formulations disappears. This also means that empiricist reductionism conceived along these lines totally denies what might be called the ontological aspect of reduction.

[*B*] *The Ontological Aspect of Reduction; the Theory of Levels*

The concept of 'reduction' in science and scientific thought has often been understood ontologically: attention focussed not on the reduction of language expressions, of predicates from one level to predicates of another, but on a reduction that assumed that the phenomena or processes of the objective world were structured hierarchically. This is the sense in which it is sometimes said that certain chemical phenomena are reduced to physical phenomena, or psychological processes to physiological, and so forth.

The concept of 'reduction' in the ontological sense has sometimes aroused suspicions: ideas crop up of a mechanistic picture of the world where all, even the most complex processes would be explained as mechanical actions, i.e. changes of the spatiotemporal and energy characteristics of certain 'elementary' objects, such as molecules or atoms. Or there is the idea of uniform 'foundation stones' of matter, which has a longstanding tradition in philosophy and the natural sciences, or an idea of uniform laws with which to operate in all spheres, among which the sole difference is quantitative. There is of course more than one version of mechanism, whether connected with the ideas of Laplace's determinism, the idea of uniform 'foundation stones' of nature, or uniform laws of nature. Mechanism has also taken on different guises in physics, biology, sociology or psychology and so on. But in any case the concept of 'reduction' has usually led to the kind of associations in which the idea of the old mechanism figures in some form or another.

Because this idea of the old mechanism, which is usually associated with the nineteenth century level of science, does not suit contemporary scientific knowledge, mechanism itself has been declared an unsatisfactory scientific standpoint. This rejection of mechanism has often been accompained by a denial of any kind of importance to reduction in science. On the grounds that each domain is unique and specific, any connections between them have sometimes been denied. Thus it is necessary to point out, in rough terms, two extreme positions:

(1) At one extreme there is mechanism and some of the later conceptions that followed in the mechanist tradition, whether in the ontological, or merely the phenomenalist respect, e.g. physicalism, the movement for unified science in some of its aspects, and others. This viewpoint is characterised by the assumption of a fundamental reduction base, represented by elements of the elementary level, by means of which it is possible to explain the phenomena and processes of higher levels. Hence it also recognises the principle whereby a scientific picture of the phenomena and processes of the higher levels can be presented by means of the conceptual and theoretical apparatus provided by the scientific analysis of the elementary level.

(2) The other extreme is represented by, for example, vitalism, gestalt psychology and other conceptions that cling to the notion that each separate domain of nature has its own principles of explanation, quite distinct from those of others, and that each has its own specific basic elements and basic laws. Hence the desire here to set up quite specific conceptual means that cannot be applied in other domains.

It is not the purpose of this work to analyse the whole vast complex of problems connected with the relations between these two extremes or the interrelations among the various domains. But these problems are connected with some important methodological questions, especially those to do with determinism, causality, scientific explanation, the nature and role of scientific laws, the question of frontiers and so forth.

It would also be possible to point to a number of dogmatic versions of the two extremes, one of the roots of dogmatism being the (scientifically, historically conditioned) impossibility of imagining any other point of view than that believed in. In connection with the actual purposes of this work we shall note just some of the partial problems out of this complex, namely those connected with the system of scientific language.[56]

The ontological aspects of reduction are best captured by the 'theory of levels', which was an offshoot of physics and came about in connection with certain difficulties in the interpretation of quantum theory, but which is of much wider significance. The theory of levels assumes that the structure of nature is hierarchical. Unlike mechanism it emphasises that the scientific analysis of the world has no absolute boundaries. As J.P. Vigier put it [158], [174], there are no a priori limitations to the development and possible progress of human cognition. No scientific theory, if tied to one of the levels, can claim to give a complete explanation of the totality of nature (which puts an end to the ambitions of the old mechanism to achieve 'complete description', 'complete explanation' and 'complete reduction'). But it is possible to extend the horizons of investigation, refine the analysis of nature and so transfer to a different level. At the same time it is assumed that – theoretically – nature contains an infinite and inexhaustible number of levels. This is why with moves towards 'lower' levels in physics, i.e. to the atomic and quantum levels, the 'subquantum' level is assumed as a possibility.

This outline of the conception of levels, which was drawn up by D. Bohm, Vigier and others, could be made more complete by the communication model of scientific cognition. According to that model the 'channel' of the observer or experimenter, or more generally the 'channel of science', it itself also connected, by its properties, threshold values, degree of accuracy and its theoretical principles, to one particular level. This is where those difficulties arise that are connected with the fact that the source and the 'channel' of the observer or experimenter (e.g. the properties of his metrical apparatus) belong to different levels. It can be shown that Heisenberg's relation of uncertainty is actually a special case of these difficulties and problems, which are of much more general significance and do not concern just the interrelations of the microlevel and macrolevel in physics. Also the theory of levels assumes no absolute or stable 'reduction base'. Any 'reduction base' is relative and is only of partial significance for the aims and tasks of a given investigation. At the same time it transpires that not even the deductive schema of reduction is satisfactory since, while it is capable of modelling what might be called 'complete reducibility', it cannot respect the fact that (a) with the constitution of formations of a 'higher' level on the basis of elements of a 'lower' level new qualities are born, and (b) the reduction of these formations to elements of the lower level can usually be done only statistically.

The conception of levels contains implicitly the idea of essential parameters (or more precisely, relatively essential parameters). Vigier formulates it as follows: In certain spheres it is possible to abstract from the infinite number of parameters necessary to the description of some matter, and it is possible to start out from a certain number of variables that play a fundamental role for the given level and describe its properties truly objectively [174, p. 100]. Similar views were expressed by Bohm [14]. This not entirely clearly formulated idea about essential parameters (and so too about essential concepts, for this concerns not only empirical predicates, but also theoretical concepts) could be supplemented along the following lines: the conception of reduction, relativised to a certain level, also clarifies why the *finitist programme in science* is so productive and expedient. Nature is infinite and inexhaustible and has to be analysed with relatively finite and ultimately restricted means, whether these are the means of measuring and experimentation or means of a theoretical order. The finiteness of these means – to return to the communication model of science – is due to the finiteness and limited nature of 'capacities', 'memories' and 'delays'. What is always important, however, is for every finite picture of an infinite and inexhaustible world to apprehend those aspects that are essential with respect to the given level and to the aims of the investigation in question.

The role of reduction in the context of the theory of levels and the finitist programme in science can be adjudged both in connection with the range of data that are accessible at the given level, and in connection with the data that are ascertainable at other levels. Hence any scientific portrayal of a given domain of nature abstracts both from the infinity of parameters at a given level, and from the infinity of parameters at other levels. At the same time the measure of how far an area under investigation can be viewed as a relatively isolated area and how far one can disregard the encroachments of other areas at the same level or the influence of the microlevel or macrolevel (relative to the level of investigation) depends both on the nature of the area in question and on the equipment of the experimenter and scientist and the aims of his work.[57]

[C] *The Semantic and Pragmatic Aspects of Reduction*

The semantic aspects of reduction are associated with the requirements that we have already formulated in the appraisal of the operationalist criterion of sense. These requirements, as we shall be showing, concern not just empirical data, but any data relevant to a given task. With respect to the problem area of reduction these requirements may be reformulated as follows:

(1) In the process of reduction the sense of the expressions that enter into the reduction procedure is preserved. This requirement, which may be described as the *requirement of invariance of sense*, is on the whole obvious intuitively. For example, if we replace predicates P_1, P_2 and P_3 by the complex predicate Q, we require that Q preserve the sense associated with the original predicates.

(2) In order to replace a given sequence of expressions, e.g. the predicates P_1, P_2 and P_3, by another expression, e.g. the predicate Q, we require to know the procedure that allows us to deduce Q on the basis of P_1, P_2 and P_3. This requirement may be called the *requirement of deducibility*.

It may be added that the requirements so formulated are more like the extreme possibilities that in any real task connected with reduction (of data, parameters, predicates, laws of science) we can only approach more or less. At the same time it is of advantage, as we have shown, to use the communication model in modelling a reduction procedure. In that model reduction is portrayed as a transmission of a message through the channel, the achievement of minimal loss being desirable. The requirement of invariance of sense therefore assumes the ideal situation, i.e. the channel operating without noise. The requirement of deducibility then assumes that the channel operates deterministically (in the sense of Laplace).

It has to be emphasised here that the requirement of invariance of sense cannot be taken abstractly, i.e. unconnected with some class of tasks, some objectives of a given investigation, communication etc. This brings us face to face with the question of *criteria of invariance of sense* with respect to the given class of tasks, or, to put it more generally, the question of criteria of the kind of preservation of sense that, under reduction, is desirable or admissible in the solution of the given class of tasks.

The principles for a solution to this problem, and so too for a solution to the admissibility or otherwise of one version of reduction, based on the results of information theory put forward by A. Perez, is given in [110]. This solution employs the communication, or more exactly the communication-decision, model of reduction and constitution procedures and exploits the facilities offered by the mathematical theory of information.

The virtue of this solution is not only that it replaces the sweeping requirement of invariance of sense by a requirement relativised to a given class of tasks, or decision problems, but also, and preeminently, that it puts forward pragmatic criteria for the preservation of sense in reduction, without the need for a separate solution with each individual decision problem.

The solution of the reduction problem based on an application of information theory may be seen as an extension of the original deductive nomological model of reduction based on both the above requirements. One example of this kind of deductive nomological model of reduction are the systems of reduction sentences set up by Carnap, or alternative solutions that are based on the fact that the result of reduction is deducible if what is given is a deductive form of the particular reduction procedure. This means that no account is taken of other than purely formal criteria of the invariance of sense. The switch from the deductive nomological, or deduction, model to a model based on the theory of information offers *objective criteria of the preservation of sense* in the process of reduction by linking these criteria to the quality of decisions taken on the basis of the reduced data. If the quality of decision-making can be measured, in terms, for instance, of the average 'loss' or average 'risk' entailed, the measure of quality can be taken as the basis for assessing the admissibility or inadmissibility of a given version of reduction. In effect this conception is a consistent application of the principle, already mentioned, that the sense of an expression (piece of data expressed by the relevant sentence) cannot be considered in isolation, but in relation to a particular task or decision problem. Hence the quality of the solution of a task or decision process is a yardstick of the preservation of sense in a reduction process.

It needs to be stressed that the significance of reduction (and constitution) procedures cannot be weighed outside the framework of a particular task or decision problem, yet it is precisely in this context that the real need for reduction becomes justified. For example, if we are to assess a case of diagnosis on which to demonstrate this kind of solution, it is obvious that the optimum diagnosis would call for the greatest possible number of tests, data, laboratory and clinical findings, specialist consultations etc. But the doctor who is making the diagnosis is hampered by the time factor (e.g. the urgency of treatment), limitations in the capacity of laboratories and clinics, the availability of only his own and his colleagues' knowledge, in short by having available only limited 'capacities', 'memories' and 'delays'. He cannot therefore make decisions without regard to the objective state of the patient and the conditions available to him. Yet his decision-making is goal-oriented, motivated by the desire to achieve the best solution, or by the expectation of the best solution.

After these comments we can proceed to review the basic features of the suggested reduction model.[58] We may do so in terms of the case of diagnosis, although the formulation of the basic elements of the model will have to be abstract enough to suit other cases of reduction as well. (The suggested model is based on concepts that are current in communication theory.)

The main thing here is what may be described as the *input space* **X** with elements **x**. With a diagnostic problem the space **X** represents the sum of all the different possible ailments or pathological states. Then there is the *output* space **Y**, with elements denoted **y**. In diagnosis **Y** is the sum of the ascertainable symptoms. (The concept of symptom need not be identified inevitably with the concept of empirical predicate. Allowance may be made for symptoms that are not 'observable' in the sense of the original homocentric conception of empiricism or operationalism. If we were to identify the space **Y** with an initial empirical space, we would have to assume the more general conception of empiricism referred to earlier in the account.)

Another element in this model of reduction procedures is what may be called the *decision space* **D**, with **d**, i.e. the elements of that space, representing individual decisions. In the classical model of message transmission this is the ascertainment (decoding, detection) of the message sent on the basis of the signal received. Therefore **D** is usually a suitably selected sub-set of **X** and may in the extreme case coincide with **X**. In diagnosis **D** is the sum of the illnesses available to the doctor (or diagnostic machine).

In addition there is what may be called the *decision functions space* **B**, each $\mathbf{b}(\mathbf{b} \in \mathbf{B})$ enabling us to assign to each $\mathbf{y} \in \mathbf{Y}$ either a particular $\mathbf{d} \in \mathbf{D}$ or a particular distribution of probabilities in **D**. Each **b** therefore represents one of the possible states of diagnosis (with respect to the sum of the means available and to the given population).

Between the input and output spaces there is assumed to be a *channel of observation* which enables us to take decisions about elements of the input space on the basis of the accessible elements of the output space, but which also causes 'noise' (e.g. the possibility of errors, imperfections of the measuring and experimental apparatus, etc.).

In order to operate with these elements we still have to assume an *a priori distribution of probabilities* $\mathbf{P_X}$ in the input space $\mathbf{X}$ (i.e. the relative incidence of diseases in the given population) and the '*two-dimensional distribution of probabilities*' $\mathbf{P_{XY}}$. In diagnosis $\mathbf{P_{XY}}$ expresses the relative frequency of the occurrence of different pairs of diseases, and symptoms.

In order to measure the quality of decisions we will also need what is called the *weight or loss function* $w(\mathbf{x}, \mathbf{d})$, which to each pair $(\mathbf{x}, \mathbf{d})$ assigns a certain loss or evaluation (weight). Intuitively it is obvious that the more adequate the decision selected, the less the loss. Thus it is desirable for the average value of this loss, or the *average risk*, to be minimal.

Decisions about the admissibility of a particular kind of reduction, e.g. reduction of the elements at the input, i.e. reduction of certain symptoms, are therefore tied to the qualification of decision-making by means of the above devices. Thus the admissibility of a particular version of reduction may be judged by the degree to which the quality of decisions is not made worse by it, i.e. the degree to which the average risk does not rise above the required level. On the other hand, if that level is exceeded, we are prompted to extend (by constitution) the range of data taken as the basis for solving the given task. Another stimulus for making such an extension – and not only an extension of the empirically or experimentally ascertainable data that may correspond to elements at the output, but also an extension of the search for new laws that may correspond to decision functions, or an extension of the theoretical concepts that may be relevant and may correspond to elements at the input – may be the desire to achieve even more adequate decisions or to enhance the quality of the solution to a given task, i.e. a further reduction of the average risk.

Therefore this model of reduction need not concern just the reduction of output elements (e.g. reduction of the empirically ascertainable data, or reduction of the empirical predicates), but also reduction of the input space, the decision space and the decision functions space (e.g. reduction of theoretical concepts, scientific laws etc.). What is essential here is that reduction of the elements of these spaces is not considered within the framework of the relations among the elements (the syntactic viewpoint), but from the point of view of a particular universe and the point of view of the quality of decision-making (the semantic and pragmatic viewpoints). In the case of diagnosis this means, for example, that the admissibility of reduction (or constitution) of certain symptoms, or tests, measurements, clinical or laboratory tests and so on, is considered with respect to the given population and the distribution of the relevant diseases or pairs of diseases, or symptoms, in that population, and with respect to the desired quality or adequacy of the decisions taken.

Thus we may say that this model does provide a criterion of the preservation of sense which by its nature embraces both the semantic and the pragmatic aspects. Of course this criterion is not absolute, but always relativised to a given range of tasks or decision problems. Thus it does not fulfil the ambitions of certain conceptions of a criterion of sense, namely of providing a universal means for deciding the sense of any language expression without respect to the context, the nature of the language, and the aims and tasks connected with the use of the given expression. At the same time, however, this model does give a sufficiently objective yardstick for assessing sense-preservation. It may be added that this is essentially a practical yardstick.

To return now to the concept of 'having a sense', we may add the following to what was said before: While this model does not provide a direct explanation of what an expression's 'having a sense' is, it does provide an objective yardstick for a decision about what 'preserving sense' or 'having the same (or approximately the same) sense' means, not absolutely, but in relation to a particular task or decision problem. It may also contribute to a refinement of the concept of 'synonymity', since it gives a more exact definition of what we have called 'marginal conditions'.

The concept of 'having a sense' as a metalanguage predicate is, as should be obvious from the account so far, a many-placed predicate, one of its arguments being the statement of the task or decision problem in respect of which the sense of a language expression is considered.

Since a language expression (just as a particular parameter in a scientific or technical task) may figure in different kinds of tasks, the relevance or irrelevance of its sense may vary with respect to different tasks. At the same time it is desirable that, on the one hand, the sense of an expression with respect to a given task be sufficiently clear-cut (making 'have a sense' a non-vague predicate) for reliable decisions to be taken, and, on the other hand, its sense should be sufficiently vague to enable a switch from one task to another and from one decision problem to another.

VII. VAGUENESS

1. VAGUENESS AND THE UN-SHARPNESS OF BOUNDARIES

In the account up to this point we have encountered the term 'vague' (or certain other similar notions as opposites of the terms 'exact' or 'clear-cut' and others); it expresses one property of certain expressions of language. In everyday speech the concept of 'vagueness' is usually used to indicate a lack of exactness or definiteness. It can easily be shown that the concept of 'vagueness' is vague in itself and that it is not always easy to reach agreement on its use.

'Vagueness' always assumes an interpretation, for although we assign it as a property to certain expressions of language (making 'vagueness' a metalanguage predicate), we assume that the property will manifest itself in particular situations. Roughly speaking, a term is vague if there are situations when there is a possibility of uncertainty as to whether or not the term, or its contradiction, may be applied.

If 'vagueness' always assumes an interpretation and the possibility of situations involving this kind of uncertainty, the question arises as to where the uncertainty resides and from what does it stem. A number of questions may be formulated to encapsulate the more important problems that arise in connection with vagueness: if vagueness is a property of expressions of language, is it possible to ascertain with anything like accuracy what these expressions are? Is the source of vagueness the nature of language, especially natural language, or the nature of the objects to which the vague expressions relate? Is it circumstances connected with the language user that are the source of the vagueness of language expressions? Is vagueness a property of an expression that is ever-present, whatever the circumstances of the expression's use, or is it a property that is merely relativised to certain circumstances? (In other words, is vagueness a metalanguage predicate, is it a one- or multi-place predicate?) Is vagueness subjective in origin, or does it have objective causes?

These questions are far from exhausting the whole range of problems that occur in any analysis of vagueness. In what follows we shall focus our attention primarily on the methodologically important questions to do with vagueness, in particular those that may be encountered in science and in using the language of science.

[*A*] *Vagueness and the Empirical*

On the face of it, vagueness is pre-eminently a property of those empirical predicates that are qualitative in character. Terms such as 'high', 'distant', 'dark', 'bald' and the like are, when compared to terms such as 'two meters high', 'without a hair on the head', vague. In some instances, terms which effectively describe qualitatively a characteristic that may also be quantified, give rise, because of the indefiniteness of the characteristic or the lack of clarity of the boundaries of the term's applicability, to difficulties that may even be formulated as paradoxes. Some of the familar classical paradoxes, such as that of the Bald Man, the Heap etc., are essentially of this kind of origin. Bareheadedness comes about by the *gradual* uprooting of the hair, but if we *gradually* uproot one hair at a time, we do not make a man bald. A heap of grain comes about by gradually adding to it grain by grain, but if we add one grain to another, we do not get a heap. The indefiniteness of a qualitative characteristic that is essentially a quantitative determiner, and the impossibility of establishing a clear-cut boundary, may together give the impression that we are dealing with the mysterious emergence of new qualities.

These and similar examples of vague concepts express the quantitatively determined change of a quality or the emergence of a new quality. Yet the quantitative determination itself is so uncertain that situations may arise where it is possible to apply both the expression for the new quality and the negation of that expression, thereby giving rise to paradoxes. At the same time, the description of the emergence of these situations (the genesis of a heap or of baldness etc.) uses expressions ('gradual uprooting of hair', 'putting one grain with another') that are likewise vague and may be prone to different interpretations in different situations.

Thus it might seem that vagueness is a concomitant phenomenon in situations where we use empirical predicates that are themselves not quantitatively determined exactly or that are difficult to quantify exactly. This relates to very many terms of all sorts which may be empirical predicates by nature and where the quantification does not play any major role. If a 'window' is an aperture in the wall of a building through which light may enter the building, is any aperture in the wall of a building a window? We would obviously be uneasy were we to have to determine the dimensions above which we would be prepared to treat an aperture as a window, the more so in that even a doll's house may have 'windows'.

It must be added that an exact quantitative statement about a characteristic will not always overcome the vagueness or the difficulties that may arise in this connection. There are, however, cases in which elimination of the difficulties by this means is desirable. This is usually done by introducing what may be described as an *interpretational norm.*

In using language in certain areas of our social life, and in science and technology, interpretational norms are an important device in the elimination of undesirable vagueness and so too of interpretational uncertainty. It is obvious that, for example, the term 'house' is a vague term. The term 'family house' is also vague, although it does provide a more definite characterisation. Therefore, in using this vague term, for instance in legal practice, there is a call for an interpretational norm, for example of the kind: By family house is understood a house that has no more than five rooms and whose habitable area is not in excess of 120 square metres. It should be noted that interpretational norms, which are usually conventional in nature, do not eliminate vagueness altogether, but do restrict it with respect to certain goals or tasks, making it compatible with the nature of those tasks.

Vagueness may occur in the case of empirical data based on quantification, as in the case of measuring. These are, of course, rather different instances of vagueness from those mentioned previously. Nevertheless, it is well known that any quantitatively defined result of measurement invariably depends on certain relations between the object measured and the metrical apparatus, or on interaction between them. Any measure or measuring equipment has only a limited discriminatory capacity. In essence, measuring is only a process of the transmission and processing of information, while 'noise' (measurement errors) may occur in the channel. Thus the procedure of measuring is a process in which uncontrollable influences may also occur, so that the outcome of the procedure has to be judged statistically. At the same time the required accuracy of measurement and the degree of accuracy afforded by the properties of the metrical apparatus depend on the aims or tasks connected with the particular act of measurement. Therefore there are also limits of measurability, which are not of course absolute, but are connected with the level of the metrical apparatus in relation to the level of the object being measured. It follows from this that even the expressions of language which express the results of measuring and therefore relate to objects, properties, relations or whatever other kind of entities that have been determined quantitatively, may also be vague. At least they may be vague in the sense that there can occur reasoned possibilities for rendering more exact the quantifications expressed by those expressions.

[*B*] *Vagueness and Theoretical Concepts and Constructions*

The concept of 'vagueness' is also applicable to certain theoretical concepts, in a number of respects moreover: One might consider the interpretation of some concepts that are the product of idealisation, relating them to the empirically discernible universe; or one might also assess the vagueness of theoretical concepts in the same way as that of empirical predicates, i.e. by analysing the situations where there is some uncertainty as to the applicability of some concept or its contradiction.

In the case of the concepts that are the product of idealisation (e.g. 'ideal gas', 'mass point', 'absolutely black body', etc.) it is possible to confront these concepts with real or empirically ascertainable entities, and this will show the differences. Taking the entirely naive approach to the confrontation we might find ourselves asking on which side does the 'imperfection' lie: Is an orange an imperfect copy of the shape of the ideal sphere, or are the configurations of spherical geometry a 'simplified' version of physical objects and their spatial properties? Questions like this might be reminiscent of some of the conceptions of Platonism concerning the perfection of the ideal world in relation to the real world which is only its 'imperfect' reflection. We take the view that to ask this kind of question without any kind of context is pointless, although this in no way alters the fact that certain theoretical concepts and the models of the real world built upon them may differ in greater or lesser degree, which may give rise to situations where there may be uncertainty as to their applicability in much the same way that we are sometimes hard put to it to say whether a given body is or is not a sphere.

This situation is also typical of some cases of the application of mathematics and mathematical concepts in the analysis of systems that are also studied empirically. This fact was expressed by Einstein [41, p. 3]: "If the theorems of mathematics relate to reality they are not reliable, and if they are reliable, they do not relate to reality". Einstein, who has in mind here reality as empirically ascertainable, is formulating the difficulties which arise in the application of theoretical concepts to the study of a reality that is apprehended empirically (or *also* empirically), and the formulation is a paradox. The heart of the problem – including that of the application of mathematical concepts – resides primarily in the fact that an interpretation in one context where vagueness does not figure, or where we disregard it, may be confused with another interpretation where vagueness cannot fail to manifest itself. To return to the example of the sphere: it can be shown that this term, when applied as an exactly defined term of spherical geometry within geometry itself, is not vague, yet its homophone as applied to bodies that are studied empirically *is* vague. The two are then easily confused. The same situation may come about with many other theoretical concepts that feature once as exactly defined and once as vague.

[C] *Vagueness and the So-called Fringe*

Bertrand Russell has said, in his thoughts on vagueness [135], that the law of tertium non datur is only justified if clear-cut and exact terms are being used; it is not justified if these terms are vague. To this he adds that all terms are in fact vague. This in turn means, as Russell emphasises, that all languages are also more or less vague. By contrast, logic assumes that the terms which it employs are exact, which leads Russell to add that logic is not applicable to life on Earth, but to an imaginary superterrestrial existence.[1]

This viewpoint calls for some refinement. Logic, especially classical logic, genuinely assumes that the principle of tertium non datur can be applied to non-logical terms. At the same time these terms are usually handled in such a way that they could be interpreted, although no concrete interpretation is given in so many words. In this situation the problems do not and cannot arise. They do so only when practical decisions on an interpretation are to be made.

Russell's viewpoint – it is useful here to consider the conception of the entire system of *Principia Mathematica* – is effectively based on the conviction that only logical terms can be exact and therefore not vague. By contrast non-logical terms are more or less vague, unless some definite interpretation is assumed. It also follows from this that only a formalised language can be completely rid of vagueness, and that only if the interpretation of non-logical terms is left out of consideration.

There is the possibility here of a viewpoint that links the concept of vagueness with the semantic, and possibly also the pragmatic, level of language: if we abstract from these levels of language and keep to just the syntactic level, we need not worry about vagueness. There are certain objections that may be raised against this view, which, while not stated explicitly by Russell, does follow from his argument.

In natural language even those expressions that correspond to the logical terms of formalised language are often vague.[2] In a formalised language these terms are naturally given exact definitions so the vagueness is eliminated.[3]

Even in the case of the non-logical terms the situation is not as hopeless as Russell's argument suggests. It is invariably possible to accept certain interpretational norms and so obviate uncertainty of interpretation. The actual forms of acceptance or introduction of interpretational norms are moreover immaterial: they may be of the nature of conventions, their competence may be limited to certain specific language users, or to certain specific tasks, and so on.

Leaving aside the problem of the vagueness of logical terms to concentrate on the question of the vagueness of non-logical terms, we may specify vagueness by ascertaining what Black [10] calls the 'fringe'.

The concept of 'fringe' can be analysed in connection with the interpretation of predicates. Using the predicate P to form the class $(\hat{x})(Px)$ and the complementary class $(\hat{x})(\sim Px)$, the concept of 'fringe' can be visualised in two ways:

(a) If we postulate a single language user XY, the class $(\hat{x})(Px)$ has a fringe if XY is not sure whether $y \in (\hat{x})(Px)$ or $y \in (\hat{x})(\sim Px)$. The domain of individuals where such uncertainty can be demonstrated may be treated as the fringe with respect to P.

(b) If we postulate a plurality of language users, the situation may arise where some consider y to be an element of the class $(\hat{x})(Px)$ and others consider it to be an element of the complement of that class. The fringe is then the domain of individuals where this kind of disunity of assignment can be demonstrated.

It is obvious that the principle of tertium non datur has not been suspended. The classes $(\hat{x})(Px)$ and $(\hat{x})(\sim Px)$ are by definition complementary, but with one or more language users there may be uncertainty over the application of the predicate or the incorporation of some element into the class made up by that predicate.

The foregoing remarks on vagueness and the fringe have concerned predicates. It would be possible to extend the idea and consider in like manner the vagueness of sentences in the context of the uncertainty of a single individual language user over the assignment of a truth value, or in the context of the disunity of decisions by a number of language users about truth value assignment.

So far in looking at 'vagueness' and the concept of 'fringe' we have assumed that the language user or users make decisions about the denotation of the relevant expressions, with some room for uncertainty in these decisions. This approach raises some questions of practical importance: How is vagueness to be measured? How can we determine whether some expressions are more exact or less vague than others? Is it possible to restrict vagueness and so increase the accuracy of some expressions? In order to provide answers to these questions we must first analyse the pragmatic, semantic and syntactic aspects of vagueness.

2. SOURCES OF VAGUENESS AND WAYS OF ANALYSING VAGUENESS

The fact that there is more than one kind of vagueness is due to the fact that there is more than one cause of it. In analysing vagueness we may point to the following major sources of it: (a) circumstances connected with the class of users of a given expression (e.g. differences in their equipment, whether psychological, somatic, cultural, linguistic or other prerequisites); (b) circumstances connected with the nature of the class of individuals designated (e.g. unclear boundaries between one quality and another, problems associated with discreteness and continuity, etc.); (c) circumstances connected with the nature of language itself (e.g. the logical structure of the particular language, the nature of the logical signs used in the particular language, etc.).

From this brief review of the major sources of vagueness it is obvious that there may also be quite varied ways of ascertaining vagueness. These ways may be divided up according to the major levels of language analysis.

[*A*] *The Pragmatic Aspects of Vagueness*

In considering the pragmatic level of language analysis we have to take into account the class of language users. We have already seen that there is no particularly systematic categorisation for pragmatic concepts. If we are to analyse 'vagueness' from the pragmatic point of view, it seems expedient to distinguish two types of pragmatic concepts: (1) pragmatic concepts such as 'pronounce', 'declare', 'use the term ...' etc., which while expressing a particular relation on the part of the language user to the language expression, express relations that are semantically irrelevant; (2) pragmatic concepts such as 'believe', 'accept', 'recognise' etc., which express a relation of the language user to the language expression which is not semantically irrelevant. Understandably, the differences between the two types of pragmatic concepts may be only one of degree in natural languages.

In analysing vagueness from the pragmatic point of view it is indispensable to consider the second type of pragmatic concept. This means taking account of a relation of the language users to the language expressions that includes within it the acceptance or rejection of a certain semantic characteristic of a given expression (e.g. the truth of a sentence, the relation of denotation etc.). We shall take as the prototype of this second kind of pragmatic concepts the concept of '*semantic decision*'.[4] By a semantic decision we mean any attitude, however displayed, of a language user to the semantic characteristics of a given expression which he understands.

A semantic decision need not necessarily be a two-value decision, e.g. in the form 'XY considers ... to be true', 'XY does not consider ... to be true' and the like. However, for simplicity's sake we shall only be considering two-value semantic decision-making.

Semantic decision-making about the application of the predicate P may be treated as a statistical decision. If Y is the set of all the individuals y that may be elements of the class $(\hat{x})(Px)$ or elements of the complement of that class, it is possible to investigate a single language user to determine:

(a) with which individuals he will decide unambiguously in favour of $y \in (\hat{x})(Px)$;
(b) with which individuals he will decide unambiguously in favour of $y \in (\hat{x})(\sim Px)$;
(c) with which individuals he will betray uncertainty over making the assignment.

Postulating a plurality of language users, we may study the proportion of the set of language users and the proportion of the set Y, and the connection between them, with respect to a given decision. The connection can be expressed in graph form by giving the figure for the set of language users along one axis and that for the set Y on the other. For example, if we consider the predicate 'be a heap', we may consider the number of grains for which a certain proportion of the set of language users will decide that they do or do not constitute a heap. In the graphic representation

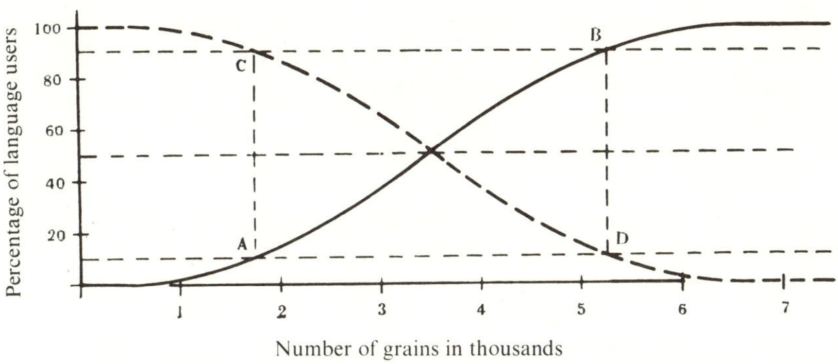

Fig. 15.

the first decision is indicated by the unbroken line, the second by the broken line. Black, who used this graph method for expressing the measure of vagueness [10, p. 44], calls the curve that emerges in this way the consistency profile for the application of the predicate *P*.

It is obvious that this consistency curve for *P* is relative and depends on the properties of the users and of the set *Y* in respect of which the applicability of *P* is being considered. It is not at all difficult to imagine the transfer from two-value decision-making to three-value decision-making, when, for example, some of the language users refuse to make a decision, or even to higher value decision-making.

In contrast to the consistency graph (Figure 15) which shows the position for a very vague predicate, we can imagine an exact predicate whose consistency curve could be presented as in Figure 16.

This graphic method, which is no more than one way of expressing the statistical approach, may also be treated as an empirical or experimental means of measuring what has been called the 'fringe'. If we wish to determine the 'fringe' on the basis of the pragmatic approach, we shall have

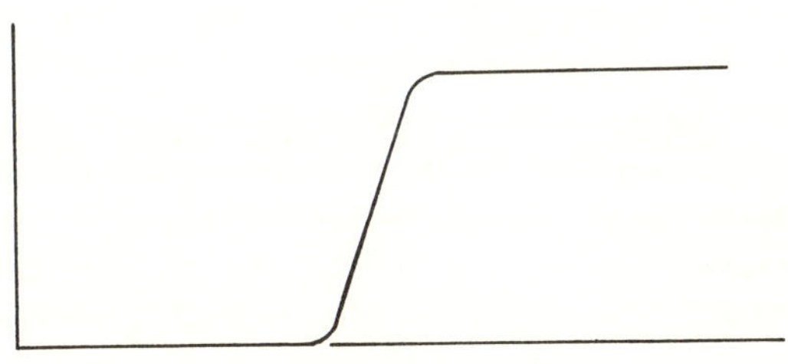

Fig. 16.

to delimit that proportion of the language users that we shall take as standard. (In the first graph we chose 90%, or 10% for the reverse decision. The curve between points A and B, or between points C and D, then gives the extent of the fringe.) We might proceed in like fashion if we wished to start from a particular proportion of the set *Y*.

The graphic representation of semantic decision may be viewed as the most elementary form of measuring vagueness and comparing the vagueness of different terms. It may of course be extended in such a way that the consistency of the application of *P* with respect to *Y* will be taken as not only dependent on the semantic decision-making of a certain class of language users, but also on a given situation, set of circumstances, in which the language user is deciding on the applicability of *P* with respect to *Y*. If it is possible to express the distribution of these situations quantitatively or graphically, it is possible also to measure the dependence of semantic decision-making on this distribution.

In this connection we might recall again the fact that semantic decisions are somehow fixed, i.e. that there are certain limits to the 'fringe' with respect to a vague term, and that these are set by the introduction of interpretational norms. For example, certain units of measurement such as 'inch', 'foot', 'ell' and the like might once have been viewed as vague terms. Before serious demands for accuracy were made, a certain distribution of semantic decisions within admissible limits was tolerated. (Even in the Middle Ages failure to abide by these limits was of course severely punishable.) This means that the vagueness of such terms was not really felt. But with the introduction of more exact units of measurement and when the needs of science and technology began to place incomparably greater demands of accuracy, interpretational norms had to be brought in to state, for example, just how many centimetres and millimetres made one 'foot'.

Seen in this light, interpretational norms represent a unification of semantic decision-making. Their aim is to ensure that all language users are bound, if they use a particular term, by a certain interpretation of that term, however that obligation may be presented.

The analysis of the concept of vagueness from the pragmatic point of view offers relatively good facilities for empirically or experimentally oriented 'measurement of meaning'. At the same time, this kind of measurement can, if it concentrates on the objectively apprehensible results of semantic decision-making, if it works with behavioural methods, provide fairly broad possibilities. For example, in the analysis of natural languages the method of analysis suggested may be used in dialectology. The same line can be taken for contrasting the exactness (in pragmatic terms) of different terms. There is even the possibility of analysing the vagueness of certain signals intended to produce reactions in 'reception mechanisms' in technology, or in reading systems etc.

[*B*] *The Semantic Aspects of Vagueness*

If vagueness is connected with differences in the semantic decision-making of a certain class of language users, it would be a mistake to see the heterogeneity of this class as the sole reason for vagueness. This is of course a major factor and one which is connected with many of the individual differences among language users. In this connection reference is usually made to the process of learning which tends towards the semantisation of terms (e.g. [122, p. 125]). Since the learning process is always based on experience, it is a stochastic process in which individual differences cannot be excluded.

Irrespective of these and other factors connected more with the pragmatic aspects of vagueness, there are also plenty of reasons for seeing the causes of vagueness in the semantic sphere. In a semantic analysis of vagueness there is no overlooking the fundamental fact that language offers the facility of an essentially *discrete mapping* of phenomena, actions or events which may (if we consider the limited possibilities of empirical, experimental or theoretical investigation) be both discrete and continuous.[5]

The discrete nature of the language mapping of extralinguistic phenomena and processes (and also the finiteness of that mapping) brings in a number of serious problems. There is no need for any lengthy proof that the principles of this discretisation are also human, being conditioned by the given level of society, by traditions, social practice, knowledge and so forth. Suffice it to recall the investigations of comparative linguistics, cultural anthropology, the research associated with the Sapir-Whorf hypothesis of linguistic relativity. While the Eskimo has a number of different terms for the different kinds of snow, the Central European has just one, be it 'snih', 'Schnee', 'śnieg', 'neige' or whatever. Thus the question might be asked how far do we, in certain circumstances, force the discreteness of language upon the respective universe. Moreover, it is not always that the universe is not also discrete; the main concern is the degree to which we prescribe for nature that conception of discreteness that is proper only to the given language and culture. According to situation these aspects will obviously be apparent in varying degree. Finally, it is one of the characteristic features of verbalism that exactness and definiteness in language are taken for exactness, definiteness and unequivocality of the denotates of individual terms.

Hence one of the causes of vagueness may well be seen to reside in the lack of absolute coincidence between the discreteness of a given universe and that of a given language, although there will always be a measure of correspondence between them. Without this correspondence language would be incapable of performing one of its basic functions, namely that of the relatively reliable transmission of messages.

The conception of discreteness in language on the one hand, and the discreteness and continuity of a given universe on the other, may differ, the nature of the differences varying from one kind of language expression to another. Considering the different kinds of language expression we may point to the following:

(a) In the case of the proper names of individuals or individual descriptions (in Russell's sense) it is generally assumed that there is no, or only minimal, vagueness. The proper name 'Water Scott' or the individual description 'the writer who is the author of Waverley' or other analogous expressions may generally be reckoned to be exact. The ultimate possibility of some vagueness is not of course excluded, but the sources of it will be factors of a pragmatic or rather semantic nature. Nevertheless it can be said that by and large we operate with proper names of individuals and with individual descriptions as if they were exact terms.[6]

(b) Vagueness can be found to play a relatively greater role in the case of general terms, and pre-eminently those that are of the nature of empirical predicates. (Moreover we assume that this applies to both one-place and multi-place predicates, and that these expressions can be interpreted both as 'classes' and 'properties'.) Insofar as the interpretation of these expressions is connected with experience, it is the statistical character of experience that is to the fore, since experience is based on a particular distribution of the values of the arguments of a given predicate.

This distribution may take on many forms. There may be cases where the borderline or fringe objects occur at both edges of the distribution, or cases where there is a continuous transition of values, the differentiation of distinct qualities being dependent also on the sensory equipment. In fact many empirical characteristics of physical, biological, social and other phenomena frequently tend to accentuate properties with extreme values (e.g. people are tall or short, fat or thin, rich or poor and so on), while the real distribution of values is densest in the middle ranges. By contrast, distinguishing the different colours of the spectrum assumes that the attention is concentrated on a particular segment, the borderline cases being the controversial ones.

(c) Vagueness cannot be eliminated even in the case of theoretical concepts. But since the meaning of many theoretical concepts depends on a particular theoretical conception which can be exactly defined, the vagueness of such terms is generally restricted. In the preceding category of expressions we associate vagueness with a kind of uncertainty of empirical interpretation or with ill-defined borderlines or with continuity within that empirical interpretation. With theoretical concepts any interpretation of this kind is more or less mediated. Yet what is fundamental is the interpretation within the limits of some theory, law or theoretical conception or hypothesis. In a sense the concept of 'ideal gas' has more accurately defined properties than the empirical concept of 'gas', for the properties of an ideal gas are defined by equation of state.

The account of the semantic aspects of vagueness has so far concerned first and foremost that which might be described as *denotational vagueness*, i.e. the lack of definition of the boundaries in semantic decision-making with respect to a given universe and the individual partial domains within it. It is worthwhile to distinguish between 'denotational vagueness' and what might be termed '*vagueness of sense*'.[7] Vagueness of sense means the ill-defined limits of that property of an expression that is preserved under any correct translation of the expression. Carnap [28] has aptly pointed out that there are possible instances when denotational vagueness[8] is quite insignificant, while vagueness of sense may have a role to play. The denotational vagueness of the term 'man' need not be very important, especially if we disregard such quesions as the origin of man, the prehistory of the human race etc. In the present day world we can visualise the clear-cut set of all men. But this denotational exactitude does not mean that the term 'man', still less the related adjective 'human', is exact as to sense. On the contrary, these terms can be treated as vague in sense, particularly because their sense assumes a definite anthropological, sociological, psychological, emotional or other conception of man.

Exactness or vagueness of sense are heavily dependent on the nature of the sense of the given expression, or on the genesis of the sense or on the category of the so-called rules of sense. The two most important kinds of rules of sense that can be distinguished are:

(1) empirical rules of sense,
(2) axiomatically formulated rules of sense.[9]

With the empirical rules of sense, which are consolidated inductively and on the basis of experience, there is much greater scope for vagueness. By contrast, in the case of postulates or axioms, which are binding in their semantisation of terms (e.g. in the form of meaning postulates), any such scope is limited or even wiped out completely. Therefore in the case of theoretical concepts whose sense is tied to some theory, or to the postulates that express the theory, it is possible, at least from this semantic point of view, to exclude vagueness completely.

3. VAGUENESS, AMBIGUITY AND DENOTATIONAL OPACITY

[*A*] *Ambiguity*

In the section on synonymity we touched on the difference between the two kinds of metalanguage predicates whose arguments are the names of expressions from the object language. In the case of the concepts 'synonymity' and 'heteronymity' it has to be assumed that we are dealing with at least two expressions. By contrast, in the case of the concept 'vagueness' we are dealing with a metalanguage predicate whose argument is the name of just one expression from the object language.

From this point of view the concepts 'ambiguity' and 'denotational opacity' share properties with the concept 'vagueness'. 'Ambiguity' differs from 'vagueness' in that with ambiguity more than one interpretation of a given expression is assumed.[10] On the other hand with vagueness we assume only one interpretation but with the possibility of uncertainty as to the definition of its limits. If an ambiguous term has more than one interpretation, it goes without saying that any or all of the interpretations may be vague. (An expression is ambiguous if there are at least two occurrences of it, each having a different interpretation. For example, in Czech the expression 'zámek' is ambiguous because there is its interpretation as a building (= chateau) and its interpretation as a device for securing a door (= lock). Any natural language has plenty of ambiguous expressions. This also means that there are at least two instances of one and the same expression that are heteronymous.

Thus it is possible to treat ambiguity as a special case of heteronymity. Since heteronymity is a relation between two distinct expressions (i.e. **Heter**(Δ_i, Δ_j), where Δ_i is different from Δ_j), the ambiguity is due to the fact that the two arguments of the metalanguage predicate 'heteronymous' are identical (i.e. $\Delta_i \underset{\text{id}}{=\!=} \Delta_j$). This also throws more light on the remark made previously to the effect that there may be ambiguity of expressions in a language in which an absolute prohibition of **Heter** (Δ_i, Δ_j) is not excluded.

For these reasons we may also use the concept of 'heteronymity' in the definition of ambiguity:[11]

$$(D_1 - \text{VII 3 A})\ \mathbf{Amb}(\Delta) = \text{df}[\mathbf{Inter}(\Delta, \Gamma_1)\cdot\mathbf{Inter}(\Delta, \Gamma_2)\cdot\mathbf{Heter}(\Gamma_1, \Gamma_2)]$$

This definition of ambiguity could be made more exact by using the concept of 'marginal conditions'. It is intuitively obvious that an expression is ambiguous if it occurs in a set of marginal conditions of which there are two sub-sets, each of which is connected with a different interpretation of the expression. It is possible, for example, for two parts of the same text to contain occurrences of the expression 'zámek' used in each of the two interpretations. It should be noted that ambiguity is felt in precisely those situations where there are different marginal conditions which are treated, however, as components of the same set of marginal conditions.

[*B*] *Denotational Opacity*

In operations using terms from natural language it is very common to encounter terms which for one reason or another defy interpretation. It is a frequent state of affairs in the study of a foreign language, when we meet a new term and do not know its translation or the definition of what it means using the terminological apparatus already known; it also happens in the assimilation of new terminology in science.

The impossibility or inability to find a term's interpretation may in principle be of one of two kinds:

(a) the total ignorance of what the term signifies and denotes, i.e. total ignorance of its sense and denotation;

(b) ignorance of what the term denotes, i.e. lack of clarity or lack of certainty as to denotation. In this case we may be able to translate this kind of term with its unclear denotation, providing another term with the same sense, while not being able to reveal the denotation itself.

From the semantic point of view we need not concern ourselves with case (a). What we in fact have there are not expressions of language, but signs that may, under certain conditions, become expressions of language.

Case (b) is of greater importance to the semantic analysis of language. In this connection writings in logic and semantics use the concept 'opacity', or more precisely denotational or referential opacity, as the opposite of 'transparency'.[12]

In confrontations of natural and formalised language notice is usually taken of two properties that somehow impede operations with natural language terms, namely vagueness and ambiguity. To these may be added a third, denotational opacity, which differs from both vagueness and ambiguity.

Keeping to interpretation from the point of view of denotation, i.e. the interpretation of meaning as a class of entities, we can distinguish the situation where the class cannot be given clear-cut boundaries, the fringe being non-empty (vagueness), from the situation where a number of different classes can be assigned. With denotational opacity there is a quite different situation again: while we assume that the relevant class exists, we are partly or completely incapable of specifying that class. This means that while we accept that a particular expression denotes 'something', we are not in a position to characterise or define that 'something' with anything like accuracy.

Denotational opacity cannot be confused with the situation where the denotation – which we are treating here as a class – is actually defined exactly, but where for one reason or another it cannot be attained empirically or cannot be imagined (for instance because the class is an infinite one). Denotational opacity may be a property not only of terms that denote classes, but also of individual names. If someone says that "Halmahera is an island", he need not be capable of stating exactly what the denotation of the individual name 'Halmahera' is, i.e. he may not be able to say that it is an island in the Moluccas, or give its exact location and so forth.

[*C*] *Extremes and Graduated Differences*

So far we have distinguished 'vagueness', 'ambiguity' and 'denotational opacity'. Each of these concepts has its conceptual opposite, giving us essentially the following pairs:

exactness – vagueness
unambiguity – ambiguity
transparency – denotational opacity.

In connection with these pairs there are a number of questions to be raised concerning primarily the roles of the concepts involved in natural language. It seems that in any process of communication it is desirable to abide by exactness, unambiguity and transparency.

As far as the maintenance of exactness is concerned, it would only be possible and completely feasible if the conception of discreteness that is proper to a particular language were perfectly matched by the discretisation of the universe. But we have to assume that the correspondence is only more or less approximate, but also that it may change and be disrupted in connection with refinements to the differentiation that we are capable of making within the objective world itself. This is why even in scientific work some operations with more or less vague terms are unavoidable. It is of course desirable that it should always be possible to improve the accuracy of the terms used in connection with the needs of science, in tasks that are connected with a particular system of terms and a system of relevant semantic rules. Hence the interpretational norms that are assumed in any terminological system, whether they are expressly formulated or merely tacitly acknowledged, cannot constitute a stable system, but a system that has to be adapted and made amenable to new tasks and needs.

Ambiguity and transparency are properties of language and have been discussed in connection with the name relation and the principles of univocality and subject matter. It should be noted again that both principles are more like ideal requirements the observance of which in some adequate measure is the guarantee of real intercommunication. At the same time, in natural language, where these requirements must always be relativised to a context, it is impossible to overlook the role played by that context. It should be noted that even terms and language expressions generally that appear to be ambiguous and denotationally opaque may, in a certain context, play the part of unambiguous and transparent terms.

Vagueness, ambiguity and denotational opacity need not therefore be dismissed as an expression of the imperfection or inferiority of natural language. The main consideration here is clearly the nature of the tasks connected with the use of language. What may appear in science, and anywhere else where exact and unambiguous interpretation matters, as a shortcoming, may, in other tasks, such as some areas of verbal art, appear as a virtue. In the first case the object is to minimalise the language user's arbitrary interpretations. This is what gives rise to what V.A.

Uspenskij [179] described as the need for a 'non-pragmatic language'. The role of interpretational norms must also be assessed in this context. In the second case what is desirable is the opportunity for the language user to select his interpretations or even seek an interpretation of his own.

Between the two extremes, i.e. of exactness, unambiguity and transparency on the one hand, and vagueness, ambiguity and denotational opacity on the other, there may be a whole scale of transitions. We have already pointed to one possibility for measuring vagueness on a statistical basis. Ambiguity and denotational opacity also lend themselves to similar ways of measuring semantic decision-making.

VIII. SEMANTICS AND SOME PROBLEMS OF ONTOLOGY

1. SEMANTICS AND ONTIC DECISION

A connection is sometimes made between problems of semantics and some of the philosophical problems of ontology. Sometimes the two are even mistaken for each other, which results in confusion and a standpoint that has rightly been called 'semantic philosophy'. The problems of the world are confused with the problems of language and its interpretation. The problems of the world, be they finding solutions to social conflicts or theoretical mappings of individual domains, cannot be solved by the means furnished by semantics. The tasks of semantics are, as Tarski put it in his criticism of such aspirations [149], much more limited and more modest. So semantics cannot replace sociology, social politics, and become an instrument of 'social therapy', as some of the adherents of general semantics would have it, but nor can it replace any of the special sciences or philosophy or the theory of cognition. An in-depth criticism of these false aspirations, which are to the detriment not only of the solutions to any of these deserving problems, but also of semantics itself, is beyond the purview of the present work.

Rejection of these false aspirations is not of course to be taken as meaning that semantics is somehow 'neutral' with regard to certain ontological or gnoseological problems or that it is entirely unconnected with them. That which we have called semantic decision cannot be identified with what has been called ontic decision, the acceptance or rejection of a particular conception of the objective world itself, i.e. of that which *may* be the universe with respect to a given language. But is there some justification for the views (expressed, for example, by Carnap in [27]) that semantic decision is simply unconnected at all with decisions about the objective world, whether we are talking about questions of the reality of the world as a whole or questions of some other kind? In our view it is not possible to deny all connection between certain semantic problems and problems of a philosophical-ontological nature. Moreover, the denial of any such connection, often linked with pronouncements to the effect that the second group of problems are 'pseudoproblems' or 'apparent problems', in fact leads to the situation where a semantic solution stands in for a philosophical solution, or where semantic decision is taken as a substitute for ontic decision.

[*A*] *The Use of Terms and Ontic Commitments*

The problem of ontic commitments[1] was first touched on by Whitehead and Russell in their discussion of the theory of descriptions (although they did not use the term, which is of a later date). Russell's theory of individual descriptions presupposes that any individual description can, if true, be translated into a sentence which will state the existence of that individual. For example, the sentence 'The highest mountain in Bohemia is in the Giant Mountains' is interpreted as: 'There exists an x, which is the highest mountain in Bohemia, and whatever is the highest mountain in Bohemia is identical with that x, and x is in the Giant Mountains'.

The interpretation of individual descriptions can be expounded in two ways. If $(\iota x)(Px)$ is an individual description, its interpretation can be expounded as:

(a) a statement that Px is satisfied by a single individual;
(b) a statement that Px is satisfied if and only if there *exists* a sole individual which satisfies Px.

The (b) version presupposes something which has been described as *ontic commitments*. The (a) version is neutral with respect to ontic commitments.

Russell inclined to view (b). By contrast, Quine was against the application of ontic commitments in the use of individual descriptions. He considers a mistaken ontic commitment to be a confusion of meaning with naming.

It is right in principle that any use of a term which has the nature of an individual description or is the name of an existing or assumed individual does not of itself entail any definite ontic commitment. In other words, semantic decision, i.e. the language user's operation by which he assigns semantic characteristics to individual expressions, or his attitude to those characteristics (e.g. acceptance or rejection of a particular semantic characteristic), cannot be identified with what could be called ontic decision, i.e. decision-making about the objective existence of an entity whose existence is independent of the language user or observer.

Russell's conception of individual descriptions, which does lean towards a recognition of some ontic commitments, is motivated by the desire to eliminate unwanted descriptions. This gives rise to the question of whether simple recognition of ontic commitments is itself sufficient as a means for the (scientific or whatever other method of) elimination of unwanted individual names or descriptions. (This is where the many examples of mythological proper names come in, e.g. 'Pegasus'.) The conception of 'ontic commitments' in the form put forward by Russell strikes us as very narrow, since it is attached only to the concept of acquaintance. This means that 'ontic commitments' are assessed – as it would appear on the face of it – immediately, without any reference to a theoretical, scientific or ideological conception of a particular universe. But if, in the elimination of unwanted individual names and descriptions, we do want to take account of the ontological aspects, we have to consider the ontological claims of the entire theoretical, scientific or ideological conception within the framework of which the given expression is used. It can be shown that even Russell's conception of 'ontic commitments' is ultimately not an immediate, but a mediated conception, since it is tied to his conception of an (objectively) existing world as a world which may be the object of immediate experience. Finally, there are to be found 'ontic commitments' in the spirit of the subjective idealist 'esse est percipi'.

The controversy over the 'ontic commitments' of individual descriptions cannot be settled by simply agreeing with either Quine, who denies any such thing as ontic commitments, or Russell, who accepts certain ontic commitments in the spirit of his conception of immediate experience – 'acquaintance'. In principle Quine is right in saying that any use of a term or language expression which is an individual description or individual name does not yet carry with it any particular ontic commitments or, to put it another way, that semantic decision cannot be identified with ontic decision. On the other hand, any theoretical, scientific or ideological conception which always makes certain ontological claims (and this applies just as much to the 'anti-ontological' conceptions – those which in so many words reject any kind of 'ontology') is not and cannot be without some influence on semantic decision. Hence ontic decision is not irrelevant to semantic decision.

We shall pass now from individual names and descriptions to predicates. For example, if we have the sentence $(\forall x)(Qx \longrightarrow Px)$ and interpret Q as 'have a specific gravity lower than 1' and P as 'float on water', the sentence can express Archimedes' Law. But it must be stressed that the existence of objects that float on water cannot be inferred from this sentence if the class $(\hat{x})(Qx)$ is an empty class. This means that the simple use of a predicate does not entail any ontic commitments any more than it allows us to infer the existence of classes created by the predicates used.

The rejection of ontic commitments with respect to the undefined use of predicates does not of course mean that all ontic decision is precluded, as would follow from the nominalist conception of that rejection. It is precisely in this connection, i.e. in connection with the interpretation of predicates and the problem of ontic commitments, that there was a revival of the ancient controversy, already known in classical and medieval times, which has been most frequently described as the conflict between Platonism[2] and nominalism. This conflict is of greatest immediacy in the interpretation of predicates of a higher type, although in principle it concerns all predicates.

The nominalist viewpoint rejects any consideration of ontic commitments in the use of predicates, especially predicates of a higher type, on the grounds that in the objective world there are no individuals to which it would be possible to assign the properties expressed by the words 'function', 'class' etc. On the other hand, the Platonist view admits the use of predicates of the higher type as expressions for classes or properties of entities sui generis. Nominalism, which does not recognise the reality of so-called universals – in the sense of the medieval 'universalia sunt nomina, sunt flatus vocis' – therefore concentrates ontic decision on the starting level, which is in logical terms connected with the domain of individuals.

Platonism and Platonic realism do not make a distinction of principle between the reality of concrete and abstract entities and therefore they do not limit ontic decision to any one level. Hence Platonism does not formulate any criterion of ontic commitments. By contrast, nominalism definitely needs some such criterion.

Quine, who holds to the nominalist view, put forward the following formulation of a criterion: 'An entity is assumed by a theory if any only if it must be counted among the values of the variables in order that the sentences affirmed in the theory be true' [117, p. 103]. Quine stresses that this is not a criterion of an 'ontic state', it is not concerned with what exists objectively, but with what is the ontic commitment of a particular conversation or communication. Thus we are not dealing with ontic commitments with respect to man as language user, but with respect to a certain communication in which there is assumed to be more than one participant. Another relativisation of this criterion is the fact that it is tied to a particular theory or conception of the universe.

Quine is undoubtedly right in suggesting that there is no point in raising the question of ontic commitments outside the framework of some theory and outside of a particular communication. This also means that in the end the responsibility for ontic commitments is shifted onto a theory or theoretical conception of the whole universe. But this 'shifting' of responsibility does not actually solve the problem of ontic commitments: there is after all a substantial difference between what is said to exist (or not exist) in the course of a conversation, what is held to be real (or not real) within the framework of some theoretical or ideological conception, and what actually exists. Quine, insofar as he does analyse the question of what exists (especially in [121]), insofar as he formulates a criterion of ontic commitments (especially in [117]), really only looks at the first version of the whole problem.

But has the second version any real point? Can the question of ontic commitments really be raised outside a particular communication, conception or theory? The point of the second version is not that it annuls any relativisation of ontic commitments, but that it points to the differences between various relativisations, requiring that relativisation be made with respect to some definitely qualified theory. If the responsibility for ontic commitments is shifted onto just any theory or conception of the universe, or onto any communication, it is easy to prove that a child who listens to tales of fairies and dwarfs and believes in their existence is just as entitled to speak of the existence of these beings as the physicist is to speak of the existence of photons.

So if Quine is right that the question of ontic commitments cannot be raised outside the framework of a theory, this in itself is not enough until we specify the demands which we make of the theory. Otherwise we shall find ourselves unavoidably justifying complete arbitrariness in matters of ontology, simply on the basis of the semantic analysis of language. The task of specifying the demands made of a theory, however qualified, is of course not a task for semantics. It is a matter for science and is determined by the contemporary standard reached by science and verified by practice. Hence the question of 'what exists' falls not within the competence of semantics and the semantic analysis of language, but within the competence of science.

[B] *Linguistic Framework and So-called External and Internal Questions*

So far we have suggested a solution to the problem of ontic commitments and the relations between semantic and ontic decision with respect to those expressions of language that have the nature of non-logical constants. In this connection we ran into the problem of nominalism and Platonism in the analysis of language (and also in the construction of a formalised language). Before passing to an analysis of the semantic aspects of the conflict between nominalism and Platonism and the controversies which are the latterday analogue of that conflict in logic and logical semantics, we will take a look at the question of whether it is possible to think in terms of the ontic commitments of an entire language system.

A similar question was asked by Carnap in his thoughts on the relation between empiricism, semantics and ontology [27]. Unlike Quine and the other nominalists, Carnap took, in a sense, a more liberal view in assessing the semantic admissibility of predicates of the higher type, or abstract entities, while at the same time denying the justification of so-called external questions, questions that are frankly ontological.

The question of the existence or reality of abstract entities is, as Carnap was at pains to emphasise, an ambiguous one. Carnap urged the strict separation of two interpretations of the question or indeed of two kinds of questions: what he called *internal questions*, which concern the existence of certain entities and possibly of a new kind of entities within the framework of a particular linguistic system, and the *external questions*, which concern the existence or reality of the system of entities as a whole. He gives the following justification for the difference: if we wish to speak of some new kind of entities, we have to bring in a system of new ways of expressing, subject to new rules. This procedure, which he does not specify further and which presumably includes the construction of both the relevant semantic and syntactic rules, is described as the construction of the *linguistic framework* for the respective new entities.

Internal questions, i.e. questions of the existence of entities within the framework of a given language system, are viewed by Carnap as legitimate. Such questions can be resolved either by purely logical or by empirical methods. The question of the reality or objective existence of any object, if we treat this as an internal question, is then a matter of incorporating the object in the system of objects and events that can be admitted as real by the rules of the relevant linguistic framework. In the case of, for example, empirically apprehensible objects, the linguistic framework is constructed as an empiricist language, the truth or falsehood of sentences in the language having to be verified empirically. Thus it is possible to answer even such questions as those concerning the existence or reality of centaurs, unicorns and so forth. (In the negative, naturally.) Carnap calls this kind of linguistic system factual. By contrast a system like that of natural numbers, such as Peano's axiomatics (in which the terms '0', 'number' and 'successor' are primitive terms), is a system of a logical nature. In this system we can answer the question whether there exists a number which is the successor of two different numbers. (The answer is connected with one of Peano's five postulates whereby no two different numbers may have the same successor.) This means that even here the question of the existence of a particular number makes sense and can be answered meaningfully, the answer being based on logical inference, i.e. following from the postulates of Peano's system of natural numbers.

As against these internal questions, whose legitimacy Carnap fully recognises, he has the external questions, which he sees as problematic. These are questions about the existence of the entire material world, the existence of numbers in the traditional ontological sense and so forth. Carnap adds the comment that this kind of question comes neither from the scientist nor the man in the street, but from the mouths of philosophers. Questions of this kind, which may be answered either in the realist spirit (and, in the case of abstract entities, in Platonic idealism) or in the spirit of subjective idealism, are viewed by Carnap as questions of dubious formulation which lead to insoluble controversies. On the other hand 'being real' in the scientific sense means being an element of some system and therefore this predicate cannot be applied to the system as a whole.

The basic argument for the assertion that questions about the reality of the world as a whole are pseudoproblems is the assumption that sentences of a given system can be tested either on an empirical basis, in the case of a factual linguistic framework or on a logical basis, in the case of a logically constructed linguistic framework. By contrast the thesis on the reality of the objective, or as Carnap calls it the 'material', world cannot be formulated either in a factual linguistic system for it to be verifiable by empirical means, or in a logical system for it to be tested on a logical basis.

This view of Carnap's, which dates from 1950, does not differ greatly, then, from the early conceptions of the Vienna Circle, the period of the 'Scheinprobleme' and the principle of tolerance. There are several objections of principle which must be raised against it.

(1) Firstly, Carnap is obviously considering only that which we have called semantic decision and is trying to deny the sense of any kind of ontic decision. The so-called internal questions of the linguistic framework, whether this be an empirically or a logically constructed framework, are connected with semantic decision.

(2) Carnap's view leads to an absolutisation of the boundary between

internal and external questions. This boundary is variable, both in the empirical and in the logical aspect. That this is so in the empirical aspect is perfectly plain from the thoughts about the discovery of new entities whose names have to be introduced into the relevant language system. That much is admitted by Carnap himself. But in the logical aspect one can point to the possibility of 'external questions' about a logical system (such as questions about its completeness or consistency) and these can only be answered if some higher or more complex system is assumed.

(3) The sharp contraposition of internal and external questions in Carnap is also connected with the fact that he considers ontological problems to be, always and under all circumstances, external questions. Of course this is characteristic of only traditional ontology, whether of the objective idealist, i.e. Platonist, type or of the naive realist type. The question of the reality of the objective world in the conception of scientific philosophy is not, however, a question of traditional ontology, but, as Lenin showed [79], primarily a question of gnoseology. In other words, the property 'be real' is not ascribed to the world as a whole outside the framework of any system, but within the system: objective world – man.[3]

(4) Claiming that the problems of science are connected solely and exclusively with internal questions, Carnap is underestimating the fact that the systems of entities studied by science are also ordered hierarchically. There is then sense in thinking of one system as an element of another, just as there is sense in thinking of one conglomerate of such systems within the framework of the system objective world – man.

Having made these critical comments we can now ask whether there is any legitimate reason for separating external and internal questions. In our day-to-day life, if we ask certain questions, we do not usually appreciate the unstated or assumed context, the framework, within which the question is justified and can or cannot be answered. If I ask what the time is because I do not have a watch on me, I am satisfied by the answer of the man who had simply glanced at his own watch. I do not concern myself with the fact that the question is only legitimate if I take into account its incorporation into some system, such as the system of time standards, Central European Time and so on. This means that, in naive terms, one can ask questions which do not have the nature of internal questions, i.e. questions that are asked within the framework of some system. So we may say that the questions with which science is concerned do have the nature of internal questions. Thus far we can agree with Carnap. But in this case the contraposition of internal and external questions is not a contraposition of the scientific and the philosophical-ontological approaches, but of the scientific and the naive approach, the latter usually being described as the 'common sense' approach. To say that scientifically legitimate questions are those which have the nature of internal questions does not mean either that (a) all scientifically legitimate questions which are internal questions can be answered, or (b) there do not exist questions which, with respect to a given system, have the nature of external questions. In other words, it is impossible to deny the relativisation of questions to a particular system, nor is it possible to deny that there are questions which cannot be answered within the framework of one system while they can be answered within another framework.

2. NOMINALISM, PLATONISM AND SEMANTICS

[A] *The Reification of Abstract Entities and the Problem of Nominalism*

The conflict between nominalist and Platonist conceptions in the context of the interpretation of expressions of language, expecially of abstract entities, is an age-old matter. There is some justification for the assertion that in modern logic, mathematics and logical semantics it has revived, although at a rather higher level, as the controversy over the reality of the so-called universals. Progress has undoubtedly been made in the sense that what medieval philosophy called 'universalia' can now be described with greater accuracy and also classified on the basis of, for example, the differentiation of predicates and the theory of types.

The main thing to be noted is that the terms 'abstract entities' and 'universalia' are very unclear. There are two possible interpretations:

(a) They may mean all expressions which can be formed on the basis of predicates, irrespective of what predicates these are or of what type they are. From this point of view any classes or abstracta are expressions which may be the subject of disputes. For these expressions we can use the term 'universal'.

(b) They may mean expressions which can be formed on the basis of theoretical predicates and predicates of the higher type, that is classes and abstracta in a rather narrower sense than in case (a). For these expressions we can use the term 'abstract entity'.

If we ask after the reality of that which corresponds to 'universals' or 'abstract entities', what we have in mind may be either the first (the broader) or the second (the narrower) interpretation. It must be added that in the original versions of the controversies between nominalism and Platonism the two were not usually distinguished. The nominalists would answer the question in the negative, without explicitly stating which interpretation was involved. They would protest against what they usually called the reification or hypostatisation of abstract entities. They protested against the attribution of concrete reference to expressions for abstract entities. This does not mean, of course, that they spoke against any kind of use of expressions for abstract entities. If that were so they would have to have thrown out the whole of classical mathematics. They merely objected against the assignment to these expressions of something that could be described as 'real' or having the nature of 'material existence'.

The resistance against reification or hypostatisation of abstract entities is also connected with the requirement that entities should not be multiplied, i.e. with Occam's razor. Hypostatising or reifying the properties of objects in a way similar to the objects themselves means duplicating the world. The nominalist argument runs approximately as follows: There can be no doubt of the existence of certain 'round things', but a protest must be made against the existence of the object 'roundness'. Recognition of its existence means that the entities which we call 'round things' are being duplicated by the new entity we call 'roundness'.

The nominalist criticism extends across two fronts: on the one hand against the Platonising reification of abstract entities, against the objective idealism of the Platonic type, and on the other hand against naive realism, which shapes its image of the world of real things around the image of language. This is an undoubtedly positive feature of the nominalist conception in the analysis of semantic and ontic decision.

However, there are several objections to the nominalist conception as here described.

(1) If nominalism protests against the hypostatisation of abstract entities, it must be said that in fact it hypostatises entities of the initial level itself (entities connected with expressions of the zero type, the system of individuals) as 'things'. The concept 'things' is not only itself an abstraction, but any substantialisation of entities from the initial level will prove to be problematic as soon as we refine our differentiations and discover that what was treated as a 'thing' under the first approach becomes, under the new approach, an aggregate of other 'things', that is a class or property of other objects.[4]

(2) Other objections against nominalism come from phenomenalist positions. Against a nominalism which is willing to concede the reality

only of individuals we can object that we never perceive individuals, but only their properties or qualities; we do not perceive a 'tree', but the green of its leaves, the shape of its trunk etc., and we merely assume that the tree exists. The controversy over the epistemological priority of property or individual is resolved by nominalism in favour of the individual and by phenomenalism in favour of the property.[5] We believe that there can be no blanket solution to the conflict and that if stated in general terms and without a framework of conditions, however defined, the very question of the epistemological priority of property or individual is pointless. If we do define these conditions with something like accuracy, for example in terms of the communicational model of scientific investigation, what matters is whether the input elements are directly accessible to the observer or accessible by means of some other channel, or whether they can be reconstructed on the basis of data ascertained at the output.[6]

In connection with the problem of the epistemological priority of property or individual, or thing, it is worth noting that which could be described as the transitions between things and properties.[7] If we alter the properties of what we have called the 'observer channel' or the 'channel of science', it might become possible to investigate as a class or property something that it had been previously possible to treat as a thing, i.e. a relatively independent individual entity.

This last remark is also connected with the problem of how to specify individuals. It might be asked whether there is any way of establishing criteria for the specification of individuals. In principle there are two kinds of criteria: (a) internal criteria, which are connected with the logical structure of the nominalist system; (b) external criteria, which are connected with what Goodman [49] has called the 'non-logical basis' of such a system. It is symptomatic that internal criteria can be drawn up with greater exactitude. We have already quoted Quine's formulation for 'ontic commitments', which determines what entities are admissible as values of variables and hence as individuals.[8] More complicated is the problem of external criteria. Here a major part is played by whether we take as basis the theory of similarity or the theory of identification. (These problems will be dealt with at a later stage.)

[B] *'Praeter Necessitatem'*

One of the basic arguments of nominalism was the so-called Occam's razor, which forbade the duplication or multiplication of entities. But the original version of Occam's razor has an extra tag: 'Entia non sunt multiplicanda *praeter necessitatem*'. This raises the question of whether there might not be compelling reasons which would justify the 'multiplication of entities'. To be more precise, has Occam's original 'praeter necessitatem' any particular justification in the problems of logical semantics?

The answers to the question of what might be that compelling argument which admits of the 'multiplication of entities' are many and various. In principle they fall into two kinds.

(1) There is no avoiding the admission that abstract entities are a quite separate sphere which is (a) independent of individuals, (b) independent of the 'subject', 'observer' or 'user' of the language. This is the view adopted by Platonism and Platonic realism. In contemporary logical and mathematical thinking platonic realism has a kind of analogue in the conception of *logicism*, supported in the main by Frege, in part by Russell (in his early stage) and Whitehead; some of the views of Church and Carnap also come close to the logicist view. We might recall Frege's conception of the sense of a name (where 'name' may relate to both objects and abstract entities) as an abstract entity which is not subjective, but objective. This logicist conception is to be found in a more 'temperate' form in Church and Carnap, who are concerned with what is 'intersubjectively' stable and unchanging, as in translation.

(2) It may be admitted that abstract entities are a separate sphere which is (a) relatively independent of individuals, but (b) not independent of the 'subject', 'observer' or 'user' of the language. This sphere is, on the contrary, viewed as the product of the logico-constructive activity of the 'language user'. This is the stand taken by *conceptualism*. In contemporary logical and mathematical thinking it is largely matched by *intuitionism*, supported by L. Brouwer, H. Weyl, A. Heyting and others, or by the mathematico-logical *constructivism* subscribed to chiefly by A.A. Markov and other Soviet mathematicians.

In both answers special attention is required by point (a), i.e. the fact that abstract entities are independent of individuals. It must be stressed that (1a) and (2a) cannot be identified: with Platonism it is independence in the ontological sense. In contrast to nominalist-phenomenalist disputes over the epistemological priority of individual or property, in the nominalist-Platonist controversy it is a question of the ontological priority of individual or abstract entity (or, to put it in traditional philosophical terminology, it is the dispute over the ontological priority of universals or individuals). In scrupulous Platonism the individual is secondary both ontologically and genetically, being the product of the intersection of universals and abstract entities.[9]

Conceptualism does not share the view of the ontological priority of universals. In this respect one has to agree with it. The idea of the ontological priority of universals is wrong not only because the problem is mis-stated, being based on the mechanical separation of individuals from universals or abstract entities, but also because it can lead on to some dangerous consequences.[10]

Conceptualism also does not see it as a shortcoming of principle that the abstract entities in a given language system are related both to those which are treated in the given universe and with respect to the characteristic properties of the 'language-user channel' as individuals, and to other abstract entities. In this respect, conceptualism comes relatively closer to the intuitive conception of the interpretation of language expressions.

The 'necessity' presupposed by the formulation of Occam's razor may be viewed as more a linguistic than an ontological necessity. From that point of view it is possible to concur in the conceptualist view which gives the language user a certain right to create abstract entities (concepts) as long as these are useful not only for expressing the elements of the given universe, but also for expressing the properties of those elements, and so forth. Therefore from the semantic point of view Occam's razor is fully justified inasmuch as it forbids the multiplication of the elements in the given universe, whether these be empirically apprehensible elements or constructive elements that are the outcome of abstraction. But Occam's razor also has its other aspect, which can be described as the requirement of terminological or linguistic economy.

One advantage of conceptualism or constructivism over extreme nominalism is also the fact that it does facilitate the distinction of both aspects, whereas nominalism is notable for pursuing the programme of terminological or linguistic economy. This latter requirement does have some justification: in using language, and this applies equally in the case of scientific language, there is always the possibility of situations occurring where different terms ultimately have the same meaning. Here the requirement of the non-proliferation of names is only justified if different terms do genuinely relate to the same properties of an object. The very history of scientific thinking tells us in any case that scientists have often worked on one and the same problem using different terminological apparatus, giving rise to the illusion that two or more different problems were involved.

[C] *The Problem of Similarity and Identity*

The controversy between the nominalist and the Platonist, and, where relevant, the conceptualist viewpoints is closely connected with the problem of the discriminability and aggregation of individuals. For the nominalist the world is 'composed' of individuals. If two entities are different, they cannot be 'composed' of the same individuals.[11] If we consider the differentiation of individuals into groups, which might be called the aggregating of individuals, we cannot avoid the question of the basis on which this aggregation is carried out. Adherents of nominalism find the *similarity* of individuals an adequate principle for aggregation. Since the nominalist's starting point for the construction of predicates is the world of individuals, similarity plays a telling part in the construction.

A role analogous to that played by similarity in the nominalist system is played in the realist system by the *identity of properties*, the *identity of classes*. It goes without saying that it makes a difference if the system is built up on the basis of the ontological priority of properties, or abstract entities generally, or, which is typical of the phenomenalist conception, on the basis of epistemological priority.

In the nominalist system, where the starting point is the sphere of individuals, the identification of individuals is not precluded. But for identification there has to be distinguishability. Then that which cannot be distinguished from something else can be identified with it. This procedure has usually been described as the *method of the identification of the indistinguishable*.

The distinguishing process presupposes a sphere of individuals and a system of criteria for distinctions. Let us assume that the sphere of individuals is made up of physical objects. These objects are values of the individual variables $x, y, \ldots$. We assume next a (finite[12]) system of distinguishing criteria which can be expressed as a system of many-place predicates. These are predicates that will enable us to state differences between individuals, e.g. 'is heavier than', 'is longer than', 'is harder than', 'has a higher value than' etc. They are then relations which are irreflexive, transitive and asymmetrical and make possible the complete decomposition of the sphere of individuals.

The requirement that the distinguishing criteria should be finite is of considerable importance to the construction of the nominalist system. It also shows the close connection between nominalism and finitism.

It now needs to be stressed that the distinguishing criteria figure as predicates of the first type, which in concrete situations may make their application quite difficult or even restrict it.

With identification, allowance must be made for the possibility that the separate distinguishing criteria are interdependent. Another important problem is the relevance of each separate criterion to a particular task or class of tasks. It is therefore always expedient to give thought to the *selection* of adequate distinguishing criteria with respect to a given task and the demands made on the quality of its solution. So in selecting distinguishing criteria it is an advantage to use the methods which we suggested in the discussion of reduction and constitution.

Assuming that the distinguishing criteria are independent of one another and that we are using two-place predicates of the properties quoted, the procedure for identifying the indistinguishable can be expressed as follows:

$$\begin{aligned}(\forall x)(\forall y)\ [(\sim A_1 x, y \cdot \sim A_1 y, x \cdot \\ \cdot \sim A_2 x, y \cdot \sim A_2 y, x \cdot \ldots \\ \cdot \sim A_n x, y \cdot \sim A_n y, x) \longrightarrow (x \underset{\text{id}}{=\!=} y)]\end{aligned}$$

This can also be expressed thus: Given a (finite) system of distinguishing criteria, the individuals which are indistinguishable with respect to that system may be considered identical.

There is another possible version of the identification process. It is intuitively obvious that in investigating the distinctness of individuals we also assume the use of some kind of 'yardstick' in addition to the distinguishing criteria. Examples of this are distinguishing the lengths of objects against an indicated dimension, comparing the weights of scale weights etc. This means that usually we do not compare individuals directly when deciding on their identity, but compare the individuals with these 'yardsticks'.[13]

Taking $A_1, A_2, \ldots A_n$ as distinguishing criteria and $a_1, a_2, \ldots a_n$ as the 'yardsticks' that correspond to them, we can record the process of identification as follows:

$$\begin{aligned}(\forall x)(\forall y)\ [(\sim A_1 x, a_1 \longrightarrow \sim A_1 y, a_1)\cdot(\sim A_1 a_1, x \longrightarrow \sim A_1 a_1, y)\cdot \\ \cdot(\sim A_2 x, a_2 \longrightarrow \sim A_2 y, a_2)\cdot(\sim A_2 a_2, x \longrightarrow \sim A_2 a_2, y)\cdot \\ \cdot\ldots \\ \cdot(\sim A_n x, a_n \longrightarrow \sim A_n y, a_n)\cdot(\sim A_n a_n, x \longrightarrow \sim A_n a_n, y) \\ \longrightarrow (x \underset{\text{id}}{=\!=} y)]\end{aligned}$$

The fact that the method of identifying the indistinguishable can be applied only under the assumption that the system of distinguishing criteria is finite is essential to the conception of nominalism. But there is another side to it, which was pointed out by Hao Wang [160]. He stressed that for these reasons any finitist Platonist system can easily be nominalised.

We have a different situation when the system is constructed on the basis of Platonism.[14] First and foremost it should be noted that here the attention is focussed not on individuals, but on classes or properties. Therefore what is decisive is the identification of properties or of classes, which is easy to achieve. So in this kind of system identification creates no major difficulties. On the other hand, the extreme nominalist may object that '1 = 1' is not true.[15] Objections of this kind are only legitimate if the scope of the distinguishing criteria has not been exactly defined. In everyday practical tasks and also in those scientific tasks that are connected with the need for distinguishing and identifying, the scope of the criteria must be stated exactly both with respect to the nature of the given task and with respect to the desired quality of its solution. In any case it is perfectly obvious intuitively that we would not ask the butcher to weigh meat on pharmaceutical scales or the pharmacist to use the butcher's scale. Accordingly the problem of distinguishability and identification cannot be taken absolutely, but only with respect to a given system of distinguishing criteria.

If the point of departure for constructing a nominalist system is the differentiation of individuals, we may also consider the fact that two or more individuals may resemble each other. The concept of 'resemblance' is fundamental to the nominalist system, since it can be viewed as the starting point for the formation of concepts.

'Resemblance' can be treated as a many-place predicate of the object language with individuals as its arguments. It can be constituted in two ways:

(a) Two objects resemble each other if they cannot be distinguished in at least one characteristic. Thus to define 'resemblance' we can use the concept of 'distinguishing criteria'. If we have at our disposal a finite system of distinguishing criteria and if a distinguishing criterion is an element of that system, the weakest conditions for resemblance can be defined as follows:

$$(\mathrm{D}_1 - \mathrm{VIII\ 2\ C})\quad Res(x, y) = \mathrm{df} \sim A_i x, y \cdot \sim A_i y, x$$

(b) The intuitive conception of 'similarity' seems to be better satisfied by a different definition. Two objects are similar if we can say that they have at least one property in common. This presupposes that we have at our disposal a corpus of distinguishable properties which can be expressed in logical terms by means of one-place predicates. If $\boldsymbol{F}$ is the set of properties and Q_i one of them, resemblance could be defined as follows:

$$(\mathrm{D}_2 - \mathrm{VIII\,2\,C}) \quad Res(x, y) = \mathrm{df}(\exists Q_i)(Q_i x \equiv Q_i y)$$

From the strict nominalist point of view there are objections to both these definitions. The introduction of the concept of 'resemblance' effectively assumes that we have available a system of distinguishing criteria or a set of properties which can be assigned to individuals. But this conflicts with the nominalist conception of the epistemological and ontological priority of individuals and the derived nature of properties.[16]

If the concept of 'property' is to be secondary, we have to start from that of 'resemblance': Having discovered that in a given universe there exist individuals which are similar, we can create an expression for the common property, or we can create a corresponding distinguishing criterion. This means that for the strict nominalist the starting point is the discovery that the sentence

$$(\exists x)(\exists y)\, Res(x, y)$$

is true, which is then the stimulus for constituting the relevant property.

This discussion of 'resemblance' in connection with the construction of the nominalist system brings us to the justified conclusion that either we admit that 'resemblance' is a derived concept, in which case we have to admit a kind of priority to some system of distinguishing criteria or some set of distinguishable properties (which is at variance with the ambitions of scrupulous nominalism), or we treat properties and distinguishing criteria as derived, though in this case we are forced to concede the priority of one particular *property*, i.e. resemblance. This conclusion demonstrates the difficulties to which the extreme nominalist view can lead.

It is clear that in practical tasks we always work with a relatively a priori collection of distinguishable properties or distinguishing criteria and that these collections can be broadened or cut back as required and according to the demands of quality made on the solution. Moreover the establishment of a resemblance may be the inspiration for the formation of a new concept,[17] new concepts not necessarily being of the nature of empirical predicates. Thus the concept of 'resemblance' needs to be understood in the broadest sense – as also meaning 'abstract' or 'structural' resemblance.

There are yet other objections to the extreme nominalist conception as based on 'resemblance'. In a confrontation of the diversity of individuals in a given universe a great number of different kinds of resemblance can be revealed. But because nominalism ties its conception of identification to individuals, the spirit of nominalism is offended by any consideration of the identity of two different resemblances, i.e. any consideration of '$Res_i \underset{\mathrm{id}}{=\!=} Res_j$'.[18]

By excluding in principle consideration of the categorisation of different resemblances, we are, in practical terms, facilitating the constitution of a vast number of different resemblances combinatorially. This is an important fact with respect to the nominalist programme of terminological economy, which is connected with the formulation of Occam's razor. On the one hand the extreme nominalist view does not allow the multiplication of entities and reduces the 'necessity' referred to in the formulation to a minimum; but on the other hand the same viewpoint – without it always being an unavoidable necessity – actually imposes the multiplication of entities by excluding the possibility of $Res_i \underset{\mathrm{id}}{=\!=} Res_j$.

These critical comments addressed to the nominalist interpretation of 'resemblance' do not entitle us to conclude that the construction of nominalist language systems is a pointless exercise. Such systems may well be useful where we can make do with a language of the first order the values of whose variables are a sphere of individuals which can be exactly specified and in respect of which a finite system of distinguishing criteria can be applied. It follows from the above analysis that situations may arise where these assumptions are not enough, which leads to difficulties and lays excessive demands on the logical structure of the language. In this context we can agree with the conclusions reached by Carnap [27, p. 221]:

> "The acceptance or rejection of abstract linguistic forms, just as the acceptance or rejection of any other linguistic forms in any branch of science, will finally be decided by their efficiency as instruments, the ratio of the results achieved to the amount and complexity of the efforts required. To decree dogmatic prohibitions of certain linguistic forms instead of testing them by their success or failure in practical use, is worse than futile; it is positively harmful because it may obstruct scientific progress."

3. ANALYTICAL AND SYNTHETIC ASPECTS IN THE LANGUAGE OF SCIENCE

[*A*] *The Traditional Problems and Tasks of Semantics*

The problem of the differentiation of analytical and synthetic sentences is only connected with the ontological aspects of the semantic analysis of language insofar as synthetic sentences are usually described as sentences which presuppose some knowledge of the universe, the 'factual state of affairs', while analytical sentences contain no such assumption. (We shall show in due course that this characterisation of analytical sentences is not entirely proper, or does not apply to all kinds of analytical sentences.) However the problem of analytical and synthetic sentences is connected with other semantic issues, notably the method of state description and meaning postulates (as demonstrated by Carnap [25]), the problem of synonymity (this was emphasised particularly by Quine [121]), the differentiation of logical and non-logical signs, and also with the various reasons for semantic decision-making.

In a sense this problem lies outside the realm of questions of logical semantics and is closely linked to questions of the relations between logic and empiricism, logical necessity and contingency, apriorism and aposteriorism, and to the distinction of the empirical and theoretical levels of knowledge and so forth. These other aspects really do go beyond the scope of this work and the following analysis of the problem of analytical and synthetic sentences therefore cannot be taken as complete. We will only be mentioning these and other questions of methodology where they are linked with semantics.

The division of sentences or, as most commonly conceived in traditional logic, 'propositions', into analytical and synthetic has a long tradition in philosophy. In particular, reference is commonly made to the traditions of Kant's and Leibniz's philosophies, since certain contemporary differences of opinion are analogous to the differences between the conceptions of Kant and Leibniz. Kant's conception is limited by the traditional framework of the subject-predicate interpretation of propositions and underlines the difference between the analytical and the a priori, the synthetic and the a posteriori. Leibniz's conception is not constrained by this framework, does not work with the differences assumed by Kant between the analytical and the a priori, and associates the concepts of analyticity and syntheticity with the nature of truth. So although the solution of this problem in contemporary logico-semantic investigation is more Kantian in terminology, conceptually it is closer to the traditions of Leibniz's philosophy.[20]

In the spirit of these philosophical traditions we find even in contemporary literature repetitions of the old question: Is the sentence 'a square is an equilateral rectangle' analytical or synthetic? Is the sentence 'all spinsters are unmarried' analytical or synthetic?

It must be stressed that the questions conceived in this way and endlessly repeating examples about spinsters and bachelors who are unmarried, or about man who is an intelligent being, cannot be handled by speculation if we look at these as isolated and taken out of their language system and without paying due respect to the syntactic and semantic characteristics of the language. It is here that logical semantics has made a major contribution to the solution of these questions. The contribution can be summarised in a few points:

(1) The concepts 'analyticity' and 'syntheticity' as applied to the sentences of a language only have a sense if the entire system of the language is taken into account, including its logical (syntactic) structure and its semantics.

(2) The concepts 'analyticity' and 'syntheticity' are metalanguage characterisations. Hence the predicate 'be analytically true (in $\mathscr{L}$)' is a predicate of the semantic metalanguage with respect to $\mathscr{L}$.

This contribution of modern logic and logical semantics, while looking uncontroversial, has not led to any unequivocal solution. If the definition of analyticity depends on the semantic rules of the given language, one source of conflict may be the breadth of the sense in which these semantic rules are to be understood. We could, for example, restrict ourselves just to the rules which concern the logical terms of the language, but we might also respect the rules concerning the non-logical terms and the semantically relevant relations between them.

Another controversial matter is the possibility of a clear-cut and unambiguous divide between analytical and synthetic sentences. It was Quine in particular who opposed such a divide, the conviction that it is possible to distinguish unequivocally 'truth based solely on meaning and independent of facts', i.e. analytical truth, from truth that is dependent on facts, earning from him the label of a dogma of modern empiricism [121].

The stand adopted by Quine in 1951 led to a far-reaching debate. Leading logicians have participated in the debate and it is still not closed.[21] However we judge Quine's arguments, there is no question that his ideas have stimulated a more exact conception of analytical sentences and a tighter differentiation of the analytical and the synthetic, as well as clarifying the analyticity of sentences containing non-logical constants.[22]

[*B*] *Degrees of Analyticity*

The literature of logic contains many definitions of 'analytical sentence'.[23] It is not our present intention to analyse the various definitions, because in previous sections we have already given the definition of L-truth and analyticity as based on the method of state description and the definition of analytical truth as based on the concept of interpretation. We shall therefore quote a few examples to demonstrate what may be described as 'degrees of analyticity'. Some of these degrees give no cause for doubts, but others are controversial in that the sentences involved could be called analytical or synthetic.

(1) Analytical sentences which do not contain non-logical constants. These sentences are judged to be analytically true merely on the basis of the rules which concern the logical terms of the given language and ignoring the rules concerning non-logical constants. Sentences of this kind may contain sentence variables, e.g.

$$S_i \vee \sim S_i$$

(2) Analytical sentences which do contain non-logical constants but with which it is not necessary to take account of the relations between those constants. These sentences are likewise judged to be analytically true on the basis of the rules which concern the logical terms, since we need not take into account any semantically relevant relations of the non-logical constants; this could be because, for example, there is only one non-logical constant:

$$(\forall x)(P_1 x \vee \sim P_1 x).$$

(3) Sentences which are judged to be analytically true on the same basis as those under (1) and (2), but also on the basis of the semantically relevant relations of the non-logical constants. These relations may be presupposed in the form of meaning postulates.

This category includes the traditional examples: 'A square is an equilateral rectangle', 'Spinsters are unmarried women' and the like, but only with the reservation that we are treating these sentences as a verbal transcription of the formalised sentences of some language whose syntactic and semantic rules have been exactly determined.

(4) Sentences which are judged to be analytically true from definitions or synonymity. Ajdukiewicz [2] quotes the example of this kind of definition: 'A centimetre is one hundredth of a metre'. It goes without saying that something which is denoted by the term centimetre only satisfies the expression '*x* is one hundredth of a metre' if what we have in mind is the usual meaning of the term 'metre'.

(5) Sentences which are judged to be analytically true from the scientifically ascertained and verified general relations of objects, which are denoted by certain non-logical terms and have been shown to be unchanging in the given universe. One example is the formulation of Newton's principle of force – force is equal to the product of mass and acceleration.

This is by no means a complete review of the degrees of analyticity; each kind can be further differentiated by, for example, separating analyticity on the basis of definition from analyticity on the basis of synonymity and so on. What matters, however, is that in each of these cases we have reasons for the stated semantic decision, i.e. for describing the sentences quoted as analytical. A further question which arises is whether these are the only reasons for this semantic decision and if there are not here several types of reasons.

The first question is answered in the negative by Quine [121]. He is prepared to tolerate the concept of 'logical truth', which in the above scheme of things can only apply to degrees (1) and (2). In the case of the other degrees Quine refuses to consider their analyticity, and for a number of reasons. In the case of analyticity based on semantic rules connected with non-logical constants, and their mutual relations, there is no guarantee of substitutability salva veritate. In the case of analyticity based on definitions or synonymity, he quotes the case of propositional attitudes, which, as we have already said, make it difficult to use definitions or synonymity as a guarantee of analyticity.

It is particularly to the credit of Quine's criticism that he pointed out the heterogeneity of the reasons for a sentence's being characterised as analytical. This is why contemporary literature distinguishes two kinds of analytical sentence: analytical sentences in the narrower sense of the word (E.D. Smirnova [177]), syntactic analytical sentences (K. Ajdukiewicz [2]), formally analytical sentences (W. Stegmüller [144]) on the one hand, and analytical sentences in the broader sense of the word, semantic analytical sentences or materially analytical sentences on the other.

The analytical truth of the former group of sentences coincides with logical truth and is based on the rules connected with logical terms. Analytical sentences in the broader sence, which must be distinguished from logically true sentences,[24] are based on the semantics of the logical and non-logical terms of the given language.[25] A sentence is analytical which is a meaning postulate of the given language or can be derived from meaning postulates.

This differentiation of two kinds of analytical sentences also reveals the weak points of Quine's argument. Quine takes individual sentences of a natural language, or the partially or completely formalised sentences of the same language, and asks that it should be possible to decide whether they are analytically true. We have already stated that the concept of analytical truth, which is a metalanguage predicate, must be relativised to the semantic rules of the language, and these may be understood in a more or less broad sense. The rules themselves are not homogeneous. In natural languages these rules may be assumed only implicitly.

It is obvious from this that the applicability of the concept of 'analyticity' in the strict sense of the word is tied to languages in which these rules have been explicitly formulated. If they can only be reconstructed hypothetically – which is the common case in the analysis of natural languages and individual sentences extracted from them – the relevant statement of analyticity only makes sense with respect to the reconstruction. If language users are found who, for one reason or another, are unwilling to recognise that kind of hypothetical reconstruction, the statement of analyticity becomes problematical.

The matter at issue is then the scope, and also the nature and origin, of the semantic rules on the basis of which decisions are made about the analyticity of sentences. Carnap, as we have seen, espouses the broader conception of analytical sentences. He points out that analytical sentences are those whose truth can only be decided on the basis of semantic rules, with no consideration given to extralinguistic facts [23, p. 10]. But it is easily proved that the semantic rules on which is based any decision about the truth of sentences of degrees (3), (4) and (5) are themselves formulated on the basis of stable relations between extralinguistic facts. Therefore it cannot be said that decisions about the truth of these sentences are made without any relation to those facts.[26]

The main thing is of course the fact that this relation is, in the case of analytical truth, always mediated, the role of mediator being played by the respective semantic rules.[27] If a language system contains non-logical constants, these rules may take the form of meaning postulates. This also means that the constitution of meaning postulates broadens the original concept of L-truth with regard to the semantically relevant relations of the non-logical constants.

The relativisation of the concept of an analytically true sentence with respect to the rules of a given object language that has been enriched by meaning postulates can be met by another objection, which has already been stated indirectly by Quine in his discussion of the relation between modality and propositional attitudes. If a language user is unaware, or only partially aware, of the meaning postulates, is he entitled to decide on the analytical truth of some sentence if he has been able to convince himself in some other way, e.g. empirically or by knowing a particular theory, of the stability of the relations between the objects which are denoted by the respective non-logical constants? In response to this question, which is simultaneously a kind of objection, Carnap replied [25] that it was badly phrased. The problem of analyticity is not one of investigating relations between objects in the universe, but a problem of a particular property of expressions of language. If we accept the relevant meaning postulates, then these postulates and the sentences that can be derived from them may be described as analytical. This can be expressed another way using our terminology: the problem of analyticity is a problem of semantic decision, not of ontic decision. At the same time, of course, semantic decision may be based (e.g. genetically) on ontic decision, but the two cannot be confused. Analyticity is a property of language or, to be more precise, of certain expressions of language, not a property of extralinguistic entities.

[C] *Some Methodological Problems*

The statement that the problem of analyticity is one of semantic and not ontic decision-making is of fundamental significance to the differentiation of analytical and synthetic sentences. It means, for instance, that if we change the particular set of rules, e.g. a system of meaning postulates, we can change the boundary between the analytical and the synthetic. Nevertheless it is worthwhile mentioning some of the 'ontic commitments' that arise in connection with the problem of analyticity.

Let us take the analytical sentences of degree (2), i.e. those having the form $(\forall x)(P_1 x \vee \sim P_1 x)$. Let us assume, to use Russell's old example, that P_1 is the one-place predicate 'be bald' and that x is an individual variable in the domain X (all people). Then if, staying with Russell's example, we take the sentence 'the present king of France is bald or the present king of France is not bald', the question arises of whether or not this is an analytically true sentence. It might seem that a decision on analyticity can be inferred from the analyticity of the sentence $(\forall x)(P_1 x \vee \sim P_1 x)$. But this cannot be accepted without reservation. It is necessary to leave out of consideration individual constants with zero denotation. This means that a sentence which can be inferred from $(\forall x)(P_1 x \vee \sim P_1 x)$ where the general quantifier relates to the domain X, is only analytically true if:

(a) the individual under consideration is an element of domain X,

(b) the individual exists.

Similar conditions can be formulated for the other degrees of analytical sentences. It can be objected, however, that a language whose syntactic and semantic rules have been explicitly formulated is hardly likely to contain terms which denote nothing. All the examples quoted in logical literature have to do with mythology, by and large, and are formulated in natural language. But these examples, which operate with terms like 'Pegasus', 'centaur', 'unicorn' and others, are not the main issue. What is really important is that even in the language of science there have occurred, and may yet occur, terms which are found ex post to have zero denotation. There is no possible guarantee that even contemporary scientific thought is free of 'phlogistons', 'motive forces' and the like. Therefore the above conditions and others like them do have their significance even in the analysis of the language of science.

Another methodologically interesting problem connected with analytically true sentences is the role of 'language conventions'[28] and their relation to scientifically established relations. A telling example of this kind of relation is quoted by M. Przełecki [112]. Let us compare the semantic decision taken in respect of the two different sentences:

(1) A gram is the mass of one cubic centimetre of water at 4°C.

(2) All quantities of water having a volume of one cubic centimetre and a temperature of 4°C have the same mass.

Intuitively speaking, sentence (2) is synthetic, since it can be taken as the result of experimentation and measurement. By contrast, sentence (1), which is by nature a definition, amounts to a kind of language convention which, from the point of view of logical semantics, has the character of a meaning postulate. But since we consider to be analytical any sentence which follows logically from meaning postulates, we can also take sentence (2) as analytical. This rise to 'analyticityhood' can be accounted for in various ways. Either we bind the analyticity of a sentences to postulates which have the nature of definitions, include synonymity etc., and then, if the definition or synonymity is not formulated, we are not entitled to speak of analyticity;[29] or we link analyticity with truth, which is determined solely on the basis of meanings, irrespective of extralinguistic facts, and then of course we are certainly not entitled to speak of the analyticity of sentence (2). The first solution matches the view held by Ajdukiewicz, while the second corresponds to Carnap's orthodox conception. We have already shown that the viewpoint which says that analytical truth has absolutely no connection with extralinguistic facts is scarcely tenable and so we consider the first solution to be the more productive. At the same time, the fact of 'hidden synonymity' and 'hidden' semantically momentous relations generally among terms with which science has operated or operates, is frequently encountered in the evolution of the language of science. We can make the general assertion that in science many new terms are introduced without all their inter-relations, or even their relations to terms that are already current, being exactly fixed in advance. In this way the evolution of scientific knowledge leads not only to a broadening of the system of facts, laws, hypotheses etc., but also to the extension and supplementation of the system of the language of science by the introduction of new terms, new semantic rules and meaning postulates. And yet the language of science, even of the natural sciences, is not as a rule completely formalised. It is therefore a major task of logical semantics to formulate exactly the mutual semantical relations of both existing and new non-logical terms in the language of science. Until this task is complete, no formalisation or axiomatisation of a scientific system is possible.

After this analysis of analytically true sentences there need be no surprise in the conclusion that in the evolution of the language of science there is the possibility of both situations where certain sentences have acquired 'analyticityhood' having been previously regarded as synthetic, and situations where a sentence which has been regarded – on the basis, say, of definition or some assumed synonymity – as analytical turns out to be synthetic.[30]

For these reasons the question of whether analytically true sentences can be strictly circumscribed in the language of science cannot be answered in one way and one way only. There are two answers, and they depend on whether the question is being asked from the historical point of view or from the point of view of the current logical and semantic structure of the language.

(1) In historical terms it is obvious that the boundary is mobile. This also proves that analyticity – provided we mean analyticity in the broader sense of the word, or semantic analyticity – simply cannot be cut off from any extra-linguistic facts.

(2) In terms of the current logical and semantic structure of the language, the drawing of a line between analytical and synthetic sentences makes sense. This boundary line is of course always relative, being tied to the syntactic and semantic rules of the language, provided they are accurately formulated. (If that is not the case, the distinction loses its point.) This also means that changes of these rules result in a change of the boundary line between analytical and synthetic sentences.

IX. AN OUTLINE OF THE EVALUATION OF THE RESULTS OF SCIENTIFIC ACTIVITY IN TERMS OF SEMANTIC INFORMATION

1. THE SCOPE FOR EVALUATING SCIENTIFIC RESULTS

The problem of appraising the results of scientific activity, evaluating their theoretical and practical relevance within science itself and in every sphere of human endeavour that puts scientific results to work, would appear to be as old as science itself. In recent years and decades the problem's relevance has grown vastly thanks to a number of now familiar phenomena which can be summarised under a number of heads: the huge (in certain situations even exponential) growth in scientific output, the so-called information explosion, which is at the same time an explosion of redundance, the expansion of science into practically every sphere of living, and others. For these reasons it has become a matter of some considerable urgency to express and evaluate quantitatively and qualitatively all that can be distinguished at the output of contemporary science as a highly complex and heterogeneous system of activities. This also explains the effort put into the quantitative analysis of the impact of publications, problems of citation, copyright and other such problems, and the effort that goes into the economic evaluation of results at the output in connection with the economic evaluation of everything that may be considered as input elements of scientific and research set-ups, usually taken in terms of particular disciplines, institutions, countries or otherwise defined groupings.

In this chapter we shall try to outline a somewhat different set of possibilities for evaluating some of the results of scientific endeavour. These possibilities are offered through means that rest on the contemporary theory of semantic information. It must be emphasised that they have their limitations and no claim is here made that they might replace those means for evaluating the results of scientific activity that have been used hitherto.

The application of the means afforded by semantic information to the evaluation of scientific results is confined here to tasks involving systematisation, i.e. tasks in which we operate with both general and singular findings. At the same time it has to be borne in mind that in all scientific and research activity (and of course in all manner of practical tasks as well) we have to operate with data of two kinds and of two different origins:

(1) data acquired gradually as a task is being resolved, i.e. data a posteriori;

(2) the data that were available before work on the task was begun, i.e. data a priori.

(It should be noted that within this distinction the difference between the two kinds of data is only relative.)

Furthermore the complexity and challenge of many tasks increases with the increase in the range of a priori data, i.e. the data which the investigator must have at his command – he may have acquired it through study, through long-term successful practice, or through borrowing data from other successful investigators working in the same general area. Very often the quality with which a task is solved depends precisely on what we call erudition, experience, prior knowledge, namely the range of a priori data available to the investigator.

Tasks of the systematising type comprise only a part, although generally the major part of the set of tasks characterised as above, i.e. those in which we operate with both a priori and a posteriori data. The term 'systematisation' is, as we have already shown, a blanket term for procedures in which we operate on the one hand with knowledge of a general kind, expressed as a rule in scientific laws, hypotheses, general instructions, widely valid norms, algorithms etc., and on the other hand with concrete empirical knowledge acquired in the course of resolving a given task. This is what sometimes leads to talk of subsumption tasks, since concrete knowledge has to be subsumed under particular laws, and also 'covering laws' which are capable of embracing a certain class of empirically distinguishable facts. The terms 'subsumption' and 'covering law' are not entirely accurate or clear, however, and we shall therefore not be making any further use of them.

To make the term 'tasks of the systematising type' easier to grasp, we shall try to illustrate it by a few examples. In classifying objects, i.e. fitting them into a certain finite set of classes, we need to know the signs by which to distinguish objects and the signs by which we assign the objects to classes. This in turn assumes that we have at our disposal certain general rules governing the assignment of objects. These rules may be a product of empirical experience, the outcome of certain scientific hypotheses and laws, but also possibly a result of particular conventions. What is essential about them is that they are general in character. In solving the traffic situation at a given cross-roads, we need to establish certain concrete empirical data – the lay-out of the crossing, the road-signs present etc., and then resolve the situation on the basis of the relevant general rules. In many other practical tasks, such as those of economic management, we likewise reach decisions on the basis of both on-the-spot data and data regarding the object to be managed, and also on the basis of certain conformity patterns, rules and possibly norms that must be respected.

In scientific activity, typical tasks of the systematising type are preeminently the procedures of: scientific explanation, scientific prognosis, diagnosis, tasks of classification and others. For example, if we are to forecast a future state of an object under observation, we need to have available both data on the ascertained properties of the object in a given initial situation, and the relevant laws or hypotheses that relate to the evolution of an object of the given class. In making a diagnosis a doctor has to consider not only the data of the patient's symptoms, but also the medical knowledge that he has acquired from study and years of practice.

In the further analysis of the data that feature in tasks of the systematising type we shall assume that we have at our disposal, or are in a position to obtain, the following data:

(1) a posteriori data of an empirical nature that have the character of singular statements;

(2) a priori data that express certain scientific laws, hypotheses or general rules and have the character of nomological statements;

(3) data that characterise the goal of the given task, the procedures involved, and possibly the demands placed on the quality of the solution.

The characteristic feature of tasks of the systematising type is that what is sought is a solution to a particular task, or an ultimate decision to be made on the basis of two groups of data: a priori and a posteriori data – data of general bearing and data of an empirical nature. This means that it is vital to take the data under (1) and (2) in close conjunction, which is what is sometimes referred to rather unsatisfactorily as 'subsumption'. With respect to the given aim of a task, the issue is one of selecting the appropriate relevant a priori and a posteriori data, and of respecting the conjunction between them. The latter assumption is sometimes expressed as the indispensability of a link between the theoretical and empirical components of the data brought to the solution of a given task.

Tasks of the systematising type are thus based on the application of certain general knowledge – scientific laws and hypotheses, or general rules, norms, algorithms etc., respective to a concrete instance: resolving a particular traffic situation at a cross-roads assumes the ascertainment of the most important data and the application of the relevant parts of the highway code. Predicting the future state of a given heavenly body requires the ascertainment of the relevant parameters in the situation as it stands and the calculation of the future state from the laws of celestial mechanics. Making a diagnosis and finding the proper therapy assumes the ascertainment of the patient's symptoms and then taking a decision on the basis of the relevant medical knowledge.

Tasks of the systematising type may be characterised overall as follows: they are decision processes regularly associated with a particular goal or goals and certain demands on the quality of the solution (the decision at the cross-roads will not be expected to lead to a collision, the prediction will be expected to be sufficiently reliable, the diagnosis must be the start to effective treatment); within each such process, moreover, decisions are taken on the basis of both general knowledge, principles, norms or directions and data directly connected with the given situation. If the investigator knows the objective of the given task, he must proceed to ascertain the relevant a posteriori data while at the same time selecting the relevant a priori data to ensure that the solution guarantees that the objective will be reached.

2. BRILLOUIN'S ATTEMPT AT AN INFORMATIONAL EVALUATION OF SCIENTIFIC LAWS

In tasks of the systematising type the problem is to choose the right general facts relevant to the empirical data ascertained. The potential for an informational evaluation of scientific laws vis-à-vis established empirical or informational data has been pointed out by L. Brillouin [17]. Preeminently a specialist in problems of physics, Brillouin formulated the problem thus: What is the quantity of information contained in the laws of physics with respect to results found experimentally? Clearly, the relevance of physical laws to experimental findings is greater, the greater the amount of information contained in the laws.

Brillouin's approach to the problem as conceived is fairly straightforward: he sees a law of physics as a summary of experimental results. A physical law has been reached by our having observed a given physical system, measured certain variables and characterised a particular domain of variation for each variable. Allowance is also made for the time-scale and the degree of accuracy of the observations. Thus the information embodied in a law of physics increases with the increasing accuracy of the data, i.e. with the decrease in observational errors, and with every increase in the domain over which observations are made. If thought through consistently, this conception leads to the construction of entities that are hypothetical and whose practical realisation is hard to visualise; one such entity would be the domain of all observations of the pressure, volume and thermal ratios of all gases, if we take Boyle's law as representative of a law of physics (as does Brillouin)

$$f(p, v, T) = \frac{pv}{T} = R = \text{a constant}$$

In Brillouin's notation, a physical law in its general form looks as follows

$$f(x_1, x_2, \ldots x_n) = C = \text{a constant}$$

where x_i is a variable that is the object of observation.

It is further assumed that for each variable x_i there is a specified domain of variation X_i, the smallest interval connected with mistakes in observation $\delta x_i = \alpha$, and that the overall range of mistakes δf does not exceed the value Φ. Brillouin then assumes two forms of calculating by means of the information contained in a physical law: In the first case he considers what he describes as the 'total a priori space of variations F', which depends on all prior knowledge. In the second case what interests us is the information difference between two forms of a law – f_1 with an extent of errors Φ_1 and f_2 with Φ_2.

The first case corresponds to

$$\mathbf{I} = C.N.(\mathbf{ln}.F - \mathbf{ln}\,\Phi),$$

while the second corresponds to

$$\Delta\mathbf{I} = C.N.(\mathbf{ln}\,\Phi_1 - \mathbf{ln}\,\Phi_2),$$

where C is a constant, **ln** is a natural logarithm and N is determined thus:

$$N = \frac{x_1 x_2 \ldots x_n}{\alpha_1 \alpha_2 \ldots \alpha_n}$$

This conception, which treats a scientific law as an aggregate of experimental or observational results, has a real sense primarily in the second form, which calculates with the informational shift $\Delta\mathbf{I}$, which is due to the reduction of errors of observation or experimentation. Brillouin himself exemplifies this case by having Boyle's law replaced by van der Waals' formula.

As for the first form of calculating the information contained in a scientific law, one might speak of 'absolute information' (the term used by Brillouin himself). Since Brillouin's conception assumes that a scientific law must always be assessed on the results of observation, measurement and experimentation, it becomes necessary to postulate an 'absolute observer or experimenter', who would have properties corresponding to Laplace's 'absolute intelligence'. Clearly, the construction of criteria by which to assess the informational evaluation of a scientific law, where the construction is based on these assumptions, is highly problematical. Admittedly Brillouin was aware of the limitations of the conception, which is why he enlisted another concept for evaluation: the concept of 'relative value for a particular user', of which it always holds that the relative value for a particular user $\leqslant$ absolute information.

It goes without saying that the relative value for a particular user cannot be defined in the abstract: all depends here on the aims for which scientific laws and experimental results have been used, the tasks in which the data are applied, etc. Although Brillouin himself did hint in the abstract at the possibilities of such a relativisation of informational values, he went nowhere towards realising them. For these reasons we shall, in our further analysis devoted to the evaluation of data in tasks of a systematising type, begin with some assumptions that differ considerably from Brillouin's conception, especially the first form of his informational evaluation (i.e. the 'absolute information' of his terminology), which leads to the inevitable construction of problematical hypothetical entities. Moreover, we shall assume that scientific laws and hypotheses themselves have a much greater underlying force: a scientific law is not just the sum of experimental results, but is also an instrument of explanation and prediction; it is applied in sundry procedures both empirical and theoretical in nature, and this is so not only in the sphere of science, but also in a much wider range of applications. The measure of an evaluation will therefore depend on the quality with which all these procedures are resolved. From this point of view, as will be seen from what follows, the basic concept for assessing the semantic-informational evaluation of the results of scientific activity will not be 'absolute information', but what we shall call 'transmitted information'. Basically, we shall be moving within the following framework: In any task in which the results of science are applied, what will be assessed is the degree to which a new item of knowledge (e.g. a new hypothesis, new results of confrontation or experimentation), if added to an analysis of previously existing available knowledge, reduces any previous uncertainty associated with the solution of the task, or, in other words, how far it increases the quality of that solution.

Brillouin's conception, as we have indicated, gives no evaluation of scientific laws or any other data in connection with a task in which the data are applied. As distinct from Brillouin, we shall treat scientific laws, scientific hypotheses and any other data applied in tasks of the systematising type as statements that relate to specific entities as elements of the given universe. This means further that the immediate object of evaluation will be these very statements. This emphasis on the linguistic aspects that must be respected in any expression of the results of scientific activity, is by no means intended to mean that statements that are the object of evaluation have no objective content, or that they cannot be treated as a mapping of certain objects, facts, events etc. that have an existence independent of the language devices used and any person using those devices.

The fact that we shall be appealing to statements also justifies our referring to the semantic-informational approach to the evaluation of scientific results. The attribute 'semantic' is to be understood as follows: a) of any syntactically well-formed and meaningful statement containing the expression of data applicable to the solution of a given task or set of tasks it will be assumed that it has a particular *denotation*, i.e. objective content; b) it will be further assumed in every case that the statement is being used in a specific task, a specific decision problem, which is goal-oriented towards finding a certain solution to the task or decision problem. This second aspect means also that we are not concerned with abstract semantics, as in the spirit of Carnap's method of 'state description', but that we shall be trying to respect the use-context of the given statements – in, for instance, such typical tasks as the adequate explanation of newly discovered phenomena, the achievement of reliable predictions of future facts, the confirmation of hypotheses by the results of observation or experiment, the finding of proper diagnoses and the institution of right courses of treatment, and so forth.

3. LINGUISTIC DEVICES IN TASKS OF THE SYSTEMATISING TYPE

One requirement for the construction of reliable gauges of the semantic-informational evaluation of data that feature in tasks of the systematising type is the distinction between data and informational concepts as introduced earlier in this chapter. The data will be represented by various kinds of statements. On the other hand the informational concepts will be represented by various measures that enable us to evaluate statements or ordered groups of statements according to the degree to which the statements are able – with regard to the given task or type of task (explanation, prediction, diagnosis, i.e. tasks of the systematising type) – to transmit or communicate information, or, putting it another way, to shift the entropy threshold, lower the level of uncertainty, increase the quality of the particular decision process (e.g. explanation, prediction, diagnosis), or lower the level of risk associated with these processes. This also means that the construction of informational concepts to facilitate a particular evaluation of data will assume a degree of relativisation with respect to what we have characterised as the 'goal complex'. For example, in a case of evaluating individual tests and their results and the symptoms found, in a diagnosis, the actual evaluation depends on the given diagnostic task. Analogously it applies that the evaluation of individual hypotheses and laws of science, as well as the results of observation and experiment, will depend on the viewpoint from which the particular application is made, e.g. the viewpoint of establishing how far the data can provide adequate accuracy of predictions, qualitatively adequate decisions etc.

Since any semantic-informational evaluation will be tied to a system of data, i.e. statements expressing the results of cognitive activity, it is necessary to characterise the language in which these statements are made. For the purposes of what follows we shall assume that results of cognitive activity are expressed in a language of a type of first order predicate calculus, the vocabulary of which contains some extra-logical terms that are of the nature of predicates:

(a) the set of predicates β that denote certain practical operations, such as instructions, therapies in medicine, elementary procedures in production, etc.;

(b) the set of empirical (observational) predicates γ that denote empirico-experimental components of scientific activity, for example properties and relations that can be ascertained by means of observation, measurement, experiment and testing;

(c) the set of theoretical (auxiliary) predicates δ that denote theoretical (sometimes also new explanational or auxiliary) entities in the given domain.

We shall be assuming that any results of cognitive activity are expressed in statements that operate with elements from one, two or all three of these sets. (For example, results of observation will be expressible only by means of elements of γ, while theoretical statements, scientific laws and hypotheses will operate with elements from δ, but possibly also with elements from the other two sets.)

Statements that express the results of scientific activity can be classified from a number of points of view. For present purposes we shall introduce the following:

From the point of view of the form of a statement a line must be drawn between *singular statements* which assign elements of classes β, γ and δ direct to individual objects of the given universe, and *general statements*, which assign elements of classes β, γ and δ to a particular set of objects of the given universe. It will be assumed that general statements are *nomological* in character (sometimes also described as law-like statements). This means that we exclude 'random generalisations', the application of which in tasks of the systematising type is problematical.

In agreement with the previously maintained division of data we shall make three types of data:

(1) *A priori data*, which are generally available before a particular task is embarked upon, or a problem addressed. This will normally mean the by now familiar scientific laws and hypotheses, or the equally familiar algorithms, instructions, empirical generalisations etc. In the following account these data will be represented by the class of nomological statements

$$\{\mathbf{h}_1, \mathbf{h}_2, \ldots, \mathbf{h}_n\}.$$

(2) *A posteriori data*, which generally express empirical or experimental results obtained in the course of resolving a given task or decision problem. In the following account these data will be represented by the class of singular statements

$$\{\mathbf{e}_1, \mathbf{e}_2, \ldots, \mathbf{e}_n\}.$$

(3) Data expressing what we have described as the *goal complex*. These are data that fix the solution of the given task or decision problem, possibly including demands on the quality of the solution, the effectiveness of the task and so forth. In other words, the goal complex defines the end for which the data under (1) and (2) might be useful. In the following

account these data will be represented by the class of singular statements

$$\{\mathbf{i}_1, \mathbf{i}_2, \ldots, \mathbf{i}_k\}.$$

The aggregate of variously interlinked data of types (1), (2) and (3) constitute the *information system.* The scope of the information system depends on the nature and demands of the tasks in which the three types of data apply.

For a better understanding of the significance of the division of data, and so too of the concept of 'information system' with which we operate in solving tasks, we shall indicate a few typical examples. If we are to make a prediction about the future or expected states of a given dynamic system, then $\{\mathbf{h}_1, \mathbf{h}_2, \ldots, \mathbf{h}_n\}$ constitutes the system of hypotheses, $\{\mathbf{e}_1, \mathbf{e}_2, \ldots, \mathbf{e}_n\}$ the system of primary data ascertained, and $\{\mathbf{i}_1, \mathbf{i}_2, \ldots, \mathbf{i}_k\}$ are statements on the future or expected states of the dynamic system in question. In a case of diagnosis, an analogous triple is made up of statements expressing medical knowledge and medical theory, statements on tests performed and symptoms ascertained, and statements on the assumed diagnosis. One can also imagine a task of managing a technological process in which the a priori data are the system of technical hypotheses and general directions, the a posteriori data express the parameters found to apply in the process, and the goal complex is the system or sequence of concrete directions or decisions.

It remains to explain how the three types of data that make up the information system are related to the elements of the sets β, γ and δ, i.e. the sets that make up the system of scientific concepts. The nomological statements $\{\mathbf{h}_1, \mathbf{h}_2, \ldots, \mathbf{h}_n\}$ will be assumed to operate with elements of all three sets. A complete theory of a given domain is therefore the class of all nomological statements of the type

$$\mathbf{h}(\beta \cup \gamma \cup \delta)$$

In other words, a priori data operate with elements of all three sets, while what are of particular importance are those components of the theory that fix the dependencies between the individual elements of the individual sets.

From the point of view of the actual elements of the sets β, γ and δ with which we operate, the whole of what may be characterised as a complete theory of a given domain may be split into the following components:

(I) The basic component of a complete theory of a given domain is usually taken to be the system of nomological statements that operate solely with elements of the class δ. This (sometimes described as the Campbell component of a theory, or theory in the narrow sense) may be of the nature of postulates or implicit definitions of the elements of set δ, or it may be constructed axiomatically.

(II) The next component of a complete theory is the so-called observational component, i.e. all the nomological statements that operate solely with elements of set γ. These are usually the so-called empirical generalisations, i.e. empirically and experimentally substantiated statements that fix the general dependencies between empirically and experimentally discriminable properties and relations of the objects in the given universe.

(III) A major component of any theory that operates with theoretical concepts and relates to a certain empirical domain are the so-called correspondence rules. These are statements that fix the dependencies of the elements of sets γ and δ. Correspondence rules may vary in character, taking the form of, for example, explicit definitions, reduction theorems, 'symptom theorems', etc. One example of a correspondence rule is the statement that determines the dependence of the theoretical concept (an element of set δ) 'specific gravity' on the empirical concepts (elements of set γ) 'weight' and 'volume'.

(IV) If a complete theory also contains practical instructions or other such operations, it also contains a so-called operational component. The operational component of a theory is a system of statements that fix the dependencies between elements of sets γ and δ and elements of set β. In medical theory these would be statements that assign certain courses of therapy to certain illnesses (elements of set δ), or certain symptoms (elements of set γ).

This list of the components of what we have described as the complete theory of a given domain is not to be understood as saying that any scientific theory must include every component. Certain mathematical theories that operate only with constructive objects (elements of set δ), without any explicitly stated empirical interpretation, are limited to component (I). By contrast, theories that relate to a given empirically or experimentally distinct domain usually include components (I), (II) and (III), or just (I) and (III) or just (II) and (III). (The last case refers to theories that operate solely with elements of set γ.) Only those theories that contain ties between particular theoretical and empirical findings and certain practical operations have the operational component.

It must further be emphasised that the quality of a scientific theory is in no way governed by whether it includes more or less components. The classification as presented is important primarily from the methodological point of view, and particularly so if we seek to clarify what statements constitute what we have described as a priori data.

A posteriori data usually take the form

$$\mathbf{e}(\gamma),$$

i.e. they operate with elements of set γ.

The goal complex however may operate either with elements of set β, if it is the practical aspects that are to the fore (e.g. practical decisions in control processes), or solely with elements of class γ, if it is the theoretical aspects that are to the fore (e.g. in the case of a scientific explanation of a given phenomenon), or indeed elements of both sets (e.g. in diagnosis where not only is the disease to be identified, but an appropriate course of treatment to be found). Accordingly statements expressing the goal complex take the form

$$\mathbf{i}(\beta \cup \gamma).$$

Under the informational system presented here, the most elementary form of a (determinist) solution to a task may be expressed thus

$$\mathbf{h}(\beta \cup \gamma \cup \delta) \longrightarrow [\mathbf{e}(\gamma) \longrightarrow \mathbf{i}(\beta \cup \delta)].$$

This schema corresponds to Hempel's well-known deductive nomological model of explanation and prediction [57] and the diagnostic model put forward by Ledley and Lusted [77]. It should be added that in the schema material implication is used to map determinist dependencies. In the case of probabilistic dependencies it would be desirable to replace material implications by conditional probabilities.

So far we have given a review of the fundamental statement components with which we shall be working in the construction of various measures for semantic informational evaluation. In the construction of these measures we shall start with the means afforded by the contemporary semantic theory of information, in particular the concept of 'transmitted information'. The line taken may be defended in the following terms: We are usually not interested in the actual information carried by a particular nomological statement or a statement expressing the results of observation or experiment in themselves, but in the degree to which prior uncertainty is reduced simply by our having the statements at our disposal. In other words, what is at the focus of our attention is the shift in uncertainty that may be achieved by taking into consideration certain statements whether amounting to a priori or a posteriori data. The concept of 'transmitted information' may then become the starting-point for the construction of the basic measures of semantic informational evaluation.

4. THE CONCEPT OF 'DECISION BASE' AND THE EVALUATION OF A DECISION BASE

In characterising tasks of the systematising type we have shown that they require us to operate with two types of data simultaneously – on the one hand those of a general nature, usually of the nature of nomological statements, and on the other hand those of the nature of singular statements concerning empirically or experimentally ascertained facts.

In tasks of the systematising type we have to decide certain statements on the basis of data, some of which were available before the task was undertaken, while others were acquired immediately before, or in the course of, work on the task. In the analysis that follows we shall concentrate on those tasks of the systematising type where the a priori data have the character of nomological statements, and the a posteriori data the character of singular empirically decidable statements. This narrowing of the field of vision is important methodologically and is motivated by the choice of examples of typical tasks of the systematising type (scientific explanation, prediction, classification, diagnosis). Of course it is equally possible to visualise tasks of the systematising type in which the a priori data have the character of general norms, general injunctions and the like. Such would be the case in the application of legal norms to a specific case, in a judicial decision on guilt etc.

An ordered pair of a priori and a posteriori data, where the former are represented by a nomological statement or a class of nomological statements **h** and latter by a singular statement or class of singular statements **e**, will be denoted

$$\langle \mathbf{h}, \mathbf{e} \rangle$$

and described as the *decision base*. It should be noted that the concept of 'decision base' is always relativised to a given **i**. It should be further noted that both components of the ordered pair should be seen as non-empty classes of statements. It follows that the concept of 'decision' base corresponds to what is described in models of explanation procedures as the explanans.

In order to be able to offer a satisfactory construction of the measure of the semantic informational evaluation of the pair $\langle \mathbf{h}, \mathbf{e} \rangle$ with respect to **i**, some indication must be given of what is expected of the pair: the task of this pair is to ensure maximum certainty about the assumed **i**. In other words, the pair fulfils its purpose with proportionately greater reliability as it reduces our intial uncertainty in respect of that to which **i** refers. These intuitive demands on what should be expected of a reliable measure of the semantic informational evaluation of $\langle \mathbf{h}, \mathbf{e} \rangle$ with respect to **i** is expressed in the present analysis by a system of postulates. If we are capable of expressing our original uncertainty connected with that to which **i** refers, and if the reduction of our original uncertainty is due to our having the pair $\langle \mathbf{h}, \mathbf{e} \rangle$ additionally available, then we may say that the measure of evaluation of the decision base $\langle \mathbf{h}, \mathbf{e} \rangle$ will be proportionately greater as the difference is greater between the original uncertainty and the uncertainty that is conditional upon our having the pair $\langle \mathbf{h}, \mathbf{e} \rangle$ at our disposal. (Uncertainty so defined will henceforth be referred to simply as conditional uncertainty.)

The measure of the semantic informational evaluation of the pair $\langle \mathbf{h}, \mathbf{e} \rangle$ will be described as its *deciding power*. The concept of 'deciding power' will be treated as a quantitative concept for which we introduce the notation

$$\Delta(\mathbf{h}, \mathbf{e}/\mathbf{i})$$

From the foregoing account of the intuitive starting-point it ought to be clear that we shall demand that $\Delta(\mathbf{h}, \mathbf{e}/\mathbf{i})$ increase if there is an increased difference between the original uncertainty of that to which **i** refers and the conditional uncertainty with respect to **i** with the extra availability of the pair $\langle \mathbf{h}, \mathbf{e} \rangle$, and that it decrease if that difference decreases.

We shall further introduce the sign $\mathbf{U}(\mathbf{i})$ as the sign of the uncertainty of that to which **i** refers, and the sign $\mathbf{U}(\mathbf{i}/\mathbf{h}\cdot\mathbf{e})$ as the sign of conditional uncertainty if we have the pair $\langle \mathbf{h}, \mathbf{e} \rangle$ additionally at our disposal. In accordance with these demands for the satisfactory construction of $\Delta(\mathbf{h}, \mathbf{e}/\mathbf{i})$ it is clear that $\Delta(\mathbf{h}, \mathbf{e}/\mathbf{i}) = \mathbf{max}\Delta$ if and only if $\mathbf{h}.\mathbf{e} \underset{L}{\rightarrow} \mathbf{i}$. In other words, if **i** follows logically from $\mathbf{h}\cdot\mathbf{e}$, then **i** has been ascertained unequivocally via the decision base and nothing uncertain therefore remains. It remains to mention another extreme situation when nothing uncertain remains and when the deciding power might also seem therefore to be at maximum: this is the case where **i** is a tautological statement. Clearly, however, such instances must be excluded from further consideration. Likewise we must exclude all cases where other component statements are also tautological. For these reasons the analysis will be restricted by the following convention:

All component statements with which we operate shall be treated as meaningful synthetic statements.

In agreement with our account of the demands that must be met by a satisfactory measure of $\Delta(\mathbf{h}, \mathbf{e}/\mathbf{i})$, it is now possible to formulate postulates that fix the properties of $\Delta(\mathbf{h}\cdot\mathbf{e}/\mathbf{i})$. These postulates are clearly a modification of those for the semantic informational evaluation of nomological statements as introduced by J. Pietarinen [111]. Equally the definition of $\Delta(\mathbf{h}, \mathbf{e}/\mathbf{i})$ is a modification of the definition of systemising power.

These postulates that ought to be met by a satisfactory measure of deciding power may be recorded as follows:

(R 1) $\Delta(\mathbf{h}\cdot\mathbf{e}/\mathbf{i}) = f\,[\mathbf{U}(\mathbf{i}), \mathbf{U}(\mathbf{i}/\mathbf{h}\cdot\mathbf{e})]$

(R 2) $\Delta(\mathbf{h}\cdot\mathbf{e}/\mathbf{i}) \underset{<}{\geq} 0$ iff $\mathbf{U}(\mathbf{i}/\mathbf{h}\cdot\mathbf{e}) \lesseqgtr \mathbf{U}(\mathbf{i})$

(R 3) $\Delta(\mathbf{h}\cdot\mathbf{e}/\mathbf{i}) = \mathbf{max}\,\Delta$ iff $\mathbf{U}(\mathbf{i}/\mathbf{h}\cdot\mathbf{e}) = \mathbf{min}\,\mathbf{U} = 0$

(R 4) $\Delta(\mathbf{h}_j\cdot\mathbf{e}/\mathbf{i}) \geqslant \Delta(\mathbf{h}_k\cdot\mathbf{e}/\mathbf{i})$ iff $\mathbf{U}(\mathbf{i}/\mathbf{h}_j\cdot\mathbf{e} \leqslant \mathbf{U}(\mathbf{i}/\mathbf{h}_k\cdot\mathbf{e})$

(R 5) $\Delta(\mathbf{h}\cdot\mathbf{e}_j/\mathbf{i}) \geqslant \Delta(\mathbf{h}\cdot\mathbf{e}_k/\mathbf{i})$ iff $\mathbf{U}(\mathbf{i}/\mathbf{h}, \mathbf{e}_k) \leqslant \mathbf{U}(\mathbf{i}/\mathbf{h}, \mathbf{e}_k)$

(R 6) f is a linear function of its second argument

(It is therefore obvious that on the basis of (R4) and (R5) the values of this function increase if the values of its second argument decrease.)

(R 7) $\Delta(\mathbf{h}\cdot\mathbf{e}/\mathbf{i}) = \mathbf{min}\,\Delta = 0$ iff $\mathbf{U}(\mathbf{i}/\mathbf{h}\cdot\mathbf{e}) = \mathbf{U}(\mathbf{i})$

The intuitive sense of these postulates is perfectly apparent: the measure of deciding power is treated as a linear function of two arguments, i.e. the original uncertainty $\mathbf{U}(\mathbf{i})$ and the conditional uncertainty $\mathbf{U}(\mathbf{i}/\mathbf{h}\cdot\mathbf{e})$. The greater or lesser the difference between the original and the conditional uncertainty, the greater or lesser is the measure of deciding power. If the conditional uncertainty is reduced to zero by our having $\langle \mathbf{h}, \mathbf{e} \rangle$ at our disposal, the deciding power is maximal. (It should be noted that conditional uncertainty must be reduced to zero only by our possession of $\langle \mathbf{h}, \mathbf{e} \rangle$; by the convention introduced above we have excluded the possibility whereby $\mathbf{i}$ might be a tautological statement.) On the basis of (R4) and (R5) it is obvious that higher deciding power may be achieved in one of two ways: either by using new, more adequate nomological statements, i.e. scientific laws and hypotheses, or by carrying out new observations, experiments or other analogous empirical operations (or by a combination of both ways). This also matches the current practice of applying the theoretical and empirical components of knowledge to the solution of tasks of the systematising type. For example, in diagnosis it may be achieved either by broadening the base of adequate medical knowledge through bringing in specialists, or by making further tests and observations.

On the basis of the given postulates, which must be met by any satisfactory measure of deciding power, it becomes possible to define $\Delta(\mathbf{h}\cdot\mathbf{e}/\mathbf{i})$ thus:

$$\text{(D 1)} \qquad \Delta(\mathbf{h}\cdot\mathbf{e}/\mathbf{i}) = \mathrm{df}\,\frac{\mathrm{U}(\mathbf{i}) - \mathrm{U}(\mathbf{i}/\mathbf{h}\cdot\mathbf{e})}{\mathrm{U}(\mathbf{i})}$$

Quantification of deciding power on the basis of this definition depends on how far it is also possible to quantify uncertainty, or conditional uncertainty. Naturally, different conceptions of uncertainty may be selected. In the following discussion, uncertainty will be considered as a measure that is reduced proportionately as the number of excluded alternative possibilities grows. It will follow in due course that this conception corresponds to those initial intuitive principles that gave rise to the semantic theory of information. We have already pointed out that there are other different, but conceivable, conceptions of uncertainty: one could be associated with the quantification of gains and losses, or with the quantification of risk or mean risk, or with other modes of determining the quality of an intended goal complex. Thus the conception of uncertainty on which this work is based is by no means obligatory or the only one possible. Rather it is the intention to indicate one of the simplest ways of interpreting the concept.

As for the conditions that must be met by the most elementary measure of uncertainty and the measure of conditional uncertainty, they are analogous to the conditions given previously, in terms of which it is desirable that the measure of uncertainty be greater than or equal to zero, zero uncertainty, i.e. absolute certainty, being reserved for the extreme cases where no doubts exist regarding that to which a given statement refers. There are two logical possibilities: either the given statement follows logically from absolutely reliable premises, or the statement is tautological. Since the latter case has been excluded from consideration, the conditions for the measure of uncertainty may be formulated thus:

$$(\mathbf{u}1) \qquad \mathrm{U}(\mathbf{i}/\mathbf{h}\cdot\mathbf{e}) \geqslant 0$$

$$(\mathbf{u}2) \qquad \text{if } \mathbf{h}\cdot\mathbf{e} \underset{\mathrm{L}}{\longrightarrow} \mathbf{i}, \text{ then } \mathrm{U}(\mathbf{i}/\mathbf{h}\cdot\mathbf{e}) = 0$$

There is a further obvious condition: if $\mathbf{h} \cdot \mathbf{e}$ is tautological, then it can in no way contribute to a reduction in the original uncertainty, and therefore:

(**u**3) if $\mathbf{h} \cdot \mathbf{e}$ is tautological, then $\mathbf{U}(\mathbf{i}/\mathbf{h} \cdot \mathbf{e}) = \mathbf{U}(\mathbf{i})$

Yet another condition relates to maximum conditional uncertainty: a convention may be adopted whereby the maximum value that may be acquired by conditional uncertainty is the same as the original uncertainty. This corresponds to

(**u**4) if $\sim(\mathbf{h} \cdot \mathbf{e}) \underset{L}{\longrightarrow} \mathbf{i}$, then $\mathbf{U}(\mathbf{i}/\mathbf{h} \cdot \mathbf{e}) = \mathbf{U}(\mathbf{i})$

If it is possible to assign to each statement component the probability measure **p**, which satisfies the ordinary axioms of probability theory, it holds that

$$\mathbf{U}(\mathbf{i}) = 1 - \mathbf{p}(\mathbf{i})$$

$$\mathbf{U}(\mathbf{i}/\mathbf{h} \cdot \mathbf{e}) = \mathbf{p}[(\mathbf{h} \cdot \mathbf{e}) \vee \mathbf{i}] - \mathbf{p}(\mathbf{i}) = \mathbf{p}(\mathbf{h} \cdot \mathbf{e}) - \mathbf{p}(\mathbf{h} \cdot \mathbf{e} \cdot \mathbf{i})$$

From the conception of simple and conditional uncertainty as here described it is obvious that in the case of simple uncertainty $\mathbf{U}(\mathbf{i})$ is identified with the 'content measure' $\mathbf{cont}(\mathbf{i})$ introduced by Bar-Hillel and Carnap [6]. If however we also elected for $\mathbf{U}(\mathbf{i}/\mathbf{h} \cdot \mathbf{e}) = \mathbf{cont}(\mathbf{i}/\mathbf{h} \cdot \mathbf{e})$, we would be breaking the requirement that maximum uncertainty be equal to initial uncertainty.

This conception clearly rests on the possibility of assigning probability characteristics to the statement components with which we operate. Making this kind of assignment may give rise to some misgivings in the case of nomological statements. From the abstractly theoretical point of view there is no doubt that it would be hard to demand that any measure of probability used be equally competent for both a priori and a posteriori components. For these reasons it is expedient, for practical applications, to enlist a convention whereby the measure of probability assigned to nomological statements will be considered as the measure of confirmation by all the presently accessible evidence.

For the purposes of any application it is essential to show some of the important properties of $\Delta(\mathbf{h} \cdot \mathbf{e}/\mathbf{i})$ that stem from the postulates and conditions given above: If a given decision base $\langle \mathbf{h}, \mathbf{e} \rangle$ reduces the initial uncertainty $\mathbf{U}(\mathbf{i})$ to zero, i.e.

$$\text{if } \mathbf{h} \cdot \mathbf{e} \underset{L}{\longrightarrow} \mathbf{i}, \text{ then } \Delta(\mathbf{h} \cdot \mathbf{e}/\mathbf{i}) = \max \Delta = 1$$

This situation, as suggested previously, corresponds to the unequivocally determinist cases that can be mapped by Hempel's deductive-nomological model of explanation and prediction. In other words, the given decision base $\langle \mathbf{h}, \mathbf{e} \rangle$ is so powerful that it leaves no uncertainty with regard to **i** or, which is the same thing, it affords an unambiguous determination of **i**. If on the contrary the deciding power is not capable of reducing the original uncertainty $\mathbf{U}(\mathbf{i})$, it is irrelevant with regard to **i**.

In concrete situations we operate as a rule with a fair number of nomological statements (scientific laws, hypotheses etc.). For these reasons it is worthwhile investigating the conditions under which it is possible to aggregate nomological statements in such a way that the measure $\Delta(\mathbf{h}\cdot\mathbf{e}/\mathbf{i})$ is additive.

This measure is additive if the following conditions are respected:

[1] the aggregating of a (finite) class of nomological statements $\{\mathbf{h}_1, \mathbf{h}_2, \dots \mathbf{h}_n\}$ must be understood as a conjunctive combination with respect to the same **e**, i.e. with respect to the same circle of a posteriori data;

[2] the individual elements of the class $\{\mathbf{h}_1, \mathbf{h}_2, \dots \mathbf{h}_n\}$ are logically disjunct.

Under these conditions it holds that

$$\Delta(\mathbf{h}_1\cdot \mathbf{h}_2 \cdot \dots \mathbf{h}_n \mathbf{e}/\mathbf{i}) = \Delta(\mathbf{h}_1\cdot\mathbf{e}/\mathbf{i}) + \Delta(\mathbf{h}_2\cdot\mathbf{e}/\mathbf{i}) + \Delta(\mathbf{h}_n\cdot\mathbf{e}/\mathbf{i})$$

Like conditions apply in the case of situations where we aggregate a number of a posteriori data with respect to one and the same nomological statement.

In concrete tasks it is not usually possible to operate with an excessively large number of a priori and a posteriori data for such reasons as the amount of time it would take, the unwarranted effort it would demand, or the sheer unobtainable capacity demanded from the computer. Neverthe less it remains desirable that the planned or expected goal complex be achieved with the quality required. From this point of view it becomes possible to indicate a solution for some of the tasks associated with the admissible reduction of the elements of the class $\{\mathbf{h}_1, \mathbf{h}_2, \dots \mathbf{h}_n\}$.

The optimum decision base may be set up on the basis of the following considerations: If we have at our disposal a class of nomological statements with respect to the same **e** and the same **i**, then the optimum decision base with respect to **i** and assuming the realisation of **e** must be the minimal sub-class of the given class of nomological statements that guarantees the maximalisation of the measure of deciding power with respect to the same **i** and assuming the realisation of **e**.

It is further obvious that we can exclude from the decision base those statement components that are not capable of increasing deciding power with respect to **i**. If we have at our disposal a class $\{\mathbf{h}_1, \mathbf{h}_2, \dots \mathbf{h}_n\}$ and a class $\{\mathbf{e}_1, \mathbf{e}_2, \dots \mathbf{e}_m\}$ with respect to the same **i**, we may exclude from both classes those elements of which it holds that

$$\Delta(\mathbf{h}_j \cdot \mathbf{e}/\mathbf{i}) = 0 \quad \text{or}$$
$$\Delta(\mathbf{h} \cdot \mathbf{e}_j/\mathbf{i}) = 0$$

Pairs like $\langle \mathbf{h}_j, \mathbf{e} \rangle$, or $\langle \mathbf{h}, \mathbf{e}_j \rangle$, for which the given relation applies, may be described as *systematisingly irrelevant* with respect to **i**. On the basis of this characteristic it becomes possible to formulate a *rule for the reduction of the decision base:*

If $\langle \mathbf{h}_1, \mathbf{h}_2, \dots \mathbf{h}_n, \mathbf{e}_1, \mathbf{e}_2, \dots \mathbf{e}_m \rangle$ is taken as the decision base with respect to **i**, all components that are systematisingly irrelevant with respect to **i** may be excluded from that decision base.

In practice situations may be encountered where the decision base known to date is insufficient with respect to a certain **i**. These are situations where, for example, existing hypotheses and empirical data are insufficiently reliable for making predictions or diagnoses. In situations like this we have to select procedures that are the reverse of reduction, i.e. procedures that lead to a broadening of the existing decision base sufficent to increase the measure of deciding power. In general terms, all steps that broaden the decision base and lead to an increase in the measure of deciding power can be treated as plausible. But in practice we content ourselves with only a broadening of the decision base that guarantees that the measure of deciding power reaches a certain conventionally set value.

On the basis of the foregoing we may introduce the concept of 'adequate decision base' : $\langle \mathbf{h}_1, \mathbf{h}_2, \dots \mathbf{h}_n, \mathbf{e}_1, \mathbf{e}_2, \dots \mathbf{e}_m \rangle$ is the adequate decision base with respect to **i** if

$$\Delta(\mathbf{h}_1 \cdot \mathbf{h}_2 \cdot \dots \mathbf{h}_n \cdot \mathbf{e}_1 \cdot \mathbf{e}_2 \cdot \dots \mathbf{e}_m/\mathbf{i}) > \pi,$$

where π is the conventionally set value for the measure of adequate deciding power with respect to **i**. (It is usually to be desired that π does not deviate greatly from 1.)

In fixing the optimum decision base, or the adequate decision base as the case may be, it is necessary to respect a certain demand that is analogous to the demand of total evidence for inductive and probability logics. The demand may be formulated as follows: In fixing the optimum decision base, or adequate decision base, with respect to a given **i**, account must be taken of all the statement components that are not systematisingly irrelevant with respect to **i**.

5. THE RELEVANCE OF A POSTERIORI DATA

In solving tasks of the systematising type we often find that within a given task or set of tasks there is a mass of different a posteriori data that has been gathered, this having come about because various different empirical and experimental operations have been performed. These a posteriori data need then to be collated, evaluated and assessed. Collation and evaluation can only be based on the same or similar assumptions as those on the basis of which the above-mentioned measures of the semantic informational evaluation of data were constructed. Naturally, the assessment of empirical and experimental operations may also be based on other criteria, such as criteria of technical and economic feasibility in any assessment of the cost of the operations. Such other possibilities will not be considered in the present analysis.

As devices for the semantic informational evaluation of data we shall introduce the terms '*empirical theoretical relevance*' and '*test relevance*'. The former enables us to compare two a posteriori data with respect to a certain **h** and **i**, while the latter enables us to express the relevance of a particular a posteriori datum with respect to a certain nomological statement or class of nomological statements **h**.

The concept of 'empirical theoretical relevance' may be brought in on the basis of the following considerations: Let us assume that we are in a position – with deciding a certain **i** as the goal in mind – to carry out two complexes of empirical or experimental operations the results of which are contained in the statements or classes of statements $\mathbf{e}_j$ and $\mathbf{e}_k$. In addition we have at our disposal a complex of a priori data expressed in the nomological statement or class of nomological statements **h**. If we are to choose between the complexes $\mathbf{e}_j$ and $\mathbf{e}_k$, we shall obviously prefer $\mathbf{e}_j$ to $\mathbf{e}_k$ if

$$\Delta(\mathbf{h}\cdot\mathbf{e}_j/\mathbf{i}) > \Delta(\mathbf{h}\cdot\mathbf{e}_k/\mathbf{i}) \text{ or if}$$
$$\mathbf{U}(\mathbf{i}/\mathbf{h}\cdot\mathbf{e}_j) < \mathbf{U}(\mathbf{i}/\mathbf{h}\cdot\mathbf{e}_k)$$

Because $0 \leqslant \Delta(\mathbf{h}\cdot\mathbf{e}/\mathbf{i}) \leqslant 1$, while $0 \leqslant \mathbf{U}(\mathbf{i}/\mathbf{h}\cdot\mathbf{e}) \leqslant \mathbf{U}(\mathbf{i})$, it is more expedient to have as the basis of the quantitative concept of 'empirical theoretical relevance' the measure of deciding power. The complex of empirical or experimental operations whose results are $\mathbf{e}_j$ becomes positively relevant with respect to $\langle \mathbf{h}, \mathbf{e}_k, \mathbf{i} \rangle$ if

$$\Delta(\mathbf{h}\cdot\mathbf{e}_j/\mathbf{i}) > \Delta(\mathbf{h}\cdot\mathbf{e}_k/\mathbf{i}),$$

or it is negatively relevant if

$$\Delta(\mathbf{h}\cdot\mathbf{e}_j/\mathbf{i}) < \Delta(\mathbf{h}\cdot\mathbf{e}_k/\mathbf{i}),$$

or finally it is equally relevant if

$$\Delta(\mathbf{h}\cdot\mathbf{e}_j/\mathbf{i}) = \Delta(\mathbf{h}\cdot\mathbf{e}_k/\mathbf{i}).$$

The proceeding as described naturally enables us to collate and order only two or more complexes $\mathbf{e}_j, \mathbf{e}_h, \ldots$. Any such comparison is only useful if we need to decide which complex of empirical or experimental operations to prefer in a given task. In a case where the two complexes are equally relevant, i.e. offer the same deciding power, we may begin to look at other criteria, such as costs, time-consumption etc.

The concept of 'empirical theoretical relevance' may also be treated as a quantitative concept. Assuming that we may avail ourselves of **h** and that is our aim to decide **i**, we may, for the measure of empirical theoretical relevance of the complex $\mathbf{e}_j$ with respect to the complex $\mathbf{e}_k$, introduce the sign

$$\Sigma(\mathbf{e}_j, \mathbf{e}_k/\mathbf{h}\cdot\mathbf{i}),$$

and then two possibilities offer themselves for quantification: One possibility corresponds to Carnap's concept of relevance quotient, and the other to his relevance measure [24, p. 356, 360]. The former assumes that quantification is based on the quotient of both measures of deciding power, so that in the case of the positive relevance of $\mathbf{e}_j$ with respect to $\langle \mathbf{h}, \mathbf{e}_k, \mathbf{i} \rangle$ the quotient of empirical theoretical relevance is greater than 1, in the case of its negative relevance it is less than 1, and in the case of equal relevance it is equal to 1. In our definition of the measure of empirical theoretical relevance we shall start out from the second possibility, i.e. we shall be operating with the difference between the two deciding powers. This second way of introducing the measure of empirical theoretical relevance makes it possible that if

$$0 \leqslant \Delta(\mathbf{h}\cdot\mathbf{e}/\mathbf{i}) \leqslant 1$$

and if we use the above mentioned measures of original and conditional uncertainty, then

$$-1 \leqslant \Sigma(\mathbf{e}_j, \mathbf{e}_k/\mathbf{h}\cdot\mathbf{i}) \leqslant 1$$

The measure of empirical theoretical relevance that meets these requirements is then defined as follows:

$$\text{(D 2)} \qquad \Sigma(\mathbf{e}_j, \mathbf{e}_k/\mathbf{h}\cdot\mathbf{i}) = \text{df } \Delta(\mathbf{h}\cdot\mathbf{e}_j/\mathbf{i}) - \Delta(\mathbf{h}\cdot\mathbf{e}_k/\mathbf{i}) = \frac{\mathbf{U}(\mathbf{i}/\mathbf{h}\cdot\mathbf{e}_k) - \mathbf{U}(\mathbf{i}/\mathbf{h}\cdot\mathbf{e}_j)}{\mathbf{U}(\mathbf{i})}$$

The measure of empirical theoretical relevance defined thus enables us to determine how far deciding power with respect to **i** is increased or diminished by replacing complex $\mathbf{e}_k$ by complex $\mathbf{e}_j$ in the decision base $\langle \mathbf{h}, \mathbf{e}_k \rangle$. This definition always shows the relativised measure of empirical theoretical relevance. For example this means that if the decision base $\langle \mathbf{h}, \mathbf{e}_k \rangle$ functions with respect to **i** on the basis of the deductive nomological model, i.e. if it holds that

$$\mathbf{h}\cdot\mathbf{e}_k \underset{\mathrm{L}}{\longrightarrow} \mathbf{i},$$

than any $\mathbf{e}_j$ is negatively relevant or at least equally relevant with respect to $\langle \mathbf{h}, \mathbf{e}_k, \mathbf{i} \rangle$. For these reasons the given definition can be applied first and foremost in comparisons of two or more complexes of empirical or experimental operations within the framework of a given task and a given corpus of a priori data.

In addition to comparing two complexes of empirical or experimental operations within the framework of a given task and a given corpus of a priori data there are yet other situations that can occur during the solution of tasks of the systematising type. And these are still situations that assume the addition of complex $\mathbf{e}_j$ to an existing complex $\mathbf{e}_k$, with the same goal complex and the same corpus of a priori data. We may be interested in how far, in such situations, deciding power is increased or reduced if in respect of **i** we move from base $\langle \mathbf{h}, \mathbf{e}_k \rangle$ to base $\langle \mathbf{h}, \mathbf{e}_j\cdot\mathbf{e}_k \rangle$. This problem is analogous to what Carnap described as the problem of the relevant measure of two observations and their connection [24, p. 365].

From the purely formal point of view it might seem that deciding power will always increase, or at least remain the same, if we switch from decision base $\langle \mathbf{h}, \mathbf{e}_k \rangle$ to $\langle \mathbf{h}, \mathbf{e}_j\cdot\mathbf{e}_k \rangle$, and therefore that

$$\Delta(\mathbf{h}\cdot\mathbf{e}_j\cdot\mathbf{e}_k/\mathbf{i}) \geqslant \Delta(\mathbf{h}\cdot\mathbf{e}_k/\mathbf{i}).$$

However, in actual empirical and experimental procedures in science what may also happen is that newly acquired results cast more or less doubt on earlier findings. The extreme case is the situation where the results of a subsequent complex of empirical or experimental operations $\mathbf{e}_j$ completely refute the previous findings of $\mathbf{e}_k$, because, for example,

$$\mathbf{e}_j \equiv \sim \mathbf{e}_k .$$

In this case the deciding power of base $\langle \mathbf{h}, \mathbf{e}_j \cdot \mathbf{e}_k \rangle$ with respect to $\mathbf{i}$ is reduced to the lowest value, i.e. zero.

In solving tasks of the systematising type there of course exist no universally valid rules to determine the relation between empirical and experimental operations carried out previously or subsequently. For example, one approach might assume that later findings are more trustworthy than earlier findings, particularly if the later ones suspend those made earlier. This approach corresponds to the principle that *actio posterior derogat priori.* In other cases account has to be taken of the reliability and discriminatory power of measuring and experimental apparatus. And finally allowance must be made for the fact that different, or even contradictory, results of empirical and experimental operations in fact relate to different phases in the evolution of the system under investigation – as in the case of a patient's disease where some symptoms present themselves at some phases while being absent at others. For these reasons a posteriori data have to be assessed against consideration of the different relations between them, and only in terms of those relations can one admit aggregation of the data or otherwise. In other words, if we switch from decision base $\langle \mathbf{h}, \mathbf{e}_k \rangle$ to decision base $\langle \mathbf{h}, \mathbf{e}_j \cdot \mathbf{e}_k \rangle$, we have to respect the nature of the actual relations subsisting between $\mathbf{e}_j$ and $\mathbf{e}_k$.

The measure of empirical theoretical relevance $\Sigma(\mathbf{e}_j, \mathbf{e}_k/\mathbf{h} \cdot \mathbf{i})$ characterises the relations between two complexes of a posteriori data with respect to a certain $\mathbf{h}$ and $\mathbf{i}$, i.e. a certain complex of a priori data and the goal complex. However in many tasks this kind of evaluation of a posteriori data is found wanting because of the need to clarify just which data ought to be selected with respect to a given complex of a priori data. It is then a question of the empirical or experimental operations that are relevant with respect to the corpus of scientific laws and hypotheses with which we shall be operating in a given task. To assess how right a choice of empirical or experimental operations has been with respect to a given $\mathbf{h}$ we may resort to another concept which may be characterised as the *test relevance* of a given **e** with respect to a given **h**. This then means assessing how far certain a posteriori data are adequate with respect to a given complex of a priori data.

The measure of test relevance, which we shall denote

$$\mathscr{T}(\mathbf{e}/\mathbf{h}),$$

may be treated as a special case of the measure of transmitted information, i.e. a measure of how far a given complex of empirical or experimental operations whose outcome is **e** is relevant with respect to that to which the complex of a priori data **h** relates.

The measure of test relevance $\mathscr{T}(\mathbf{e}/\mathbf{h})$ thus corresponds to the concept introduced by R. Hilpinen [59] for the information supplied by observation with respect to a given hypothesis. Thus to establish the properties of $\mathscr{T}(\mathbf{e}/\mathbf{h})$ we select the same postulates as for the concept of transmitted information. This means that

(R1) if $\mathbf{e}_j \underset{L}{\equiv} \mathbf{e}_k$ and $\mathbf{h}_i \underset{L}{\equiv} \mathbf{h}_g$, then

$\mathscr{T}(\mathbf{e}_j/\mathbf{h}_i) = \mathscr{T}(\mathbf{e}_k/\mathbf{h}_g)$ and

$\mathscr{T}(\mathbf{e}_k/\mathbf{h}_i) = \mathscr{T}(\mathbf{e}_j/\mathbf{h}_g)$

(R2) $0 \leqslant \mathscr{T}(\mathbf{e}/\mathbf{h}) \leqslant 1$

(R3) if $\mathbf{e} \underset{L}{\longrightarrow} \mathbf{h}$, then $\mathscr{T}(\mathbf{e}/\mathbf{h}) = \mathbf{max}\,\mathscr{T} = 1$

(R4) if $\mathbf{e}_j \vee \mathbf{e}_k \vee \mathbf{h}$ is logically true, then

$\mathscr{T}(\mathbf{e}_j \cdot \mathbf{e}_k/\mathbf{h}) = \mathscr{T}(\mathbf{e}_j/\mathbf{h}) + \mathscr{T}(\mathbf{e}_k/\mathbf{h})$

(R5) $\mathscr{T}(\mathbf{e}_j \vee \mathbf{e}_k/\mathbf{h}) = \mathscr{T}(\mathbf{e}_j/\mathbf{h})\,\mathscr{T}(\mathbf{e}_k/\mathbf{h} \vee \mathbf{e}_j)$, or

$\mathscr{T}(\mathbf{e}_j \vee \mathbf{e}_k/\mathbf{h}) = \mathscr{T}(\mathbf{e}_k/\mathbf{h})\,\mathscr{T}(\mathbf{e}_j/\mathbf{h} \vee \mathbf{e}_k)$ respectively.

The interpretation of these postulates also matches the interpretation mentioned of the concept of transmitted information.

Let us further note that by (R2) the minimum value of $\mathscr{T}(\mathbf{e}/\mathbf{h})$ is 0. This minimum value is obtained when either

(a) **e** is a tautological statement (and cannot therefore furnish any information about anything), or

(b) the a posteriori data **e** are so 'weak' that negation of them logically implies **h**.

Thus it holds that:

if **e** is a tautological statement, or if

$$\sim \mathbf{e} \underset{L}{\longrightarrow} \mathbf{h}, \text{ then } \mathscr{T}(\mathbf{e}/\mathbf{h}) = \mathbf{min}\,\mathscr{T} = 0.$$

These required properties of the measure of test relevance fulfil the definition that is analogous to the definition of the measure of transmitted information:

$$\text{(D 3)} \qquad (\mathbf{e}/\mathbf{h}) = \mathrm{df}\ \frac{\mathbf{cont}(\mathbf{e} \vee \mathbf{h})}{\mathbf{cont}(\mathbf{h})}$$

It is clear that the definiens operates with the content measure. If it is possible to determine this content measure by the probability measure **p**, it holds that

$$\mathscr{T}(\mathbf{e}/\mathbf{h}) = \frac{1 - \mathbf{p}(\mathbf{e} \vee \mathbf{h})}{1 - \mathbf{p}(\mathbf{h})} = \mathbf{p}(\sim \mathbf{e}/\sim \mathbf{h}).$$

The measure of test relevance will be seen to disregard the goal complex **i**, being concerned only with the aggregate of a priori and a posteriori data. Hence this measure is applied especially in assessing the suitability of the choice of this or that complex of empirical or experimental operations that could well be applied in the light of the given complex of a priori data. It is then a kind of preliminary evaluation that does not yet take into account the particular goal complex in which the two kinds of data are to be used. The measure of test relevance is therefore an aid to selecting those empirically or experimentally discriminable facts that may serve as evidence in confirming or testing given a priori data.

6. EVALUATION OF THE GOAL COMPLEX AND THE CONCEPT OF 'EPISTEMIC GAIN'

All the foregoing discussion of means for the semantic informational evaluation of data in tasks of the systematising type has been mainly to do with the decision base and its components, i.e. a priori and a posteriori data. It is however also necessary to evaluate what we have described as the goal complex, i.e. data which govern how a given task will be tackled. This includes the evaluation of any practical decision taken on the given decision base, assessment of any prognostic statement made on the basis of a particular analysis of the a priori and a posteriori data, appraisal of any diagnostic decision taken on the basis of available medical hypotheses and the empirically or experimentally ascertained symptoms, and others. It is further necessary to describe in general the principle on which our evaluation of the goal complex **i** is to be based: in evaluating the goal complex, for instance a particular scientific prediction, chosen decision etc., we cannot take into consideration data which are not available at the time. We may only consider what we have at our disposal *hic et nunc*, i.e. the decision base $\langle \mathbf{h}, \mathbf{e} \rangle$. This means that the evaluation of a particular $\mathbf{i}$ is invariably relativised to a particular decision base. Therefore any evaluation may change if we move from, say, base $\langle \mathbf{h}, \mathbf{j} \rangle$ to the informationally richer base $\langle \mathbf{h}', \mathbf{j}' \rangle$. The principle that says that any immediate evaluation of a certain selected decision may only be made on the basis of data available hic et nunc does not then mean that the evaluation cannot be changed in the future if new data are acquired. It remains true of course that any future evaluation will only be possible on the basis of data available at that future date.

The measure of the semantic informational evaluation of the goal complex $\mathbf{i}$ on the basis of $\langle \mathbf{h}, \mathbf{e} \rangle$, which we denote

$$\Pi(\mathbf{i}/\mathbf{h}\cdot\mathbf{e})$$

may be treated as the measure that expresses conditional *credibility* of $\mathbf{i}$ on the basis of $\langle \mathbf{h}, \mathbf{e} \rangle$, or something that may be described as the measure of the *nomologico-empirical support* of $\mathbf{i}$ on the basis of the decision base $\langle \mathbf{h}, \mathbf{e} \rangle$.

A satisfactory measure of the credibility of the goal complex $\mathbf{i}$ on the basis of $\langle \mathbf{h}, \mathbf{e} \rangle$ ought to meet the following demands:

(R1) if $\mathbf{h}_i \underset{L}{\equiv} \mathbf{h}_g$ and $\mathbf{e}_j \underset{L}{\equiv} \mathbf{e}_k$, then
$\Pi(\mathbf{i}/\mathbf{h}_i\cdot\mathbf{e}_j) = \Pi(\mathbf{i}/\mathbf{h}_g\cdot\mathbf{e}_k)$, or
$\Pi(\mathbf{i}/\mathbf{h}_i\cdot\mathbf{e}_k) = \Pi(\mathbf{i}/\mathbf{h}_g\cdot\mathbf{e}_j)$ respectively.

(R2) if $\mathbf{h}\cdot\mathbf{e} \underset{L}{\longrightarrow} \mathbf{i}$, then $\Pi(\mathbf{i}/\mathbf{h}\cdot\mathbf{e}) = \mathbf{max}\,\Pi$

(R3) if $\mathbf{h}\cdot\mathbf{e} \underset{L}{\longrightarrow} \sim\mathbf{i}$, then $\Pi(\mathbf{i}/\mathbf{h}\cdot\mathbf{e}) = \mathbf{min}\,\Pi$

(R4) $0 \leqslant \Pi(\mathbf{i}/\mathbf{h}\cdot\mathbf{e}) \leqslant 1$

It is obvious from these demands that they answer to a concept that is capable of expressing the measure of certainty of $\mathbf{i}$ on the basis of $\langle \mathbf{h}\cdot\mathbf{e} \rangle$ and that is therefore the complement of $\mathbf{cont}\,(\mathbf{i}/\mathbf{h}\cdot\mathbf{e})$. In accordance with the method of quantification given earlier and with respect to the above postulates it becomes possible to produce the following definition of the measure of credibility, or nomologico-empirical support, of $\mathbf{i}$ on the basis of a given $\langle \mathbf{h}, \mathbf{e} \rangle$:

(D4) $\Pi(\mathbf{i}/\mathbf{h}\cdot\mathbf{e}) = \mathrm{df}\ 1 - \mathbf{cont}\,(\mathbf{i}/\mathbf{h}\cdot\mathbf{e})$

If we use the probability measure $\mathbf{p}$ to determine the content measure, we obtain

$$\Pi(\mathbf{i}/\mathbf{h}\cdot\mathbf{e}) = \mathbf{p}(\mathbf{i}/\mathbf{h}\cdot\mathbf{e})$$

The measure of credibility may then – still in accordance with the requirements, acquire the extreme values under the following situations. If the goal complex **i** is supported by the given decision base ⟨**h**, **e**⟩ so strongly that

$$\mathbf{h}\cdot\mathbf{e} \underset{L}{\longrightarrow} \mathbf{i},$$

we reach maximum credibility, equal to 1. If the goal complex **i** is supported by the given decision base so weakly that its negation follows logically from **h**·**e**, i.e. if

$$\mathbf{h}\cdot\mathbf{e} \underset{L}{\longrightarrow} \sim\mathbf{i},$$

we reach minimum credibility, equal to 0.

For the semantic informational evaluation of a certain goal complex on the basis of a given decision base it is possible to construct certain further devices. Assuming that a decision on a certain goal complex **i** using decision base ⟨**h**, **e**⟩ is being worked out in a situation where all the data are fully and unquestionably reliable, there is no need to calculate with a risk element, nor therefore with the possibility of any kind of losses. If that is not the case, i.e. if the decision is being worked out with a risk element present, it is expedient to consider what we might gain, or lose, if on the basis of ⟨**h**, **e**⟩ we decide for **i** and not for ∼**i**. It should be noted that by 'gain' and 'lose' we do not mean gain and loss in economic terms, but gain and loss in a cognitive sense. For these reasons we shall speak of *epistemic gain* and *epistemic loss*, as the case may be.

We shall first outline, as in the foregoing cases of the measures of semantic informational evaluation, the intuitive thoughts on which the construction of the measure of epistemic gain, or loss, is based. The epistemic gain that accrues from our opting for **i** on the basis of ⟨**h**, **e**⟩ grows in proportion to the growth in credibility of the goal complex **i** on the basis of ⟨**h**, **e**⟩, and in proportion to the deciding power of ⟨**h**, **e**⟩ with respect to **i**. There is the simultaneous consideration that says that epistemic gain decreases in proportion to the increase in the credibility of ∼**i**, on the basis of ⟨**h**, **e**⟩, and also in proportion to the increasing deciding power of that base with respect to ∼**i**.

On the basis of this idea it becomes possible to use the previously introduced concepts and bring in a quantitative measure of epistemic gain (or loss), which we designate

$$\Gamma(\mathbf{i}/\mathbf{h}\cdot\mathbf{e})$$

by means of the following definition:

$$\text{(D5)} \qquad (\mathbf{i}/\mathbf{h}\cdot\mathbf{e}) = \text{df}\, \Pi(\mathbf{i}/\mathbf{h}\cdot\mathbf{e})\Delta(\mathbf{h}\cdot\mathbf{e}/\mathbf{i}) - \Pi(\sim \mathbf{i}/\mathbf{h}\cdot\mathbf{e})\Delta(\mathbf{h}\cdot\mathbf{e}/\sim \mathbf{i})$$

Taking into consideration the previously given quantification, and assuming that

$$\mathbf{cont}\,(\mathbf{i}) = 1 - \mathbf{p}(\mathbf{i})$$

$$\mathbf{cont}\,(\mathbf{i}/\mathbf{h}\cdot\mathbf{e}) = 1 - \mathbf{p}(\mathbf{i}/\mathbf{h}\cdot\mathbf{e})$$

we obtain

$$\Gamma(\mathbf{i}/\mathbf{h}\cdot\mathbf{e}) = \mathbf{p}(\mathbf{i}/\mathbf{h}\cdot\mathbf{e})\mathbf{p}(\sim(\mathbf{h}\cdot\mathbf{e})/\sim \mathbf{i}) - \mathbf{p}(\sim \mathbf{i}/\mathbf{h}\cdot\mathbf{e})\mathbf{p}(\sim(\mathbf{h}\cdot\mathbf{e})/\mathbf{i})$$

Thus it holds that

$$-1 \leqslant \Gamma(\mathbf{i}/\mathbf{h}\cdot\mathbf{e}) \leqslant 1,$$

i.e. the measure in question may even acquire a negative value, which in any case answers well to the intuitive conception of gain and loss.

We shall now outline the extreme situations, i.e. those of maximum gain and maximum loss, when

$$\Gamma(\mathbf{i}/\mathbf{h}\cdot\mathbf{e}) = 1 \qquad \text{and}$$

$$\Gamma(\mathbf{i}/\mathbf{h}\cdot\mathbf{e}) = \sim 1$$

In the case that $\mathbf{h}\cdot\mathbf{e} \underset{L}{\longrightarrow} \mathbf{i}$ applies and we have opted for **i** and not for $\sim \mathbf{i}$, we obtain

$$\Pi(\mathbf{i}/\mathbf{h}\cdot\mathbf{e}) = 1$$

$$\Delta(\mathbf{h}\cdot\mathbf{e}/\mathbf{i}) = 1$$

$$\Pi(\sim \mathbf{i}/\mathbf{h}\cdot\mathbf{e}) = 0$$

Even if it holds that

$$\Delta(\mathbf{h}\cdot\mathbf{e}/\sim \mathbf{i}) \geqslant 0,$$

we obtain

$$\Gamma(\mathbf{i}/\mathbf{h}\cdot\mathbf{e}) = \mathbf{max}\,\Gamma = 1$$

In the case that $\mathbf{h}\cdot\mathbf{e} \underset{L}{\longrightarrow} \mathbf{i}$ applies, but we have opted for $\sim \mathbf{i}$ and not for **i**, we obtain, analogously,

$$\Gamma(\sim \mathbf{i}/\mathbf{h}\cdot\mathbf{e}) = \mathbf{min}\,\Gamma = -1$$

It is obvious that we will only obtain extreme values of epistemic gain in cases of the applicability of the deductive nomological model, i.e. in cases of determinist decision-making as it is usually described in decision theory. In all other cases

$$\Gamma(\mathbf{i}/\mathbf{h}\cdot\mathbf{e})$$

acquires values between 1 and -1.

This method of constructing the measure of epistemic gain or loss is based on the assumption that we decide for **i**, or $\sim \mathbf{i}$, given our decision base $\langle \mathbf{h}, \mathbf{e} \rangle$. Naturally this represents only the most elementary form of alternative decision-making. Allowances must be made for the possibility that the decision process may have to take into account a whole spectrum of possibilities, i.e. $\{\mathbf{i}_1, \mathbf{i}_2, \ldots \mathbf{i}_k\}$. But even with a broader spectrum of possibilities there may sometimes be a point in confronting an element with its complement in the same class, i.e. $\mathbf{i}_1$ with $\sim \mathbf{i}_1$, $\mathbf{i}_2$ with $\sim \mathbf{i}_2$ etc.

A few additional remarks need to be made on all the above mentioned measures of the semantic informational evaluation of data used in tasks of the systematising type. These measures represent certain fundamental ways of arriving at a semantic informational evaluation, but they may be further refined and elaborated. This applies in particular to the use of principles drawn from statistical decision theory and to the refinement of the devices employed in such evaluation. We have also outlined the basic assumptions, many of which are not easily realised, on which the construction of the basic measures rests. This applies in particular to assumptions on the quantification of those measures, most especially to our knowledge of the essential probability characteristics of the elements with which we operate. What is of particular importance is the fact that these characteristics are assigned to statements of different kinds that represent different kinds of data. This naturally calls for a refinement of the semantic starting point that would make it possible to make this assignment of probability characteristics in a uniform manner. (We touched briefly on the probability characteristics of scientific laws and hypotheses, i.e. nomological statements, in this connection earlier.) Clearly therefore, the approach that is being put forward here is by no means an easy one and it is fraught with a number of quite tricky problems.

In spite of the above constraints, and probably other ones as well, we feel justified in defending it in the face of certain possible reservations, which may be summarised in the following points:

(1) The approach is based on very strict relativisation, i.e. the consistent evaluation of all elements with respect only to other elements. This is indeed the case, but in our view there is no genuine alternative. This means that individual subsidiary data, which in our conception take the form of statements of different kinds, cannot be evaluated 'sub specie aeternitatis', but only, and invariably, with respect to the task in which the data are applied, to the degree in which they contribute to the desired quality of the solution, or to the reduction of original uncertainty, and so forth. In other words, in order to be able to assess a shift in the measure of uncertainty, a shift of levels of entropy, we must always characterise with accuracy the level of knowledge that is our starting point.

(2) The approach is based on respecting finitist principles. For any calculation we can only take into consideration a finite corpus of a priori and a posteriori data, but also the finite capacity of the means available to us – our computer technology and so forth. By contrast we know that the path of human cognition, the evolution of science, is infinite and inexhaustible. However, in order for us to be able to assess and evaluate each subsidiary and usually infinitesimal step on the way, we can only take into account subsidiary parts of the overall path, and these are always finite and limited.

(3) The construction of all the given measures of semantic informational evaluation is relativised to the state of knowledge available 'hic et nunc'. It goes without saying that it is desirable that this state of knowledge should be the highest level of scientific discovery so far achieved. For example, for a fairly complex task in diagnosis in contemporary medicine it is quite typical for the broadest range of knowledge, including the most recent findings, to be brought to bear, notably by specialists being brought in for consultations. Similarly broad expertise is also sought before a book is published or a higher qualification conferred. It is therefore perfectly desirable that the data with respect to which a relativisation is made should be associated with the highest level of knowledge so far reached. But this fact too is not a shortcoming of the approach suggested, but the only possibility open. What is of course important is that what is available 'hic et nunc' should not be obsolete. What we must have are the newest findings, for, as Karl Marx put it so pregnantly, the anatomy of man, as the ultimate in evolution, is the key to the anatomy of the monkey, and not vice versa.

X. THE SEMANTICS OF PREFERENCE ATTITUDES

1. THE ROLE OF PREFERENCE AND PREFERENCE ORDERING

Questions concerning the formal status of preference relations have been raised in a number of areas: for example, it is a known fact that applications of the means and models of statistical decision theory or game theory are based on the supposition that there exist possible subjects of a given decision process or game who are able to evaluate the alternatives from the point of view of their preference ordering. A preference schema also forms the basis for any rational decision (in the Bayesian sense). This is also the reason why, in books and papers explaining mathematical models of decision, we find various systems of axioms or postulates which must be satisfied by a rationally constructed preference relationship.[1] The construction of preference schemas is important in other domains: A good example can be found in modern prognostics. Prognostics is today required to do much more than just point out the probability distribution of possible future states; it is expected to point out the immediate, and especially the future desirability of those states. This means linking the possible or expected future states to some system of values, criteria and preferences. Sometimes it is even required that prognostics should transfer its main emphasis to precisely this sphere of values, decision criteria and preferences. Another domain in which the problems of preference and preference schemas are being taken increasingly into consideration is that of the processing, storage and retrieval of data for various purposes, i.e. information science in the narrow sense of the word.

We assume that no unique preference schema exists or can exist. In books and papers on statistical decision theory we usually find a conception of preference that describes it as a binary relation between objects of any nature which can be *compared* on the basis of a certain evaluation. It is of course necessary to distinguish preference (which may be considered as a justification for a decision) from an actual decision or choice of one or other alternative. Therefore it is sometimes stressed that preference indicates only a possible choice.[2] This assertion cannot be interpreted to mean that any decision or choice is a preference which has been materialised, since the decision might have been accidental, irrational or motivated by some other preference (unknown to the observer) and so forth.

Just as it is not possible to consider the relation between a preference or order of preferences on the one hand and a decision or any other choice on the other as a kind of equivalence relation, it is not even always possible so to consider the relation between a preference or order of preferences and some form of evaluation that might afford a justification for the given preference. Nevertheless this kind of approach is very common, even though it can lead, as we shall show, to certain difficulties. What lies at the core of the problem is the supposition that the preference and the evaluation which may give it some justification can be substituted for one another – salva veritate. In the simplest case this may take the form:

(1) $xPy[\equiv]\mathbf{w}(x) > \mathbf{w}(y)$,

where P is the preference relation, x and y are preferable entities, and the function $\mathbf{w}$ is a function that assigns certain numerical evaluations to these entities; $[\equiv]$ has the properties of an equivalence relation.

This approach is based on a simple application of the function $\mathbf{w}$ and disregards the probability measure assignable to the entities x and y. This means, for instance, that the hope that x or y may be realised is disregarded, as also is the chance that the two entities might occur, and any other kinds of probability evaluation. And yet it is well known that this is just what was required by cartesian logic (the Logique de Port Royal), which was conceived, incidentally as 'l'art de penser'. If $\mathbf{p}$ is a probability measure which must be assumed to satisfy the ordinary (Kolmogorov) axioms of probability theory, then (1) may be rewritten as

(2) $xPy[\equiv]\mathbf{w}(x)\mathbf{p}(x) > \mathbf{w}(y)\mathbf{p}(y)$

In the case that the entities x and y exhaust all possible alternatives and are thus complementary, and if $\mathbf{p}(x)$ is given the value r, then (2) has the form

(3) $xPy[\equiv]\mathbf{w}(x)r > \mathbf{w}(y)(1-r)$

It is therefore expedient to distinguish three different spheres among which relations of an equivalency type are usually interpolated:

(a) The sphere of actual decisions or choices among different alternatives. It is expedient to assume that this sphere is observable, or that choice is like behaviour which can be objectively recorded or reliably reconstructed on the basis of the ascertainable consequences of the alternatives chosen.

(b) The sphere of preferences; this sphere can be reconstructed as a system of relations between various alternative entities. As already shown, this system indicates a possible decision or choice where it is desirable that the decision or choice be not justified by preference.

(c) The sphere of evaluations, whether conceived as simple or probability weighted evaluations, or evaluations relativised to certain aims, criteria or postulates. It is deemed desirable that evaluation be conceived numerically or at least in a form which enables the entities that are the object of an evaluation to be fully ordered.

The interpolation of equivalency relations between these spheres, which are essentially different in principle, is connected with attempts to exploit certain assumed interdependencies, the simplest variants of which are (1), (2) and (3) above, or expressions containing similar interdependencies, to reconstruct a justification for the individual steps in the spheres just described. The commonest way in which a justification is reconstructed is (c) ⇒ (b) ⇒ (a). It may of course be objected that this kind of reconstruction is turned upside down, and that other forms of reconstruction, e.g. (a) ⇒ (b) ⇒ (c), are no less justified. We shall next endeavour to show that this kind of reconstruction, and especially the interpolation of equivalency relations, which makes possible a mapping of an ordering in one sphere by means of a corresponding mapping in another sphere, has certain limitations. In some instances the limitations show themselves in the fact that the relevant matrices can be interpreted in one direction only, and not in both directions as would be desirable under the assumption of relations of equivalency and symmetrical dependency.[3]

2. THE COMPARABILITY PRINCIPLE AS A PRESUPPOSITION FOR THE CONSTRUCTION OF A PREFERENCE SYSTEM

Sentences (1), (2) and (3), or any other like sentences, are an oblique expression of the requirement which is usually described in the literature as the comparability principle.[4] If we start from sentence (1), the comparability principle means that one of the following three possibilities may be realised:

$$\mathbf{w}(x) > \mathbf{w}(y),$$
$$\mathbf{w}(x) < \mathbf{w}(y),$$
$$\mathbf{w}(x) = \mathbf{w}(y).$$

It should be stressed that these three possibilities express the comparability principle *indirectly*, i.e. express the principle in the sphere of (b) by means of (c). Moreover, it is assumed here that the evaluation is understood numerically, i.e. that it can be mapped in real numbers. But this carries with it certain important difficulties, connected especially with the problems of discrimination limits, the restricted applicability of the density principle and others.

For these reasons it is more useful to express the comparability principle in sphere (b) only. For the present we shall put aside the problem of the nature of the entities which may figure as arguments in the relation xPy. (In the following discussion we shall show that the logical conception of the status of the preference relation is best satisfied by treating these entities as denotates of statements or entities analogous to these.[5]) Intuitively speaking, we may treat as comparable those entities from the class of all the entities considered $\mathscr{X}$ which have some property in common. Moreover it is desirable that the property should be assignable in varying degrees which we are capable of distinguishing. If the predicate C is one such property, there is a comparable subclass $(\hat{x})Cx$ of the class $\mathscr{X}$. At the same time C may be understood to mean, for instance, that all the elements of $(\hat{x})Cx$ are associated with certain costs or risks, afford certain advantages, are the product of some activity that is connected with certain expected results and so forth.

If C is a property which can be assigned to all the entities of the class $(\hat{x})Cx$ in some degree, it is possible to determine whether any two of the entities have that property assigned to them in a greater or lesser degree or in the same degree. Therefore if the preference relation is relativised to property C, we may construct one very elementary from of the preference system in the class $(\hat{x})Cx \subseteq \mathscr{X}$ by means of the pair $\langle P_C, I_C\rangle$, where P_C is the preference relation with respect to C and I_C is the relation of indifference with respect to C.

The properties of P_C and I_C are expressed in the following six postulates:

(R1) $(\forall x)(\forall y)(\forall z)(xP_Cy \cdot yP_Cz \longrightarrow xP_Cz)$

(R2) $(\forall x)(\forall y)\{xI_Cy \longrightarrow [\sim(xP_Cy) \vee \sim(yP_Cx)]\}$

(R3) $(\forall x)(\forall y)(xP_C y \vee xI_C y \vee yP_C x)$

(R4) $(\forall x)(xI_C x)$

(R5) $(\forall x)(\forall y)(xI_C y \longrightarrow yI_C x)$

(R6) $(\forall x)(\forall y)(\forall z)(xI_C y \cdot yI_C z \longrightarrow xI_C z)$

Postulate (R3) is the actual expression of the comparability principle. It is obvious that P_C is transitive (R1), I-reflexive (R2), I_C is reflexive (R4), symmetrical (R5) and transitive (R6). There are now many theorems which can be formulated on the basis of these postulates, e.g.

$$(\forall x)(\forall y)(\forall z)(xP_C y \cdot yI_C z \longrightarrow xP_C z),$$
$$(\forall x)(\forall y)(xP_C y \longrightarrow \sim yP_C x) \qquad \text{etc.}$$

The pair $\langle P_C, I_C \rangle$ therefore makes a quasi-series in the class $(\hat{x})Cx \subseteq \mathscr{X}$, giving the preference relation all the properties of a comparative predicate.[6]

The procedure by which we have justified the comparability principle has been based on the assumption that a preference system can be reconstructed in a class $\mathscr{X}$ or in a part of this class provided that a common property C (distinguishable in its degree) can be set up in such a way that

$$(\hat{x})Cx \subseteq \mathscr{X} \qquad \text{and}$$
$$(\hat{x})Cx \neq 0.$$

But the comparability principle may have another significance: If the common property C is not known, yet it is still possible to compare the entities of a class, comparability may suggest that there is one or more properties common to all the entities of the class. This will be recognised as the idea that lies behind Marx's well known procedure leading to the introduction of the concept of 'value'. It is also possible to introduce a two-place relation K whereby

$$xKy \equiv xPy \vee yPx \vee xIy$$

It is easily proved that this relation is symmetrical, reflexive and transitive.

Theoretically it would be possible to assume yet another interpretation of the comparability principle. If, in a given class of entities $\mathscr{X}$ which have a common property, we are capable of preferring entity x to entity y, are we justified in always expecting to be able to find a z which can be interpolated between x and y in the given preference system? In other words, can we state

$$(\forall x)(\forall y)(\exists z)(xPy \cdot xPz \cdot zPy)\,?$$

It seems that this need not always be possible. First of all it should be stressed that the class of entities might be finite, making it impossible to find a z which could be preferentially interpolated between x and y. Furthermore any preferential distinction is linked to the question of extremes of distinguishability. This means that if xPy applies, the difference of preference could be so slight that no entity could be preferentially interpolated between x and y with the certainty of its being preferentially distinguished from x and y. The only thing which can be guaranteed if some z does exist such that

$$z \in (\hat{x})Cx \qquad \text{is}$$

$$\text{(4)} \qquad xP_Cy \longrightarrow (xP_Cz \vee zP_Cy).$$

This expression, which could be called the *principle of comparability in the narrow sense*, obviously does not exclude the possibility of

$$xI_Cz \vee yI_Cz$$

In this connection we are faced with the possibility of modifying our system of postulates (R 1) – (R 6) in such a way as to incorporate the narrow sense of the comparability principle. Since the comparability principle in the narrow sense only applies if xP_Cy and if there exists some z as an element of the class $(\hat{x})Cx$, we cannot write

$$\text{(5)} \qquad (\forall x)(\forall y)(\exists z)[xP_Cy \longrightarrow (xP_Cz \vee zP_Cy)]$$

Hence (R3) cannot be replaced by expression (5) and modification along these lines will not work.

3. PREFERENCES OF THINGS AND PREFERENCES OF STATES OF AFFAIRS

So far we have been discussing the concept of preference as it appears from the logical point of view, as a two-place comparative predicate whose arguments are unspecified entities that are the individuals in a certain non-empty universe. This conception, which can be described as the preference of things, would appear to be limited in its use. Moreover, this conception of preference does not match the intuitive interpretation of preference, as the following examples will show. If we ask whether peaceful coexistence or armed conflict is the better, the preference relates to certain *states of affairs*. Moreover, preference relations can be constructed between one state of affairs and its negation, as, for instance, when we say that at a particular time it is better for it to rain than not to rain. Most of the preference relations with which we operate on a day-to-day basis seem to be more preferences of states of affairs than preferences of things as such. If we say that we prefer a semi-detached to a tower block, what we mean is that we prefer living in a semi-detached to living in a tower block.

The conception of the preference relation as preference of a state of affairs obviously changes the logical status of the concept of preference. What are taken as the arguments of preference are entities, however specified, which can be described as states of affairs or, as some authors, such as G.H. von Wright, call them, proposition-like entities [168]. This conception of the preference relation has been espoused by most authors who deal with the logic of preference. Rescher, for example, speaks of propositional preference ordering [126, p. 292]. The expression pPq then means that the state of affairs marked p is preferred to the state of affairs q. The meaning of the sign P is obviously different here from its meaning when used as a comparative predicate. The expression pPq does not express a preference between the statement of p and q, i.e. a preference between the linguistic forms of p and q, but between possible states of affairs or alternative facts as denoted by the statements. Since each possible state can be assumed either to be going to happen or not to be going to happen, we have to assume both positive and negative instances of them; for the negative instances we shall use the ordinary sign for negation $\sim$. Then the expression

$$pP \sim p$$

means that the state of affairs denoted by p is preferred to the negation of that state of affairs, denoted $\sim p$.

If the arguments of the preference relation P are to be states of affairs or however else described proposition-like entities, we also have to admit as arguments the designates of molecular expressions, such as

$$p \cdot rPq$$

$$p \cdot \sim qP \sim p \cdot q$$

$$pPq \vee r \qquad \text{etc.}$$

This fact, that as arguments of the preference relation we may take not only the designates of atomic expressions, but also those of their negations and of molecular expressions, places new demands on the concept of preference. The difficulties arising here concern above all the possibilities for comparing that which is permitted as valid in sphere (b), i.e. the sphere of preferences proper, and what is admissible as valid in sphere (c), i.e. the sphere of evaluations. If, for example, it holds that

$$pPq$$

then this obviously implies

$$\sim (qPp)$$

If $\mathbf{w}(p) = \mathbf{A}$ and $\mathbf{w}(q) = \mathbf{B}$, then

$$pPq [\equiv] \mathbf{A} > \mathbf{B}$$
$$pIq [\equiv] \mathbf{A} = \mathbf{B}$$

In most systems of preference logic[7] the following are also recognised as valid:

(6) $pPq \longrightarrow \sim qP \sim p$ or

(7) $pPq \longrightarrow (p \cdot \sim q)P(\sim p \cdot q)$ or

(8) $\sim qP \sim p \longrightarrow (p \cdot \sim q)P(\sim p \cdot q)$

This is the case in the Halldén system of preference logic [52], where preference system **A** is set out by the following axioms:

(A1) $pPq \longrightarrow \sim (qPp)$

(A2) $pPq \cdot qPr \longrightarrow pPr$

(A3) pIp

(A4) $pIq \longrightarrow qIp$

(A5) $pIq \cdot qIr \longrightarrow pIr$

(A6) $pPq \longrightarrow (p \cdot \sim q)P(\sim p \cdot q)$

(A7) $pIq \longrightarrow (p \cdot \sim q)I(\sim p \cdot q)$

It is obvious that (6), (7) and (8) are axioms or provable theorems in this preference system.

If we now write the application schema for the evaluation function $\mathbf{w}(s_i)$, where s_i is always the relevant expression, we get the table:

	$\mathbf{w}(s_i)$
p	**A**
q	**B**
$\sim p$	**C**
$\sim q$	**D**

and it should always hold that

$$\text{if } \mathbf{A} > \mathbf{B}, \text{ then } \mathbf{D} > \mathbf{C}.$$

But it is perfectly easy to imagine, for example, the following evaluation:

	$\mathbf{w}(s_i)$
p	10
q	3
$\sim p$	−2
$\sim q$	−4

which obviously does not meet the above requirement. This also means that there is a violation here of the supposed equivalence relation which we interpolate between spheres (b) and (c) in order for a decision in one sphere to be a guide for a decision in the other and vice versa.

If we now consider as arguments of a preference relation the designates of molecular expressions, the relations between spheres (b) and (c) become even more complicated. If we write the respective evaluations in the following tables:

(I)

	$\mathbf{w}(s_i)$
p	**A**
q	**B**
$\sim p$	**C**
$\sim q$	**D**

(II)

	$\mathbf{w}(s_i)$
$p \cdot q$	**a**
$p \cdot \sim q$	**b**
$\sim p \cdot q$	**c**
$\sim p \cdot \sim q$	**d**

then the acceptance of expressions (7) and (8) above as axioms or theorems of a preference system would require that

$$\text{if } \mathbf{A} > \mathbf{B}, \text{ then } \mathbf{b} > \mathbf{c} \quad \text{and}$$
$$\text{if } \mathbf{D} > \mathbf{B}, \text{ then } \mathbf{b} > \mathbf{c}.$$

It is obvious once more that there are no a priori guarantees of this.

Another problem which arises here resides in the question of whether the values of $\mathbf{w}(s_i)$ in table (I) are determined independently of or in dependence on the values of $\mathbf{w}(s_i)$ in table (II) and vice versa. Accepting that the two tables are dependency related, which of them do we have to treat as primary?

One solution to these questions has been put forward by N. Rescher [125] [126]. It can be shown, however, that this solution leads to the values of table (I) being calculable on the basis of the values in table (II), i.e. that table (II) has to be taken as primary. If the reverse is the case, the problem is insoluble, and this is also what sets limits to the acceptability of expressions (6), (7) and (8) or any others like them. (The simplest analogues of these expressions are expressions which operate with the relation of indifference, e.g.

$$pIg \longrightarrow (p\cdot \sim q)I(\sim p\cdot q) \text{ etc.)}$$

Rescher's conception also shows that one concept of preference is not enough, and he proceeds to put forward two different concepts.

The motivation behind Rescher's conception is semantic. If p, q, r ... etc. are atomic statements, there have to be certain possible states of affairs corresponding to them, and these can be understood as possible worlds in the Leibnizian sense or as the designates of admissible state descriptions in Carnap's sense. If we take the triple of statements p, q and r, considering only them and their negations, there are four possible states of affairs corresponding to p which can be expressed as a disjunction:

$$p\cdot q\cdot r \vee p\cdot q\cdot \sim r \vee p\cdot \sim q\cdot r \vee \sim p\cdot \sim q\cdot \sim r$$

On the basis of this semantic motivation Rescher then constructs his two different concepts of preference, the first denoted $P^{\#}$, the second denoted P^{*}. (It can be shown that in the case of semantically independent statements the two concepts coincide.) The preference denoted $P^{\#}$ is based on the average of the evaluations of the states involved in the statements. The expression of preference $pP^{\#}q$ is true if the average evaluation of the states involved in p is greater than the average evaluation of the states involved in q. For simplicity's sake let us assume just four states and their assigned evaluations:

possible states:		their evaluations:	
	$p\cdot q$		**a**
	$p\cdot \sim q$		**b**
	$\sim p\cdot q$		**c**
	$\sim p\cdot \sim q$		**d**

The evaluation of p, i.e. $\mathbf{w}^{\#}(p)$, is then found from the average of the state evaluations denoted $p \cdot q$ and $p \cdot \sim q$, i.e.

$$\mathbf{w}^{\#}(p) = \frac{\mathbf{a}+\mathbf{b}}{2}$$

It holds similarly that:

$$\mathbf{w}^{\#}(q) = \frac{\mathbf{a}+\mathbf{c}}{2}$$

$$\mathbf{w}^{\#}(\sim p) = \frac{\mathbf{c}+\mathbf{d}}{2}$$

$$\mathbf{w}^{\#}(\sim q) = \frac{\mathbf{b}+\mathbf{d}}{2}$$

Rescher's conception presupposes that all the possible states are equally probable. But it is useful, as in the case of the probabilisation of Carnap's logical states, and their corresponding state descriptions, to assume that different states have different probability. We shall therefore introduce a weighted evaluation, as required anyway by Cartesian logic, which is shown in the following table (in which $\mathbf{r}_1 + \mathbf{r}_2 + \mathbf{r}_3 + \mathbf{r}_4 = 1$):

possible states s_i	probability of state $\mathbf{p}(s_i)$	simple evaluation $\mathbf{w}(s_i)$	weighted evaluation $\mathbf{p}(s_i)\mathbf{w}(s_i)$
$p \cdot q$	$\mathbf{r}_1$	$\mathbf{e}_1$	$\mathbf{r}_1 \cdot \mathbf{e}_1 = \mathbf{a}$
$p \cdot \sim q$	$\mathbf{r}_2$	$\mathbf{e}_2$	$\mathbf{r}_2 \cdot \mathbf{e}_2 = \mathbf{b}$
$\sim p \cdot q$	$\mathbf{r}_3$	$\mathbf{e}_3$	$\mathbf{r}_3 \cdot \mathbf{e}_3 = \mathbf{c}$
$\sim p \cdot \sim q$	$\mathbf{r}_4$	$\mathbf{e}_4$	$\mathbf{r}_4 \cdot \mathbf{e}_4 = \mathbf{d}$

If $\mathbf{w}^{\#}(p) = \mathbf{A}$
$\mathbf{w}^{\#}(q) = \mathbf{B}$
$\mathbf{w}^{\#}(\sim p) = \mathbf{C}$
$\mathbf{w}^{\#}(\sim q) = \mathbf{D}$, it follows that

$$\mathbf{A} = \mathbf{a} + \mathbf{b}$$
$$\mathbf{B} = \mathbf{a} + \mathbf{c}$$
$$\mathbf{C} = \mathbf{c} + \mathbf{d}$$
$$\mathbf{D} = \mathbf{b} + \mathbf{d}$$

It can easily be shown that this conception of evaluation imposes no restrictions on the evaluation of possible states if this kind of evaluation is taken as starting point. This means that **A, B, C, D** can be calculated if **a, b, c, d** are given. On the other hand **a, b, c, d** are not here calculable if **A, B, C, D** are given. These constraints can be expressed by the following relations:

$$\mathbf{A} + \mathbf{C} = \mathbf{B} + \mathbf{D} \qquad \text{(i.e. } \mathbf{w}^{\#}(p) + \mathbf{w}^{\#}(\sim p) = \mathbf{w}^{\#}(q) + \mathbf{w}^{\#}(\sim q))$$
$$\mathbf{A} + \mathbf{c} = \mathbf{B} + \mathbf{b}$$
$$\mathbf{D} + \mathbf{c} = \mathbf{C} + \mathbf{b}$$

It we draw up matrices for the evaluations of the possible states and for that which corresponds to the individual statements, the matrices can only be interpreted *in one direction.* In a way this is analogous to connective operations as opposed to adjunctive operations, which allow for interpretations in both direction.[8]

The same consequences are reached in an analysis of Rescher's other concept of preference, denoted P^*. Starting from the same table which included the requirement of weighted evaluation, we can construct evaluation according to P^* by considering both the weighted evaluations of the states involved in the statements used and the weighted evaluations of the states excluded by those statements, according to the rule:

$$\mathbf{w}^*(s_i) = \mathbf{w}^{\#}(s_i) - \mathbf{w}^{\#}(\sim s_i).$$

Since s_i and $\sim s_i$ are complementary and embrace the entire universe, it also holds that

$$\mathbf{w}^*(s_i) + \mathbf{w}^*(\sim s_i) = 0.$$

The limitations relevant to P^* are then given by these relations:

$$\mathbf{w}^*(p) = \mathbf{A} = \mathbf{a} + \mathbf{b} - \mathbf{c} - \mathbf{d}$$
$$\mathbf{w}^*(q) = \mathbf{B} = \mathbf{a} - \mathbf{b} + \mathbf{c} - \mathbf{d}$$
$$\mathbf{w}^*(\sim p) = \mathbf{C} = -\mathbf{a} - \mathbf{b} + \mathbf{c} + \mathbf{d}$$
$$\mathbf{w}^*(\sim q) = \mathbf{D} = -\mathbf{a} + \mathbf{b} + \mathbf{c} - \mathbf{d} \quad \text{from which it follows that}$$
$$\mathbf{A} + \mathbf{C} = 0, \quad \text{i.e. } \mathbf{w}^*(p) + \mathbf{w}^*(\sim p) = 0$$
$$\mathbf{B} + \mathbf{D} = 0, \quad \text{i.e. } \mathbf{w}^*(q) + \mathbf{w}^*(\sim q) = 0.$$

It can easily be proved that the two concepts of preference, i.e. both $P^{\#}$ and P^*, do, if the above-mentioned connective interpretation of the relevant tables is respected, satisfy sentences (6), (7) and (8).

4. PREFERENCE 'CETERIS PARIBUS'

We have so far distinguished two concepts of preference relating to states of affairs. The first, denoted $P^{\#}$, might be described as *inclusive preference,* and the second, denoted P^*, as *differential preference*. This, however, comes nowhere near to exhausting all the conceptions of preference used in practice. Compare the following two preference statements:

(α) It is better to have a bird in the hand than two in the bush.

(β) I prefer tea to coffee with all my meals.

The preference statement (α) is an example of a *relativised preference*, while (β) is a *preference ceteris paribus*. Under relativised preference two components of states of affairs are ordered preferentially under different circumstances. With preference ceteris paribus the order of preference is given for all circumstances. There is another type of preference, which

	p	q	r	t	$\mathbf{p}(s_i)\mathbf{w}(s_i)$
s_1	1	1	1	1	$\mathbf{a}_1$
s_2	1	1	1	0	$\mathbf{a}_2$
s_3	1	1	0	1	$\mathbf{a}_3$
s_4	1	1	0	0	$\mathbf{a}_4$
s_5	1	0	1	1	$\mathbf{a}_5$
s_6	1	0	1	0	$\mathbf{a}_6$
s_7	1	0	0	1	$\mathbf{a}_7$
s_8	1	0	0	0	$\mathbf{a}_8$
s_9	0	1	1	1	$\mathbf{a}_9$
s_{10}	0	1	1	0	$\mathbf{a}_{10}$
s_{11}	0	1	0	1	$\mathbf{a}_{11}$
s_{12}	0	1	0	0	$\mathbf{a}_{12}$
s_{13}	0	0	1	1	$\mathbf{a}_{13}$
s_{14}	0	0	1	0	$\mathbf{a}_{14}$
s_{15}	0	0	0	1	$\mathbf{a}_{15}$
s_{16}	0	0	0	0	$\mathbf{a}_{16}$

can be distinguished from preference ceteris paribus and can be described as *preference under explicitly defined circumstances*, e.g.

(γ) I prefer tea to coffee with an English breakfast of bacon and eggs, cheese and toast.

In everyday speech the difference between (β) and (γ) need not always be clear and accurately distinguishable, as will be appreciated in any case from example (β) and (γ). For this reason and to give a more exact distinction of (β) and (γ) we shall take a schematic example. Let us assume that we have to ascertain the preference between p and q and that the extra components r and t have also to be respected. The possible states of affairs can then be illustrated by the schema on page 316, to which weighted evaluation has also been added.

If we denote preference ceteris paribus P_{cp} and preference under explicitly defined circumstances P_{ec}, then, using the evaluation function $\mathbf{w}^{\#}$, we get

$$pP_{cp}q[\equiv]\mathbf{w}^{\#}(s_5 \vee s_6 \vee s_7 \vee s_8) > \mathbf{w}^{\#}(s_9 \vee s_{10} \vee s_{11} \vee s_{12})$$
$$[\equiv](\mathbf{a}_5 + \mathbf{a}_6 + \mathbf{a}_7 + \mathbf{a}_8) > (\mathbf{a}_9 + \mathbf{a}_{10} + \mathbf{a}_{11} + \mathbf{a}_{12})$$

Naturally we can visualise using the differential evaluation function $\mathbf{w}^*$. But since the use of $\mathbf{w}^*$ leads to an exactly equivalent solution, there is no need to distinguish the inclusive and differential concepts of preference ceteris paribus. For if we do use $\mathbf{w}^*$, we get

$$\begin{aligned} pP_{cp}q[\equiv](&-\mathbf{a}_1 - \mathbf{a}_2 - \mathbf{a}_3 - \mathbf{a}_4 + \mathbf{a}_5 + \mathbf{a}_6 + \mathbf{a}_7 + \mathbf{a}_8 - \mathbf{a}_9 \\ &- \mathbf{a}_{10} - \mathbf{a}_{11} - \mathbf{a}_{12} - \mathbf{a}_{13} - \mathbf{a}_{14} - \mathbf{a}_{15} - \mathbf{a}_{16}) \\ &> (-\mathbf{a}_1 - \mathbf{a}_2 - \mathbf{a}_3 - \mathbf{a}_4 - \mathbf{a}_5 - \mathbf{a}_6 - \mathbf{a}_7 - \mathbf{a}_8 \\ &+ \mathbf{a}_9 + \mathbf{a}_{10} + \mathbf{a}_{11} + \mathbf{a}_{12} - \mathbf{a}_{13} - \mathbf{a}_{14} - \mathbf{a}_{15} - \mathbf{a}_{16}), \end{aligned}$$

which is equivalent to the expression above.

The concept of preference under explicitly defined circumstances assumes that other components are explicitly stated. For example, if it is expressly stated that the preferability of p to q is to be determined against the circumstances $r \cdot \sim t$, then

$$pP_{ec}q(\text{for } ec = r \cdot \sim t)[\equiv]\mathbf{w}^{\#}\,(p \cdot \sim q \cdot r \cdot \sim t) > \mathbf{w}^{\#}\,(\sim p \cdot q \cdot r \cdot \sim t)$$
$$[\equiv]\mathbf{a}_6 > \mathbf{a}_{10}$$

Similarly it holds here too that the use of $\mathbf{w}^*$ will lead to equivalent results, so again there is no need to distinguish inclusive and differential concepts of preference under explicitly defined circumstances.

Let us next look at some important properties of preference under explicitly defined circumstances. While we consider the expressions

$pP^{\#} q \cdot qP^{\#} p,$

$pP^{*}q \cdot qP^{*}p$ and

$pP_{cp}q \cdot qP_{cp}p$

to be inconsistent, we are not justified in considering as inconsistent

$pP_{ec_1}q \cdot qP_{ec_2}p,$ if, for instance, in our example

$ec_1 \equiv r \cdot \sim t$ and

$ec_2 \equiv \sim r \cdot t$ and if it holds that

$\mathbf{a}_6 > \mathbf{a}_{10}$ and simultaneously $\mathbf{a}_{11} > \mathbf{a}_7$

From the logical point of view it is important to notice that preference ceteris paribus and preference under explicitly defined circumstances, just as inclusive and differential preference, both satisfy sentences (6), (7) and (8). But the same cannot be said of relativised preference. Relativised preference appears to satisfy only (A1), therefore, denoting relativised preference P_{rel}, it holds that

$$pP_{rel}q \cdot \sim (qP_{rel}p),$$

but it is scarcely possibly to envisage relativised preference as transitive and sentences (6), (7) and (8) as satisfied. It is also natural that relativised preference is irreflexive; it holds, therefore, that

$$\sim (pP_{rel}p).$$

So it is obvious that relativised preference satisfies only quite minimal demands, making it highly problematical to treat relativised preference as a starting point for the construction of preference systems.

5. THE CONCEPT OF 'PREFERABLE STATES OF AFFAIRS' AS A QUALITATIVE CONCEPT

We have so far considered five different concepts of preference, most of which (relativised preference being the exception) have some of the basic properties of comparative concepts, i.e. they are irreflexive, asymmetrical and transitive. Since comparative concepts are usually constructed in logical methodology as a super-structure above a sphere of qualitative concepts which facilitate the assignment of a certain quality discernible in degree, we might, with preference too, consider the possibility of a concept to correspond to the qualitative concepts. But it is usual in logical systems of preference to introduce qualitative concepts on the basis of concepts of preference, or indifference as the case may be. Thus it is possible to introduce a triple of qualitative concepts:

Bp (p is preferable, i.e. a good state of affairs)

Mp (p is not preferable, i.e. a bad state of affairs) and

Lp (p is an indifferent state of affairs)

and to define

$$Bp \equiv pP \sim p$$
$$Mp \equiv \sim pPp$$
$$Lp \equiv pI \sim p$$

A similar procedure is followed by Halldén [52] and von Wright in his first monograph [167]. Von Wright therefore sees a tautology of preference calculus in the expression:

$$(9) \qquad (pP \sim p)\cdot(\sim qPq) \longrightarrow (pPq)$$

which can be interpreted by saying that if p is preferable and q is not preferable, then p must be preferred to q.

The difficulty about expression (9) becomes perfectly clear if, as in the earlier example, we put in **A**, **B**, **C**, **D** for $\mathbf{w}(p)$, $\mathbf{w}(q)$, $\mathbf{w}(\sim p)$ and $\mathbf{w}(\sim q)$. Then it ought to hold that

$$(10) \qquad [(\mathbf{A} > \mathbf{C})\cdot(\mathbf{D} > \mathbf{B})] \longrightarrow (\mathbf{A} > \mathbf{B})$$

If **A**, **B**, **C** and **D** are real numbers, we clearly have no guarantee for considering (10) a tautology. For example, if

$$\mathbf{A} = 5$$
$$\mathbf{B} = 6$$
$$\mathbf{C} = 4$$
$$\mathbf{D} = 7$$

it is obvious that the expression quoted is not valid. For this expression to be acceptable, we cannot select just any preference, but only one type, and with it the relevant constraints for **A**, **B**, **C** and **D**. For example, it can be proved that (9) is satisfied on the assumption that P is understood as inclusive preference, i.e. $P^{\#}$. Then it holds that

$$\mathbf{A}+\mathbf{C}=\mathbf{D}+\mathbf{B}$$

making (10) acceptable. Likewise differential preference P^{*} satisfies (9), since it holds that

$$\mathbf{A}+\mathbf{C}=0 \quad \text{and}$$
$$\mathbf{B}+\mathbf{D}=0.$$

On the other hand preference ceteris paribus does not satisfy (9), for if it held that

$$[(pP_{cp} \sim q)\cdot(\sim qP_{cp}q)] \longrightarrow (pP_{cp}q),$$

it ought equally to apply, if we stay now with the above example with 16 possible states arising out of the components p, q, r and t, that

$$\begin{aligned}&\{[(\mathbf{a}_1+\mathbf{a}_2+\mathbf{a}_3+\mathbf{a}_4+\mathbf{a}_5+\mathbf{a}_6+\mathbf{a}_7+\mathbf{a}_8)\\ &>(\mathbf{a}_9+\mathbf{a}_{10}+\mathbf{a}_{11}+\mathbf{a}_{12}+\mathbf{a}_{13}+\mathbf{a}_{14}+\mathbf{a}_{15}+\mathbf{a}_{16})]\\ &\cdot[(\mathbf{a}_1+\mathbf{a}_2+\mathbf{a}_3+\mathbf{a}_4+\mathbf{a}_9+\mathbf{a}_{10}+\mathbf{a}_{11}+\mathbf{a}_{12})\\ &>(\mathbf{a}_5+\mathbf{a}_6+\mathbf{a}_7+\mathbf{a}_8+\mathbf{a}_{13}+\mathbf{a}_{14}+\mathbf{a}_{15}+\mathbf{a}_{16})]\}\\ &\longrightarrow[(\mathbf{a}_5+\mathbf{a}_6+\mathbf{a}_7+\mathbf{a}_8)>(\mathbf{a}_9+\mathbf{a}_{10}+\mathbf{a}_{11}+\mathbf{a}_{12})],\end{aligned}$$

for which there are of course no a priori guarantees.

On the other hand preference under explicitly defined circumstances does satisfy (9). If in our example we choose

$$ec \equiv r\cdot \sim t$$

we obtain

$$[(\mathbf{a}_6>\mathbf{a}_{10})\cdot(\mathbf{a}_6>\mathbf{a}_{10})]\longrightarrow(\mathbf{a}_6>\mathbf{a}_{10}),$$

which is plainly a tautology.

Thus we may conclude that the respective qualitative concepts, i.e. of preferable, non preferable and indifferent state of affairs, can be constructed on the basis of the concept of preference only if certain constraints are accepted. Of the preference concepts introduced in the discussion the constraints are satisfied only by inclusive preference, differential preference and preference under explicitly defined circumstances.

At first sight sentence (9) amounts to the trivial requirement that a good state of affairs is to be preferred to a bad state of affairs. But the problem is actually more complicated because the qualitative concepts used are not primitive concepts here, but concepts defined on the basis of that of preference and requiring that certain constraints be respected. Intuitively, it looks most appropriate to choose differential preference as the point of departure, for in the case of differential preference it holds that the sum of the evaluations of the complementary states is always zero, i.e. for any s_i

$$\mathbf{w}^*(s_i) + \mathbf{w}^*(\sim s_i) = 0$$

It must be added that this method of evaluating, and so too of justifying, preference mirrors the line of thought by which in making any choice from a number of alternatives we calculate both on the aggregate of gains and the aggregate of losses. This idea of 'gain' or 'loss' need not always be necessarily thought of in the economic or financial sense. It is now possible to put forward a simplification of the above definitions of $\boldsymbol{B}$, $\boldsymbol{M}$ and $\boldsymbol{L}$:

$$\boldsymbol{B}p [=] \mathbf{w}^*(p) > 0$$
$$\boldsymbol{M}p [=] \mathbf{w}^*(p) < 0$$
$$\boldsymbol{L}p [=] \mathbf{w}^*(p) = 0$$

6. PREFERENCE AS A PROPOSITIONAL ATTITUDE

The expression pPq, i.e. the state of affairs denoted p is preferred to the state of affairs denoted q, whether preference is interpreted in any one of the above five senses or in some other way, is actually a simplification. It is in effect an impersonal and atemporal modification of the expression 'x at time t prefers p to q'. It is therefore expedient to relativise preference expressions to a subject and a time.

Since the arguments of the relation P are denotates of statements, i.e. states of affairs, it is useful to treat P (in all the above interpretations) as a *special form of propositional attitude* or an epistemic modal operator. Propositional attitudes are statements of the type

x believes that p,
x knows that p,
x doubts that p and so forth.

The theory of propositional attitudes and epistemic modal operators of the kind 'knows', 'believes', 'claims', 'doubts' etc. has been elaborated in detail in various assertion logics and logics of epistemic modality.[9] The characteristic feature of propositional attitudes is the non-extensional contextual concatenation of an operator and the proposition proper, which brings these problems closer to those of the various systems of modal logic. Unlike ordinary epistemic modal operators, which fix a certain attitude of the subject to one proposition or to the denotate of one proposition, an expression of preference involves the attitude of the subject to the denotates of two propositions or, to be more exact, to a particular preference ordering of those propositions.

As with other instances of propositional attitudes, the time specification is not without importance. With a change of the time specification the subject may make a change of preference. Thus it is not an inconsistency to have a concatenation of two different preference attitudes in the one subject: 'x, at time t, prefers p to q and x, at time t' (which differs from t), prefers q to p'. For there is always the chance that in the interval which separates t' from t there may have been a change in the subject's system of values. If we record the subject of a preference attitude and the time at which the attitude is held as indices of the sign for preference P (whichever of the above mentioned, or any other, possible interpretations of P is chosen), then the expressions

$$pP_{x,t}q \cdot qP_{x,t'}p \quad \text{if } t \neq t' \text{ or}$$
$$pP_{x,t}q \cdot qP_{y,t}p \quad \text{if } x \neq y$$

must be viewed as absolutely consistent.

If we do take preference as a special form of propositional attitude (and this is the sense in which we speak of preference attitudes), it is obvious that the subject of this attitude may be not only a physical person, but also legal persons, defined bodies of people (e.g. the electorate of a given constituency). Then there has to be available a system for making a preference attitude manifest (e.g. a decision in favour of one of the possible alternatives, a decision of the relevant body if the subject of the preference attitude is a legal person, an electoral system or a system of referendum if it is a matter of ascertaining the preference of the electorate in a given constituency, and so forth). But it is also possible to visualise as the subject of a preference attitude some constructive entity, e.g. an ideal rational being, a Bayesian robot as considered by R.C. Jeffrey [64] or whatever. A statement to the effect that subjects of this sort do not exist does not mean that it is impossible to reconstruct the preference attitudes of such constructive entities.

It is for precisely these reasons that some authors stress the point of distinguishing between 'preference' and 'preferability' (as does von Wright in his last work [168]). For the concept of 'preferability' can be understood as 'being arguably better', or as the preference attitude of an ideal rational being and so on. Hence statements about preferability, as von Wright stressed, need not, though they may, be inevitably relativised to a certain person and time. Since the preference attitudes of different persons need not coincide with the preference attitudes of ideal rational beings or with the attitudes of preferability which we shall denote pBq (p is preferable to q), the expression

$$pP_{x,t}q \cdot qBp$$

has to be considered consistent.

In an investigation of different preference attitudes and the consistency of the expressions which carry them we can attach different demands to that which the subject of a preference attitude must have at his disposal. For example, if we admit that the subject has selected his preference attitude in accordance with a system of values which differs from the ideal or rationally arguable system of values, or the system of values of a Bayesian robot (in the spirit of the principle 'de gustibus non est disputandum'), we would probably be less willing to tolerate absolute ignorance of the principles of logic, e.g. contradiction. Therefore the expression of preference

$$(p \cdot \sim p) P_x q$$

will not, in the spirit of these claims, be considered consistent. This means then that from the point of view of logical semantics we are likely to be more tolerant of a host of value systems and different subjects and less tolerant of violations of the elementary principles of logic, logical syntax and semantics that apply to the linguistic means used in expressions of preference about particular states of affairs.

CONCLUSIONS

1. The findings of an analysis of the logical status of preference have confirmed that a number of different concepts of preference with different logical properties do need to be accepted. In the present analysis we have distinguished five such concepts: inclusive preference, differential preference, preference ceteris paribus, preference under explicitly defined

circumstances and relativised preference. These different concepts of preference satisfy in different ways the principles which are usually considered tautologies (i.e. axioms or theorems) of preference calculus. (In our discussion this relates to sentences (6), (7), (8) and (9).)

2. The semantic foundations of the concept of inclusive preference are procedures which correspond to the method of state description and Carnap's concept of logical range. On the other hand the concept of differential preference corresponds to the procedures used in econometric decision-making under the supposition of zero sum, i.e. for any s_i it holds that

$$\mathbf{w}^*(s_i) + \mathbf{w}^*(\sim s_i) = 0.$$

Preference ceteris paribus assumes that the preference ordering of the two components in a complex state of affairs is considered for all possible circumstances. But because these circumstances may be very numerous, it is obviously fairly awkward to justify a preference on this basis. Preference under explicitly defined circumstances transpires to be the most exact starting point for a reliable ordering of preferences.

3. Preference can be profitably considered to be a special form of propositional attitude, the concept of preference itself – whichever of the suggested interpretations we choose to give it – behaving as an operator.

4. Different forms of preference are associated with different conditions of consistency. The minimum condition of consistency must be due recognition of the elementary rules of logical syntax and semantics.

XI. THE PROBLEM OF INFORMATIONAL SYNONYMITY

In his manifesto 'On the construction of a machine language for an information machine', the Soviet pioneer of information science V.A. Uspenskij [179] specified a number of requirements which should be met by the information language to be used in any computerised information system. One of these requirements consists in the formalisation, and hence the possibility of a machine solution, of semantic identification, i.e. of a decision that two differently expressed or recorded items of data are identical in terms of their meaning. This chapter is an attempt to meet that requirement, the concept of 'semantic identity' being explicated by the means afforded by the theory of semantic information and on the basis of the concept of 'transmitted information'. It is in this light that we introduce the concept of 'informational synonymity'. Since previous attempts to solve the problem of semantic identification have usually been based on the Leibnizian principle of identity, we shall be indicating some of the difficulties and limitations inherent to methods and criteria based on the Leibnizian principle. In principle, even the concept of 'informational synonymity' keeps to the requirements which stem from the Leibnizian principle, but in a modified form.

1. THE TRADITIONAL (LEIBNIZIAN) CRITERION OF IDENTITY AND THE PROBLEM OF SEMANTIC IDENTIFICATION

The common formulations of the Leibnizian criterion are addressed to much more general issues than merely the solution of semantic identification. This criterion regards as identical all entities, linguistic or otherwise, which have all their properties in common. If $\mathscr{X}$ denotes the class of these entities, and $\mathscr{P}$ the class of predicates as the names of the properties assignable to all the elements of $\mathscr{P}$, then in the system $\langle\mathscr{X}, \mathscr{P}\rangle$, in which elements (variables) of the class $\mathscr{X}$ are denoted by small letters $x, y, u_1, u_2 \ldots$ and elements of class $\mathscr{P}$ by capitals, P_1, P_2 etc., the Leibnizian criterion can be formulated in two ways.

(a) If we restrict the elements of class $\mathscr{P}$ to one-place predicates, we can treat as identical any two elements of class $\mathscr{X}$ (Ix, y in the notation) which have all properties in common. In other words, x and y are identical if and only if all that can be said of x can also be said of y and vice versa, i.e.

$$(1) \qquad (\forall x)(\forall y)(\forall P)[Ix, y \equiv (Px \equiv Py)]$$

This formulation of the Leibnizian criterion (which actually goes back to Aristotle's *Topics*) obviously leads to some fairly awkward problems. Even in the now classical system of Whitehead and Russell's *Principia Mathematica* we find – in connection with definition 13.01 and formula 13.101, which corresponds to (1) above – the comment that identity can be understood in a variety of ways, or that identity can be of various degrees according to how what is described as the predicative function is understood. In other words, the problem consists in the different possible interpretations of that which is expressed in the intuitive understanding as 'all that can be said of …'. For this very reason (1) is inadequate, since the expression 'all that can be said of…' may be understood in a broader sense than the class of one-place predicates only.

(b) Thus if we abandon the above restrictions, i.e. if we now treat the expression 'all that can be said of …' as also including relations to other entities, we can modify (1) in the following manner (the upper index $^{(i)}$ indicating the number of arguments of the predicate P, i.e. $P^{(i)}$ is an i-argument predicate):

$$(2) \qquad (\forall x)(\forall y)(\forall P^{(1)})(\forall P^{(2)}) \ldots (\forall P^{(n)})[Ix, y \equiv (P^{(1)}x) \equiv (P^{(1)}y). \cdot(P^{(2)}x, u \equiv P^{(2)}y, u)\cdot(P^{(3)}x, u, v \equiv P^{(3)}y, u, v)\cdot \ldots \cdot(P^{(n)}x, u, v, \ldots \equiv P^{(n)}x, u, v, \ldots)]$$

It would be possible to consider yet other versions and formulations of the Leibnizian principle, which would respect expressions of the higher type, or, in Russell's terminology, functions of a higher type, including extensional and intensional functions. But there can be no doubt that this or any other similar approach only make the criterion even less transparent and less applicable in specific tasks.[1]

For any version of the Leibnizian criterion to be made applicable there has to be a finitisation, both by a selection of the essential and decisive elements of class $\mathscr{P}$, and by a restriction of the number of arguments if version (b) is used. But then any finitist formulation conceived on these lines no longer corresponds to the intuitive requirements expressed by the words 'all properties in common', 'all that can be said of…' and so on. That being so, a change has to be made in the requirement, which is reformulated as: two elements of the class $\mathscr{X}$ ($I'x, y$ in the notation) are practically (almost etc.) identical when they have all their decisive properties in common. If $\{P_1, P_2, \ldots P_n\}$ is a finite class of decisive properties, the modified notation for version (a) takes the form:

$$(3) \qquad (\forall x)(\forall y)[I'x, y \equiv (P_1x \equiv P_1y)\cdot(P_2x \equiv P_2y)\cdot \ldots \cdot (P_nx \equiv P_ny)]$$

Version (b) could be modified analogously, account necessarily being taken of a finite number of arguments and of the elements to which relations are being considered. Since semantic identification is actually a special instance of the concept I', we are now faced with the task of finding those decisive properties capable of guaranteeing semantic identification. One special case of a way of determining these decisive properties is Leibniz's criterion of 'salva veritate', to which we shall be returning yet again.

The concept I' must be distinguished from another concept which can be described as indistinguishability and is based on the criterion of identifying the indistinguishable. Two elements of class $\mathscr{X}$ are indistinguishable ($I''x, y$ in the notation) if they coincide with respect to all distinguishing criteria. Distinguishing criteria may be considered to be comparative predicates, i.e. transitive, asymmetrical, C-connected and hence also irreflexive predicates [153], which constitute the subclass $\mathscr{A}$ of all two-place predicates (i.e. $\mathscr{A} \subset \mathscr{P}^{(2)}$). If $\mathscr{A} = \{A_1, A_2, \ldots A_n\}$, the concept of coincidence with respect to A_i is denoted C_{A_i} and is always to be understood as relativised, i.e. to some distinguishing criterion. Then C_{A_i} can be characterised as follows:

$$(4) \qquad (\forall x)(\forall y)[C_{A_i}x, y \equiv (\sim A_ix, y\cdot \sim A_ix, y)]$$

Then it follows that

$$(5) \qquad (\forall x)(\forall y)(I''x, y \equiv C_{A_1}x, y\cdot C_{A_2}x, y\cdot \ldots C_{A_n}x, y)$$

Notation (5), which may be considered a definition of 'indistinguishability', is really a formalisation of the criterion of 'identificatio indiscernibilium'. The utility of (5) in specific tasks of identification is is governed by the same consideration as that of (1) and (2): for a criterion of identifying the indistinguishable to be of use in practice, the class of distinguishing criteria must be finite and it must be manageable with the means genuinely available to us, including capacities, delays and memories.

2. THE 'SALVA VERITATE' CRITERION

If any of the above versions is to be of use as a criterion of semantic identification, we need to specify class $\mathscr{X}$ and also select those properties or distinguishing criteria which are decisive for semantic identification, i.e. certain elements of class $\mathscr{P}$, or of class $\mathscr{A}$, as the case may be. The solution of this specification is really also to be found in Leibniz, although identity is there more a relation between signs than between the denoted entities, or at least the two are not strictly separated. The specification runs: "eadem sunt quorum unum potest substitui alteri salva veritate".

If this specification is to be applicable meaningfully, it must be assumed that the elements of class $\mathscr{X}$ are words or verbal signs. Then we can treat as semantically identical those verbal signs which are

(1) mutually substitutable, and

(2) do not bring about any change in the truth value of the sentence in which a substitution is made.

It is easily proved that this formulation, which is known in the literature as the 'salva veritate' test for deciding the semantic identity (synonymity) of words, leads to the antinomies that have been described as the antinomies of synonymous names. These difficulties, which have been one of the inspirations behind logical semantics and the semantic foundations of modern mathematics and logic, have been addressed by Frege, Russell, Carnap, Quine and many other pioneers of logical semantics. The problems which arise here can be summarised overall in the following questions:

(a) What expressions are concerned in semantic identifiability (synonymity)?

(b) In what kind of context or conditions can we properly consider the substitutability of synonymous expressions?

(c) When substitution is made, what is supposed to be preserved?

The original Leibnizian versions of answers to these potential questions would have been: (a) words, (b) in a sentential context, (c) the denotation of the words and the truth value of the sentence. It was Frege who revealed the difficulties of this solution and in answer (c) replaced denotation (Bedeutung) of words by sense (Sinn). Quine showed that even this solution is not satisfactory and replaced Leibniz's 'salva veritate' by the criterion of 'salva analyticitate', i.e. the requirement of the logical equivalence of synonymous words. Carnap attempted to broaden the answer (a) to take in all expressions which can denote (he called them designates), i.e. including clauses and whole sentences, but he too ran into difficulties.

His solution, based on the concept of 'intensional isomorphism', fails to resolve the problem of the synonymity of expressions which operate with a non-extensional context, such as expressions about someone's opinions, about a person XY's conviction that..., and so on, that is expressions associated with non-extensional contextual concatenation. (This includes statements about the claims, beliefs, knowledge, preferences and other similar expressions as found among the non-classical logicians.)

For any satisfactory modification of the 'salva veritate' criterion to be applicable to the purposes considered by Uspenskij, certain conventions have to be introduced representing further modifications of answers (a), (b) and (c) above:

(a) As expressions to be considered potentially semantically identifiable we shall take the basic modes of expressing data: statements (whether written down or communicated in any other way), or classes of statements or ordered sequences of statements, i.e. texts. We therefore exclude isolated words unless they are of the nature of statements (as in the case of shortened answers to questions). These basic modes of expression will be called *data objects* and they will constitute the class $\mathscr{X}$ with the elements x, y, z etc.

(b) These data objects are *communicated* to a language user and are intended to serve him as information to help solve some task or complex of tasks or to reach certain goals. The informational role of the data objects is assessed from the point of view of the degree to which they are capable of supplying the language user with information needed in solving the particular task, i.e. the degree to which they reduce his uncertainty in solving the task, increase the quality of the solution, reduce the loss or possible or expected average risk associated with the solution of the task and so forth.

(c) Data objects are mutually substitutable under the conditions mentioned under (b) if their informational value with respect to the tasks towards whose solution they are intended to contribute remains unchanged.

3. THE CRITERION OF 'SALVA RELATIONE' AND THE CONCEPT OF 'INFORMATIONAL SYNONYMITY'

In accordance with the above specification we can formulate the following criterion of semantic identification, which has its roots in the general outline of the Leibnizian conception, but which replaces Leibniz's 'salva veritate' or Quine's 'salva analyticitate' or any other similar modification by the conservation of transmitted information (relatio): Eadem sunt quae sibi mutuo substitui possunt salva relatione. To be more exact, semantically identifiable are those data objects which are mutually substitutable with respect to a given task or complex of tasks, the measure of transmitted information provided with respect to that task or complex of tasks by the data objects remaining unchanged. This criterion can be used to introduce the concept of '*informational synonymity*'. Before passing from the intuitive discussion to some attempts to provide a more exact definition, we shall note some of the difficulties and potential objections which may arise over the concept of 'informational synonymity.'

First and foremost it must be appreciated that we are not talking about synonymity in the absolute sense, as is commonly the case in linguistically oriented conceptions of synonymity where two expressions are synonymous if they share certain semantic characteristics, notably denotation, sense, mutual substitutability in any context etc. 'Informational synonymity' is always relativised, since it is based on the concept of 'transmitted information', which is itself always relativised to certain a priori information.

The next difficulty is much more serious. If we say that two data objects are informationally synonymous, given that they are mutually substitutable with respect to a given task or complex of tasks, the measure of information transmitted with respect to that task remaining unchanged, we are apparently saying that any two linguistically and semantically (however this is understood) different objects which carry no new information for the task in hand are informationally synonymous. To overcome this difficulty we need to introduce by convention a defined measure of sufficient transmitted information which can be considered the minimum measure of transmitted information below which the measure of transmitted information must not drop. Without the introduction of sufficient transmitted information the very concept of 'informational synonymity' loses all sense. Sufficient transmitted information may be said to be such information as is capable of solving a giving task as well as is required, satisfactorily removing the initial uncertainty or ignorance, giving satisfactory answers to the questions set and so on.

Having discussed some of the intuitive sources of the concept of 'informational synonymity' we can now pass to the actual constitution of the concept. Let x and y be two different data objects which are intended to communicate to the language user certain information with respect to the task or problem situation expressed by the data object or class of objects i. We shall call the data object i the a priori data base for the given task or problem situation. The data base may contain, for example, suggested solutions to the task, certain hypotheses, algorithms or previously acquired empirical data. On the other hand the data objects or class of objects x and y might express new empirical data, extra hypotheses and instructions etc. Data object i may also be the data base for asking questions, while x and y might be answers to those questions. The role of x (and also of y) with respect to the a priori data base i, however we interpret those objects, is to reduce or eliminate the original uncertainty associated with i. What x conveys with respect to that to which i relates may be expressed by the concept of 'transmitted information'. Transmitted information expresses the degree to which the new data object x (or y) is able to reduce the original uncertainty connected with i. If we denote this original uncertainty $\mathbf{U}(i)$, and if the conditional uncertainty connected with i given the availability of the data object x is denoted $\mathbf{U}(i/x)$, the transmitted information is a function of the original and the conditional uncertainty and increases with increasing values of the difference between the original and the conditional uncertainty, i.e. the difference

$$\text{(6)} \qquad \mathbf{U}(i) - \mathbf{U}(i/x)$$

In probability and inductive logics it is common practice to calculate the measure of original and conditional uncertainty by means of the logarithmic measure **inf** or the content measure **cont** as introduced in the classic work of Bar-Hillel and Carnap [7].[2] This does not of course preclude other possible ways of calculation. The following may be considered the most elementary ways of calculating the measure of transmitted information as based on the measure of information:

(I) If we begin from the logarithmic measure of information **inf**, the information transmitted by the data object x with respect to that to which data base i relates (symbolically $\mathbf{TI}(x/i)$) can be characterised as follows:

$$\text{(7)} \qquad \mathbf{TI}(x/i) = \mathbf{transinf}(x/i) = \mathbf{inf}(i) - \mathbf{inf}(i/x)$$
$$= \log \mathbf{p}(i/x) - \log \mathbf{p}(i)$$

In this case we therefore have

$$\text{(8)} \qquad \mathbf{U}(i) = -\log \mathbf{p}(i) \qquad \text{and}$$
$$\text{(9)} \qquad \mathbf{U}(i/x) = -\log \mathbf{p}(i/x)$$

(II) If we begin with the content measure **cont**, the information transmitted by data object x with respect to that to which data base i relates can be characterised as follows:

$$(10)\qquad \mathbf{TI}(x/i) = \mathbf{transcont}\,(x/i) = \mathbf{cont}\,(i) - \mathbf{cont}\,(i \longrightarrow x) = 1 - \mathbf{p}(x \vee i)$$

In this case we have

$$(11)\qquad \mathbf{U}(i) = 1 - \mathbf{p}(i) \quad \text{and}$$

$$(12)\qquad \mathbf{U}(i/x) = \mathbf{p}(x \vee i) - \mathbf{p}(i) = \mathbf{p}(x) - \mathbf{p}(x \cdot i)$$

Both the above measures of transmitted information[3] are symmetrical with respect to x and i. These measures can be normalised and thereby the symmetry disappears. Denoting the normalised measure of transmitted information $\mathbf{TI}_i(x/i)$, we obtain, according to alternatives (I) and (II):

$$(13)(\mathrm{I})\qquad \mathbf{TI}_i(x/i) = \mathbf{transinf}_i(x/i) = \frac{\mathbf{inf}(i) - \mathbf{inf}(i/x)}{\mathbf{inf}(i)} = \frac{\mathbf{log\,p}\,(i) - \mathbf{log\,p}(i/x)}{\mathbf{log\,p}(i)}$$

$$(14)(\mathrm{II})\qquad \mathbf{TI}_i(x/i) = \mathbf{transcont}_i(x/i) = \frac{\mathbf{cont}(i) - \mathbf{cont}(i \longrightarrow x)}{\mathbf{cont}(i)} = \frac{1 - \mathbf{p}(x \vee i)}{1 - \mathbf{p}(i)} = 1 - \mathbf{p}(x)\left(\frac{1 - \mathbf{p}(i/x)}{1 - \mathbf{p}(i)}\right) = \mathbf{p}(\sim x/\sim i)^4$$

It can be proved that

$$(15)\qquad -\infty \leqslant \mathbf{transinf}_i(x/i) \leqslant 1,$$

where $\mathbf{transinf}_i(x/i) = \mathbf{max\ transinf}_i = 1$ if and only if i follows logically from x, i.e. if $x \underset{L}{\longrightarrow} i$, and $\mathbf{transinf}_i(x/i) = \mathbf{min\ transinf}_i = -\infty$ if and only if the negation of i follows logically from x, i.e. if $x \underset{L}{\longrightarrow} \sim i$.

As for the extreme values of $\mathbf{transcont}_i$, it holds that

$$(16)\qquad 0 \leqslant \mathbf{transcont}_i(x/i) \leqslant 1,$$

where $\mathbf{transcont}_i(x/i) = \mathbf{max\ transcont}_i = 1$ if and only if i follows logically from x, i.e. if $x \underset{L}{\longrightarrow} i$, and $\mathbf{transcont}_i(x/i) = \mathbf{min\ transcont}_i = 0$ if and only if i follows from the negation of x, i.e. if $\sim x \underset{L}{\longrightarrow} i$.

So far we have mentioned the most important properties of the measures of transmitted information based on the logarithmical measure **inf** and the content measure **cont**. Of the other important properties reference must be made to the conditions for additivity and for informational irrelevance. In the case of the additivity conditions for two data objects x and x' (the additivity bound on their conjunctive concatenation), it holds that, using the concept of $\mathbf{transinf}_i$:

(17) $\mathbf{transinf}_i(x \cdot x'/i) = \mathbf{transinf}_i(x/i) + \mathbf{transinf}_i(x'/i)$, if the following conditions are satisfied:

(a) $\mathbf{p}(x \cdot x') = \mathbf{p}(x)\mathbf{p}(x')$, i.e. x and x' independent with respect to the selected measure of probability $\mathbf{p}$;

(b) $\mathbf{p}(x \cdot x'/i) = \mathbf{p}(x/i)\mathbf{p}(x'/i)$, i.e. x and x' conditionally independent with respect to i and to the selected measure of probability $\mathbf{p}$.

If we use the concept $\mathbf{transcont}_i$ for the measure of transmitted information, the following additivity conditions apply:

(18) $\mathbf{transcont}_i(x \cdot x'/i) = \mathbf{transcont}_i(x/i) + \mathbf{transcont}_i(x'/i)$, but only if the following condition is met: the disjunction $x \vee x'$ is logically true.

Now the conditions for informational irrelevance (again bound to conjunctive concatenation):

(19) $\mathbf{transinf}_i(x \cdot x'/i) = \mathbf{transinf}_i(x/i)$, i.e. x' is informationally irrelevant with respect to $\mathbf{transinf}_i(x/i)$, if the following conditions are met:

(a) $\mathbf{p}(x \cdot x') = \mathbf{p}(x)\mathbf{p}(x')$,

(b) $\mathbf{p}(x \cdot x'/i) = \mathbf{p}(x/i)\mathbf{p}(x'/i)$,

(c) $\mathbf{p}(x' \cdot i) = \mathbf{p}(x')\mathbf{p}(i)$, i.e. x' and i are, furthermore, independent with respect to the selected measure of probability $\mathbf{p}$.

For $\mathbf{transcont}_i$ the following conditions of informational irrelevance apply:

(20) $\mathbf{transcont}_i(x \cdot x'/i) = \mathbf{transcont}_i(x/i)$ if the condition is met whereby the disjunction $x' \vee i$ is logically true.

This survey of the most important properties of the two stated measures of transmitted information shows some of the advantages of conception (II). In any examples chosen we shall therefore calculate with conception (II).

For a definition of 'informational synonymity' we still need to constitute the concept of 'sufficient transmitted information'. Data object x is capable of providing sufficient information with respect to i if it is capable of ensuring that the task will be solved as well as is required, satisfactorily eliminating initial uncertainty, guaranteeing a solution with a risk not in excess of a certain level, securing satisfactory answers to questions etc. The transmitted information provided by data object x with respect to that which is referred to by i is sufficient if

(21) $\mathbf{TI}_i(x/i) \geqslant \varepsilon$

where ε is the conventionally determined measure of transmitted information with respect to i. It holds that two (different) data objects are informationally synonymous with respect to i if and only if they carry the same measure of information with respect to i and if that measure is greater than or equal to ε. Denoting the 'informational synonymity' of two data objects x and y with respect to i by the sign $S_i x, y$, we may make the following definition:

(22) $S_i x, y = \mathrm{df}\ [\mathbf{TI}_i(x/i) = \mathbf{TI}_i(y/i)]\cdot[\mathbf{TI}_i(x/i) \geqslant \varepsilon]$

This definition expresses the two conditions mentioned: two data objects x and y are informationally synonymous with respect to the given data base i if and only if they are mutually substitutable without bringing a change in the measure of transmitted information and if that measure is greater than or equal to sufficient transmitted information.

The data base with respect to which informational synonymity is being considered need not be thought of as a single data object, but also as a finite class of data objects $\mathscr{I}(\mathscr{I} = \{i_1, i_2, \dots i_n\})$. It could be a case of problem situations the solution to which presupposes a sequence of different steps, or a text which is an ordered sequence of statements, etc. If the measure of transmitted information with respect to $\mathscr{I}$ as a whole cannot be calculated, but only with respect to individual elements, the informational synonymity of data objects x and y with respect to $\mathscr{I}$ can be defined as follows:

(23) $S_{\mathscr{I}} x, y = \mathrm{df}\ \{[\mathbf{TI}_{i_1}(x/i_1) = \mathbf{TI}_{i_1}(y/i_1)]\cdot$
$\cdot[\mathbf{TI}_{i_2}(x/i_2) = \mathbf{TI}_{i_2}(y/i_2)]\cdot \dots \cdot[\mathbf{TI}_{i_n}(x/i_n) = \mathbf{TI}_{i_n}(y/i_n)]\}$
and for all $i_i (i_i \in \mathscr{I})$ it holds that $\mathbf{TI}_{i_i}(x/i_i) \geqslant \varepsilon$

Both these definitions of informational synonymity relativise the concept, on the one hand to the data base, and on the other to the limit of sufficient transmitted information. This also means that two data objects x and y which are synonymous with respect to i and to ε need not be synonymous with respect to some other data base j. Existing synonymity may be disturbed if the demands made on the limits of sufficient transmitted information are changed.

Since both conceptions are based on the application of probability parameters to the data objects (the question of the suitability of this or that interpretation of the probability parameters being left open, i.e. whether as frequency probability, subjective probability, measure of expectation, measure of plausibility etc.), it needs to be shown what the assumptions $S_i x, y$ are if we express them by means of these parameters. If we use conception (I), i.e. the logarithmical measure of information, we find that two data objects x and y are informationally synonymous with respect to i if and only if it holds that

(24) (a) $\mathbf{p}(i/x) = \mathbf{p}(i/y)$ and
(b) $\mathbf{log\ p}(i/x) \leqslant \mathbf{log\ p}(i)[1 - \varepsilon]$

If we use conception (II), i.e. the measure based on **cont**, we find that two data objects x and y are informationally synonymous if and only if it holds that

(25) (a) $\mathbf{p}(i \vee x) = \mathbf{p}(i \vee y)$ and
(b) $\mathbf{p}(i \vee x) \leqslant 1 - \varepsilon[1 - \mathbf{p}(i)]$

4. INFORMATIONAL RELEVANCE AND THE CONCEPT OF 'STRICT INFORMATIONAL SYNONYMITY'

The concept of 'informational synonymity' based on a same measure of transmitted information, which is greater than or equal to sufficient transmitted information, is, in a sense, weaker than Quine's criterion of 'salva analyticitate'. This means that $S_i x, y$ does not at all imply the equivalence or logical equivalence of the data objects x and y. The only thing which is guaranteed is, in the case of the application of conception (I), equality of the conditional probabilities $\mathbf{p}(i/x) = \mathbf{p}(i/y)$, and, in the case of the application of conception (II), the equality $\mathbf{p}(i \vee x) = \mathbf{p}(i \vee y)$.

The conditions for informational synonymity can be tightened in terms of the informational relevance of data object y to x with respect to the data base i. Two data objects x and y may be considered strictly informationally synonymous with respect to i if and only if the information conveyed by x with respect to i is equal to the information conveyed by y with respect to i provided that this transmitted information is equal to or greater than sufficient transmitted information and if, moreover, the aggregation of y to x or vice versa neither increases nor decreases the level of transmitted information. In other words, two data objects will be considered strictly informationally synonymous (symbolically $\mathbf{Syn}_i x, y$) with respect to i if and only if they are informationally synonymous and if the aggregation of either one to the other does not alter the measure of transmitted information with respect to i:

$$(26) \qquad \mathbf{Syn}_i x, y = \mathrm{df}\,[\mathbf{TI}_i(x/i) = \mathbf{TI}_i(y/i) = \mathbf{TI}_i(x \cdot y/i)] \cdot [\mathbf{TI}_i(x/i) \geqslant \varepsilon]$$

The concept of 'strict informational synonymity' with respect to the data base i assumes that data object y is informationally synonymous with x and simultaneously that it is informationally irrelevant with respect to $\mathbf{TI}_i(x/i)$. The above mentioned conditions are changed in the case of 'strict informational synonymity' in such a way that (a) in (24) is replaced by:

$$(27) \qquad (\mathrm{a}')\ \mathbf{p}(i/x) = \mathbf{p}(i/y) = \mathbf{p}(i/x \cdot y)$$

and (a) in (25) is replaced by:

$$(28) \qquad (\mathrm{a}')\ \mathbf{p}(i \vee x) = \mathbf{p}(i \vee y) = \mathbf{p}[i \vee (x \cdot y)]$$

But not even the concept of 'strict informational synonymity' of two data objects x and y with respect to i is any guarantee that Quine's criterion of 'salva analyticitate', i.e. the logical equivalence of x and y, will be satisfied. But it should be noted simultaneously that the logical equivalence of x and y is equally no guarantee that x and y are informationally synonymous or strictly informationally synonymous with respect to i if it holds that

$$\mathbf{TI}_i(x/i) = \mathbf{TI}_i(y/i) \qquad \text{and}$$
$$\mathbf{TI}_i(x/i) < \varepsilon \cdot$$

These properties can be demonstrated by a schematic example in which the calculation of $\mathbf{TI}_i$ is based on conception (II). The logical range of a universe (in the sense of Carnap [24]) is mapped by a square divided into sixteen equal parts (Figure 17). The extension of the individual data objects is described in terms of the numbers of the sections:

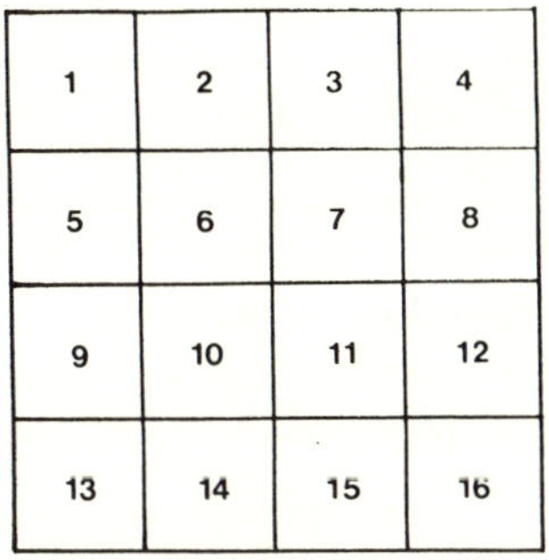

Fig. 17.

$i \ldots 4, 8, 12, 16$

$x_1 \ldots 4, 7, 8, 11, 12$

$x_2 \ldots 7, 8, 11, 12, 16$

$x_3 (\equiv x_1 \cdot x_2) \ldots 7, 8, 11, 12$

$x_4 (\equiv x_1 \vee x_2) \ldots 4, 7, 8, 11, 12, 16$

$x_5 \ldots 3, 4, 7, 8$

$x_6 \ldots 1, 2, 3, 4$

$x_7 \ldots 1, 2, 3, 4$

$\mathbf{p}(i) = \frac{1}{4}$

$\mathbf{p}(x_1) = \frac{5}{16} \qquad \mathbf{p}(i/x_1) = \frac{3}{5}$

$\mathbf{p}(x_2) = \frac{5}{16} \qquad \mathbf{p}(i/x_2) = \frac{3}{5}$

$\mathbf{p}(x_3) = \frac{1}{4} \qquad \mathbf{p}(i/x_3) = \frac{1}{2}$

$\mathbf{p}(x_4) = \frac{3}{8} \qquad \mathbf{p}(i/x_4) = \frac{2}{3}$

$\mathbf{p}(x_5) = \frac{1}{4} \qquad \mathbf{p}(i/x_5) = \frac{1}{2}$

$\mathbf{p}(x_6) = \frac{1}{4} \qquad \mathbf{p}(i/x_6) = \frac{1}{4}$

$\mathbf{p}(x_7) = \frac{1}{4} \qquad \mathbf{p}(i/x_7) = \frac{1}{4}$

If $\varepsilon = 0.8$, then

$$\mathbf{TI}_i(x_1/i) = \mathbf{TI}_i(x_2/i) = \mathbf{TI}_i(x_3/i) = \mathbf{TI}_i(x_4/i) = \mathbf{TI}_i(x_5/i) = \tfrac{5}{6}$$

This means that objects x_1, x_2, x_3, x_4 and x_5 are informationally synonymous. Using the concept ‘strictly informationally synonymous’, only objects x_1, x_2, x_3 and x_4 are involved. On the other hand objects x_6 and x_7, although equivalent, are not informationally synonymous, for $\mathbf{TI}_i(x_6/i) < \varepsilon$, nor are they strictly informationally synonymous.

XII. AN OUTLINE OF THE SEMANTIC EVALUATION OF GRAPHIC COMMUNICATION

1. INTRODUCTORY REMARKS

Contemporary cybernetic and informational technology and its peripheral equipment provides a facility for the processing and output mapping of a vast quantity of data. The speed with which this technology is capable of providing the user with huge quantities of data is a great bonus, and the data are complete and mapped with accuracy, but the operation does not always include a facility for rapid orientation in and selection of the data relevant to a given problem of management or decision or for clearly showing how the various items of data are connected. In an endeavour to solve these difficulties and snares of modern data storage, retrieval and processing, a number of trends have evolved in theoretical and technical cybernetics, including notably the following:

—investigations into the criteria and means of evaluating the relevance of data and retrieving relevant data,

—the solution of problems relating to the admissible reduction of data,

—the search for new forms of aggregating data in order to facilitate quick and easy interpretation of individual 'bundles' of data.

As these trends have evolved, data and collections of data have usually been taken in their discrete form, i.e. with the data mappable as a sequence of alphanumeric signs, or as rows, matrices etc. From the linguistic point of view they are, therefore, for the most part texts operating with verbal elements, and it is of no great importance whether these are verbal elements of natural or formalised or algorithmic languages.

Even at the early stages of the evolution of informational and cybernetic technology work began, however, with data in a non-verbal form, i.e. with the data as patterned forms, graphic patterns etc. A fairly large area of research, pattern recognition, evolved, concerned not just with the shapes produced by nature, but also with the patterned shapes that are human artefacts (e.g. written letters or other signs). But modern cybernetic technology has also inspired another sphere, which operates with patterns and graphic shapes at the output of cybernetic systems. This has given birth to a new trend in cybernetic technology, usually described as computer graphics. The concept 'graphic' is to be understood in a fairly narrow sense (since traditional high-speed printers also give a graphic output), namely as 'pattern', operating with patterns which are non-verbal in character. This covers all manner of drawing apparatus, automatic plotters, pattern displays, active and interactive displays and so forth.

Patterns and their use in information systems provide a special way of resolving problems of the selection of relevant data, the reduction or aggregation of data. In a suitable form patterns represent aggregated 'bundles' of data or a compressed form of collections of data which facilitates fairly rapid and efficient orientation in a set of data, retrieval of desirable or wanted data and easy orientation within the data retrieved. For example, if we were to convert all the data contained in a map into the traditional form of a text, the retrieval of relevant data in a text of that form would be incomparably more tedious and time-consuming than in a graphic (in the narrower sense) form. Maps, graphs, grids and other pattern forms are therefore also texts or collections of data sui generis, which, in the proper form, are quicker, easier and usually more effective to interpret. Metaphorically speaking, if graphic forms of communication in the above sense are a text, then this kind of text, unlike traditional forms of texts, can be read on more than one page at once.

It is therefore no accident that it has been found useful to operate with data and collections of data in this kind of graphic form in many areas of human endeavour, notably in technical fields, in all fields to which maps are fundamental and where great quantities of data can be localised. Even if these collections of data can be digitalised and the conversion to digital forms of recording are not a problem, the graphic form retains its advantages from the user's point of view. These facts have also inspired research into what has been variously described as 'graphic languages', 'picture languages', 'two-dimensional languages' and the like. Surveys of the problems involved, especially as regards the syntax of these languages, are to be found in [65], [100] and [128]. But this is a fairly broad range of quite heterogeneous problems which are soluble on the basis of user requirements and the technical devices employed. Any attempts to construct a universally conceived 'picture syntax' or a general syntax of graphic languages have therefore failed to achieve a solution which would be acceptable for all purposes.

In this chapter we shall point to some of the possibilities for the informational evaluation of pictures on the basis of the concept of informational synonymity introduced in the preceding chapter. This is then a possible approach to the semantics of pictures on the basis of transinformational analysis. For the syntax of pictures we shall confine ourselves to the application of a data grid structure to the conception of graphic communication put forward by J. Bertin [9].

2. GRAPHIC COMMUNICATION

We have suggested briefly some of the advantages of pictorial or graphic communication. Communication of this kind presupposes that the vehicles of information are (as opposed to sequentially organised series of alphanumeric signs in which the relevant syntactic and semantic rules are respected) *pictorial or graphic shapes* which have a meaning as a whole and can also be interpreted as a whole. (This global interpretation naturally does not exclude the possibility of partial interpretations of separate parts of a shape.)

By a shape we mean a certain organisation of constituent elements which can be produced by human activity or some technical apparatus which imitates human activity (e.g. drawing apparatus, screen display, television tube etc.). These are usually point, line or areal elements which may be variously combined together or combined with other elements such as alphanumeric signs. In some situations (such as in draftsmanship) picture shapes may be not unlike iconic models operating with elements of three-dimensional space.

Pictures or graphic shapes may be considered – and also interpreted as – an isolated manifestation of human activity, or the expressions of a graphic language which also respects that which is specific to any meaningful use of language, namely syntactic and semantic rules. Between the two types of picture use there are various graduated transitions. While the first type is more specific to the sphere of the graphic arts, where concern is with the individual works, the second type is more important to the sphere of intellectual and technically oriented communication. This differentiation of the two types of pictures is of course only relative and conditional, for even in the interpretation of isolated manifestations of human activity, as in the sphere of the graphic arts, it is not only possible, but also useful to consider 'types of writing', 'types of language' and so forth. Within this study, however, we are going to be concerned solely with the second type of graphic communication, in which the link between the signs which are utilisable and the relevant syntactic and also semantic rules is much more vital. This is also due to the fact that in certain cases, as in cases of certain maps or the graphic base material in planning documentation, some signs (operating with point, line or areal elements) and their interpretations are set up in the form of a fixed standard.

We might discuss the various possibilities for classifying pictorial or graphic shapes. One kind of graph is that of the inter-relation between two or more observed parameters, another is that of the distribution of one property from the point of view of the distribution of another (e.g. the age tree of a given population), or the representation of the distribution of certain phenomena in space by means of a map, the representation of the link between certain stages or sub-stages in time by means of a grid, and so forth. Bertin [9] distinguishes three types of graphic shapes:

(a) diagrams
(b) grids
(c) cartographic shapes.

This classification is unquestionably very rough and clearly fails to capture all the possibilities or the transitional types. What is always important, however, is the concept of 'shape' (which corresponds to the concepts of perception psychology, notably 'Gestalt', 'pattern', 'image' etc.). By shape Bertin means a significant perceivable form which we are capable of registering in a minimum time of perception. This can also be understood as meaning that the perception of a shape may reduce the time necessary for the transmission of information and at the same time ensure the transmission of a relatively large amount of information in a relatively short time.

Thus perception of a shape in graphic communication makes it possible to register and interpret in a relatively short time the essential components of the information tendered and to pass according to need from these essential components to the more detailed components. The shape expresses a complex whole which can, in its essential components, be perceived and interpreted as a whole, that whole being a fairly inextensive and yet informationally copious set of data.

If apprehension and interpretation in graphic communication are made possible by a shape as a whole, in subsequent interpretation of a graphic communication other (more detailed) data may be assigned to the whole. This means that graphic communication can be interpreted as the kind of concatenation in the set of data whereby to one superior item of data other (more detailed) data are gradually attached. In the theory of sets of data this method of concatenating a superior datum (expressing in complex form the meaning of the shape) and more detailed data corresponds to the 'master' type of data sets with a grid structure.

The master item (denoted M in the diagrams to follow) can be understood as expressing the basic characteristic of the pictured state of affairs in the given universe, this characteristic being open to rapid and complex visual perception. In graphs this could be, for instance, the nature of the distribution or occurrence of the phenomena displayed, or on a map it could be the basic features or the types of objects incorporated. The master datum is also what the user who is duly equipped with a knowledge of the graphic devices employed 'reads' first before he proceeds to read more of the graphic form used. From reading M it is possible to pass to reading the more detailed data ($D_1, D_2, \ldots, D_n$). We speak of a grid structure in this connection because it is possible to move from any detailed datum back to the master datum or to any other detailed datum.

This method of interpreting graphic communication presupposes that the data $M, D_1, D_2, \ldots, D_n$ are understood as sentences or (relatively not very large) sets of sentences. This also matches the thesis that the various forms of graphic communication are texts sui generis. Whereas traditional forms of text, operating with alphanumeric signs, are based as a rule on the sequential ordering of partial data (mapped as sentences or sets of sentences), the sequential ordering of a master datum and detailed data is only one of the possible interpretations of a graphic communication. The most important of the many possible forms of interpretation (including a digital transcription) of a graphic communication are the following orderings:

I. One of the possible forms of concatenation of a master datum and the detailed data is sequential, as expressed in Figure 18. The feedback from the last datum in the sequence of detail data to the master datum is not absolutely indispensable. This method of concatenation is applicable anywhere where it is essential to proceed through an ordered sequence of detail data, as in determining the contours of some feature on a map. The same method is used in automated digital processes where the digitiser is tracing, for instance, the course of an isoline or contours and the

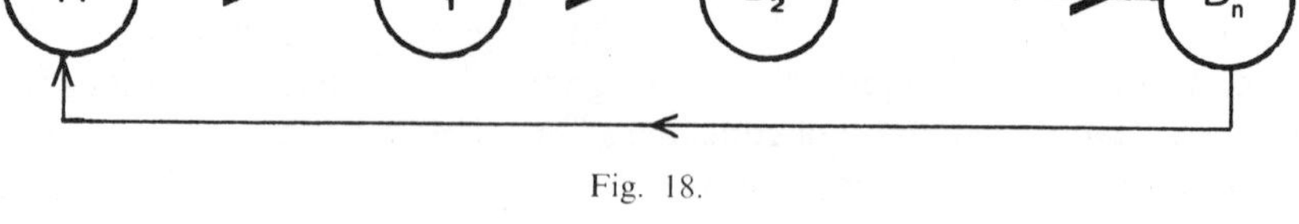

Fig. 18.

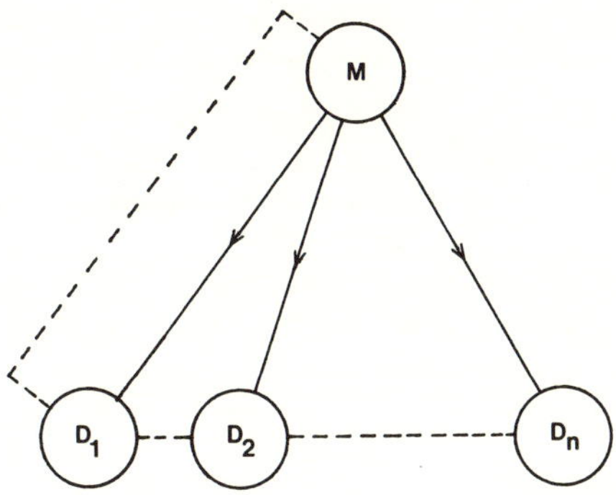

Fig. 19.

like. In digitalisation by means of a digitiser it usually does not matter whether the detail data are acquired in the direction D_1 to D_n or vice versa. The same goes for automated drawing on drawing apparatus, where the sequential ordering of the detail data is effectively mapped on the understanding that only the whole set of these data will give the overall shape.

II. Another form of concatenation of a master datum and detail data in the interpretation of a graphic communication is the kind of ordering which enables us to pass from the master to any of the detail data and also to return, if need be, to the master datum (see Figure 19). In this case there is no need to 'read in detail', i.e. go through all the detail data, in the pictorial or graphic form. This applies to situations, such as map-reading, where only certain parts are of importance to the user, who can make do with just a selected part of the pictorial or graphic 'text'. At the same time what is important here is that that part, i.e. some selected subset of the set $\{D_1, D_2, \ldots, D_n\}$, is, with the master datum M, meaningful and can therefore satisfy the user's need for information. This method of reading a pictorial or graphic form corresponds to the search for entries in a catalogue or dictionary, the selection of requisite data from a register and so forth, when it is neither necessary nor useful, for the purpose in hand, to read the entire catalogue, dictionary or other similar text.

III. A third type of concatenation of master and detail data in the interpretation of a graphic communication is the tree ordering in which movement is along the separate branches. This is illustrated in Figure 20. The tree graph presupposes that the data in the graphic communication are gradually specified in extra levels. In specifying an object on a map

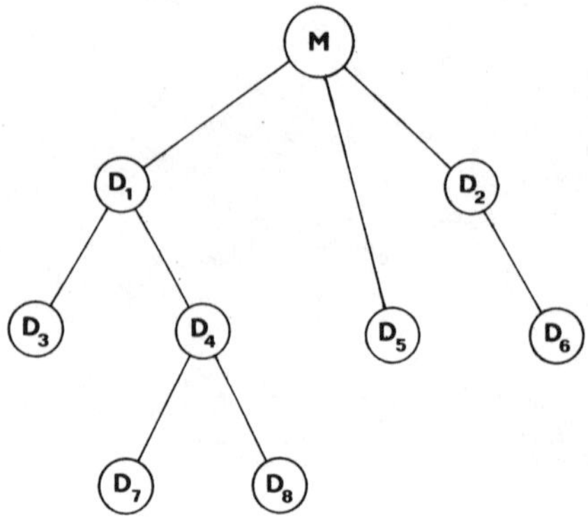

Fig. 20.

its extra characteristics are gradually specified according to the hierarchy of levels of the characteristics, which could also be portrayed in a matrix. Gradual elimination may be applied in, for instance, specifying an object's exact localisation, or in gradually applying more and more details in an interpretation of a pictorial shape, and so on.

These three types of interpretation (the list is by no means exhaustive and gives only the commonest types as they occur mainly in the interpretation of map-like data bases) can also be treated as *metalanguage transcriptions* of what is being communicated by the graphic or pictorial devices. If we take a communication expressed by graphic or pictorial means as a concrete communication in an object language, the metalanguage analysis of that communication can be made on several levels:

(a) the descriptive level,
(b) the syntactic (generative) level,
(c) the semantic level.

So it can be said that the above types of interpretation of graphic communication represent the metalanguage transcription at the semantic level, the metalanguage being a language of the natural or formalised type, operating with sequentially ordered alphanumeric signs. In information systems which assume a parallelism between the graphic and the digital forms of the output, this kind of transcription in a semantic metalanguage can be viewed as a transcription in a conversion language which is the common semantic metalanguage for both versions of the output data.

By the *descriptive level* of metalanguage transcription of a graphic communication we mean the conception of a graphic communication of certain elementary elements. The forms of this organisation are the systems of descriptive attributes or relations assignable to the individual elements, to pairs, triples of elements, etc. We consider to be elementary elements signs which operate with:

—point elements,
—line elements,
—areal elements,

or signs operating with combinations of these. The selection of elementary elements and descriptive attributes or relations depends on the nature of the graphic language used. In the case, for instance, of automatic drawing apparatus, the elements are the signs which the equipment is capable of drawing. The descriptive attributes are then, for example, the linking of two point elements by a line segment or a curve of stated properties, the properties of line elements, contiguity, overlapping, contact etc. of areal elements and so on.

The *syntactic level* of metalanguage analysis of a graphic communication is the conception of the graphic communication as the realisation of certain syntactic (generative) rules on the system of a given graphic language. Some authors, such as R. Narasimhan [101], speak in this connection of the generative level and generative description. This clearly amounts to an application of Chomsky's conception of generative grammars to the sphere of graphic communication. Accepting that a graphic language (**GL**) is made up of a system of elementary signs **X**, a system of their attributes and relations **P**, a system of formation rules **F** and a system of transformational rules **T**, the graphic language can then be defined as follows:

$$\mathbf{GL} = \{\mathbf{X}, \mathbf{P}, \mathbf{T}, \mathbf{F}\}.$$

An elementary form of generative rules corresponding to this structure of a graphic language has been put forward by Narasimhan [101].

3. THE SEMANTICS OF A PICTURE SHAPE

The semantic analysis of a picture shape can be considered from a number of angles. One way would be to consider the shape's denotation, i.e. its relation to a particular state of affairs in the relevant universe, or its sense, i.e. the meaning invariant which remains preserved however the shape is used, etc. But this method of semantic analysis, which goes back to Frege's differentiation of sense and denotation, is closely bound to purely verbal communication and is only of limited significance to graphic communication. In the following discussion we shall understand the semantics of a picture shape to mean the interpretation of the shape which is based on a transcription of the given picture or graphic shape, or pattern, into a semantic metalanguage, which is verbal in character and can be understood as a sequentially ordered text.

This conception of the semantics of a picture shape is especially useful in information systems in which it is expedient to allow for parallel data in graphic and digital form, both forms of data being retrievable from the same data base. In these cases and others like them the semantic metalanguage functions as a conversion language for the two forms of data selection. Since selections from the data base can be made in the graphic or the digital form, it is extremely worthwhile comparing the advantages and disadvantages of both forms from the informational point of view. Insofar as we are going to refer to the informational point of view or the informational evaluation of data in a graphic or digital form, what we shall have in mind is the informational evaluation which is based on a transinformational analysis.

Transinformational analysis does not assume the informational evaluation of a given datum or collection of data as such, but a method of informational evaluation which takes into account the degree to which the given data or collections of data enrich the user or addressee of the data. This means that in a transinformational analysis we consider both the a priori informational equipment of the user, e.g. a given set of tasks, problem situations etc., and the degree to which the data communicated enhance that a priori equipment, or, in other words, the degree to which they reduce original uncertainty or the original level of entropy.

From this point of view it is possible to construe the semantic analysis both of data presented in graphic form, and data or collections of data communicated in digital form. It has to be assumed, meanwhile, that the data presented in the form of a pictorial or graphic shape and the data in digital form are synonymous for the user with respect to the same a priori equipment at his disposal. Synonymity here must not be understood as absolute synonymity, but as synonymity relativised to the user and his a priori informational equipment. These demands are met by the concept of 'informational synonymity'.

In information systems we are quite often faced with the situation where we need to judge whether the graphic and the digital forms of presenting data are synonymous. So, for example, the graphic representation of a terrain in the form of a map or map schema ought to be informationally synonymous with the system of sentences which express the primary and derived (extrapolated) data acquired by measurement. It is similarly desirable for the graphic representation of the evolution of certain indices over a certain interval to be informationally synonymous with the collection of data expressing the size of those indices at contemplated moments within the given interval. Also in automated information systems or at the output of cybernetic installations for data processing it is possible to allow for both sequentially ordered sets of data as produced by the high-speed printers of the installations and for a graphic representation of the same set of data in the form of a graphic or pictorial pattern.

The semantic analysis of a pattern can therefore be usually drawn up like the semantic analysis of a text, or like text semantics. As in a sequentially ordered text the interpretation of each separate sentence is not usually complete without some regard for the interpretation of the preceding sentences in the text, so too the interpretation of each separate sign or sign-complex in a pattern requires consideration of the whole pattern. In text semantics it is usually stressed that the ordered sequence of the sentences in the text gradually fixes the universe, the concrete connections or the so-called semantic localisation. If we were to take a sentence out of context without knowing the process of semantic localisation pursued in previous sentences, we would not be able to appreciate fully the meaning of the sentence within the text as a whole. Similar arguments apply to the individual signs or sign-complexes within a picture shape in graphic communication. But there is here one very important difference: in a sequentially ordered text the process of semantic localisation is usually laborious and requires the whole or almost the whole of the preceding text to be gone through, whereas in graphic communication the analogous process in usually easier and quicker. To put it schematically: the semantic localisation of a sign or sign-complex which expresses detail datum D_i is usually achieved by a return from D_i to the master datum M, i.e. the datum which expresses the basic and rapidly perceptible characteristics of the state of affairs in the given universe.

4. INFORMATIONAL SYNONYMITY AND THE INFORMATIONAL EVALUATION OF A PICTURE SHAPE

In order to construct criteria of informational evaluation of a picture shape, we have to assume that the text represented by a sequentially ordered set of sentences and the text sui generis represented by the picture shape are informationally synonymous (in the sense of Chapter XI). Informationally synonymous are those data, sets of data, texts or whatever other communications which are mutually substitutable with respect to a given task or complex of tasks, the measure of the transmitted information (transinformation), which the communications provide with respect to the task or tasks, being preserved. Again it must be stressed that the concept of 'informational synonymity' as so described is always relativised to a particular a priori informational equipment containing the specification of the task or complex of tasks, or possibly even including what is already known about the task.

In addition to the ordinary symbols of logic we shall, in the following analysis, use these signs: picture shape will be denoted g and be treated as a text sui generis which is an organisation (e.g. in the form of the graph schemas mentioned) of a master datum $\boldsymbol{M}$ and detail data $D_1, D_2, \ldots, D_n$, i.e.

$$g = \mathbf{F}(M, D_1, D_2, \ldots, D_n).$$

Data or a collection of data in digital form will be denoted s and will be considered to be a sequentially ordered text, i.e. a sequence of individual sentences, i.e.

$$s = \{S_1, S_2, \ldots S_n\}.$$

The a priori informational equipment will be denoted i, which can also be considered to be a class of sentences which specify the task or complex of tasks or any other data which are available a priori, i.e.

$$i = \{I_1, I_2, \ldots I_n\}.$$

Two data or sets of data are informationally synonymous with respect to the given a priori equipment if and only if the measure of transinformation of both is the same with respect to i. But since the measure of transinformation may be the same if it is zero or very low, allowance must also be made for the fact it must reach a sufficient level with respect to i, which level is set by the demands of quality placed on the solution of the task or complex of tasks. For example, if we are to solve some town-planning problems in a limited part of a city, we are going to need the relevant maps as a basis to fix the planimetry and relief of that area. The same bases could also be expressed in digital form. In order to confront the two forms, the base material must in both cases be adequate to the task.

A small-scale general map showing the area and the corresponding digital expression may well convey the same information, but that information may not be sufficient. Sufficient transinformation with respect to i will be taken to be such information transmitted with respect to i as is cable of ensuring that the task will be solved to the required standard, that the original uncertainty will be satisfactorily eliminated, that the solution will be guaranteed to be associated with a risk or inaccuracy that is not in excess of a stated level, and so forth. This means that 'sufficient transinformation' is determined by convention. Transinformation furnished with respect to i is sufficient if

$$\mathbf{TI}_i(s/i) \geqslant \varepsilon$$

where ε is established by a convention with respect to the demands of the type suggested.

On the basis of these assumptions we can define the informational synonymity of picture shape g and set of (digital) data s with respect to i (denoted by the sign $Syn_i g, s$) in accordance with Chapter XI as follows:

$$Syn_i g, s = \text{df}[\mathbf{TI}_i(g/i) = \mathbf{TI}_i(s/i)] \cdot [\mathbf{TI}_i \cdot (s/i) \geqslant \varepsilon]$$

This definition expresses the conditions which are stipulated for informational synonymity: picture shape g and data s are informationally synonymous with respect to i if and only if they are mutually substitutable, the measure of transinformation with respect to i being preserved and that measure being greater than or equal to sufficient transinformation.

The above definition of informational synonymity relativises the concept 'synonymity of a picture shape and a set of data in digital form' on the one hand to the a priori informational equipment i, and on the other to a conventionally selected measure of sufficient transmitted information. This means, for instance, that if g and s are informationally synonymous with respect to i, they need not be synonymous with respect to a different task or to a different a priori informational equipment j. Violation of existing informational synonymity may come about if the demands made of the limits of sufficient transinformation are altered.

Insofar as the definition of 'informational synonymity' is based on the content measure of semantic information **cont**, it can be shown that

$$Syn_i g, s \equiv [\mathbf{p}(g \vee i) = \mathbf{p}(s \vee i)] \cdot [\mathbf{p}(s \vee i) \leqslant 1 - \varepsilon(1 - \mathbf{p}(i))]$$

The construction of the probability measure **p** may also be justified on the basis of so-called algebra of events, in which individual events or state of affairs are considered as subsets of the universe to which the picture shapes or data sets used relate.

The concept of 'informational synonymity' can serve as a foundation for the informational evaluation of picture shapes (relativised to the given a priori informational equipment, including the specification of the task or complex of tasks) but provided the shapes can be converted to data in digital form. If the conversion, usually made by the relevant digitisation, is reliable enough for us to assume that the picture shape and the digitised set of data really are informationally synonymous, the information value of the picture shape is determined mediatedly by means of the information value of the synonymous data set, i.e.

$$Syn_i(g, s) \longrightarrow \mathbf{TI}_i(g/i) = \mathbf{TI}_i(s/i)$$

5. INFORMATIONAL SYNONYMITY AND THE TIME FACTOR

If the picture shape g and the data set s are informationally synonymous, i.e. convey the same information with respect to a given a priori informational equipment and task i, this does not yet mean that g and the synonymous s are equally useful from the user point of view. For the user it is not just the quality and scope of the transmitted information which is important, but also certain other factors in the transmission. Of these other factors the time factor is the most significant.

The time factor is particularly significant from the point of view of the use of the information in management and decision processes. In processes of this kind some importance attaches to the time needed for an adequate interpretation of the data as transmitted in graphic or digital form. One of the specific features of graphic communication is the opportunity it offers of registering and interpreting the essential components of a piece of transmitted information in a fairly short time and of acquiring fairly quickly and in a compressed form these essential components in such a way that they can be gradually supplemented by the detail data.

If g is a picture shape and

$$g - \mathbf{F}(M, D_1, D_2, \ldots, D_n),$$

and s is a set of data in digital form and

$$s = \{S_1, S_2, \ldots, S_m\},$$

the interpretation of both texts takes place in a time t. This means that the time needed to interpret g, i.e. t_g, and the time needed to interpret s, i.e. t_s, have to be confronted and built into the informational evaluation of $\mathbf{TI}_i(g/i)$ and $\mathbf{TI}_i(s/i)$. If $Syn_i g, s$ and if moreover $t_g \leqslant t_s$, there is no doubt that it is better from the user point of view to base the management or decision process on the interpretation of the picture shape.

The advantages of graphic communication in terms of the time factor need not be confined to the reduction of the time necessary for the interpretation of g or s. Even more important is the fact that the two interpretations are usually made in different ways. It is obvious even intuitively that in reading a sequentially ordered text it is going to take some time before we get to the 'heart of the matter', to an appreciation of the 'merits of the case'. Usually this means going through a long line of data which can be read and interpreted only by stages and in a fixed order. So the interpretation proceeds from S_1 to S_2, S_3 etc. all the way to S_m, the progression from S_1 to S_m requiring us to read and interpret many sentences which are not always essential for revealing the merits of the case.

By contrast, in reading and interpreting a graphic shape, we start with what has been called the master datum and proceed straight to the most suitable reading and interpretation of detail data, which can in any case be surveyed practically at a glance. This is shown schematically in Figure 21. Under this procedure it is possible to omit many of the detail data as required.

The interpretation procedure can be expressed as a gradual reducing of uncertainty or of the original level of entropy. This also means that the process of information transmission, the measure of which we have expressed by means of the measure of transinformation $\mathbf{TI}_i$, always takes place as a sequence of steps, and so in time. Thus in the transmission of the information, $\mathbf{TI}_i(s/i)$ or $\mathbf{TI}_i(g/i)$ are not realised at once, instantaneously, but as a sequence of batches or quanta of information.

Since we have defined $\mathbf{TI}_i(g/i)$ and $\mathbf{TI}_i(s/i)$ as

$$\frac{\mathrm{U}(i) - \mathrm{U}(i/g)}{\mathrm{U}(i)} \quad \text{and} \quad \frac{\mathrm{U}(i) \rightarrow \mathrm{U}(i/s)}{\mathrm{U}(i)} \text{ respectively,}$$

it is possible to view $\mathbf{TI}_i(g/i)$ or $\mathbf{TI}_i(s/i)$ merely as a result which is arrived at in time t_g or t_s respectively. In fact the measure of original uncertainty

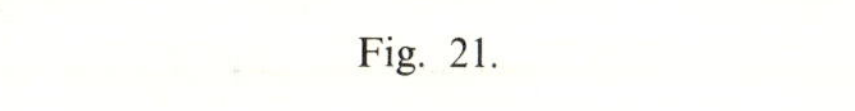

Fig. 21.

or original level of entropy are reduced gradually, i.e. as a sequence of intermediate reductions, e.g. as the sequence:

$$\frac{\mathbf{U}(i)-\mathbf{U}(i/g_0)}{\mathbf{U}(i)}, \frac{\mathbf{U}(i)-\mathbf{U}(i/g_0 \cdot g_1)}{\mathbf{U}(i)}, \ldots \frac{\mathbf{U}(i)-\mathbf{U}(i/g_0 \cdot g_1 \ldots g_n)}{\mathbf{U}(i)}$$

where $g_0, g_1, \ldots g_n$ are the gradually read and interpreted elements of the overall picture shape. The process of gradually reading and interpreting the individual sentences of a set of digital data can be expressed analogously.

The advantages of graphic communication are especially apparent from the fact that this sequential reduction of the original uncertainty can be organised so as to give a major reduction of the original uncertainty even in the early stages. This can be expressed as in Figure 22, where the

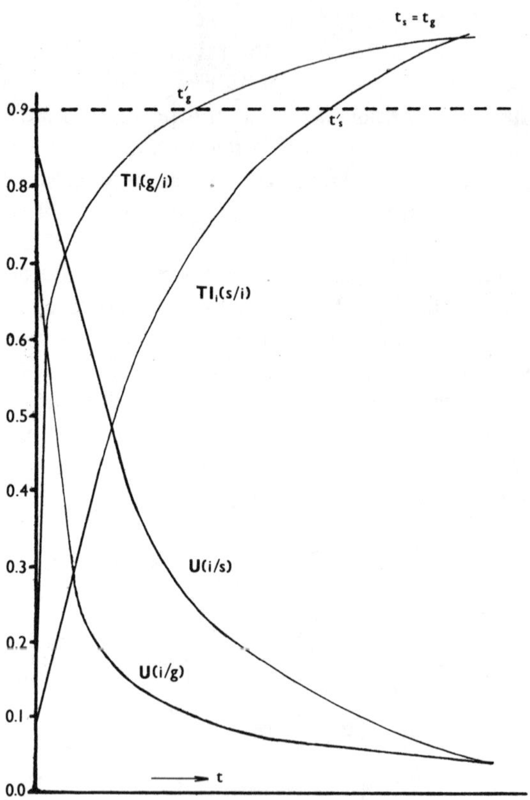

Fig. 22

x axis represents the time rhythm and the y axis the measure of uncertainty and its gradual reduction.

This figure shows that under the gradual interpretation of a graphic communication the reduction of uncertainty proceeds quite rapidly at first, since a major reduction is ensured by the master datum. By contrast under the gradual interpretation of digital data in the form of a sequentially ordered text, the reduction of the original uncertainty usually proceeds very slowly at first. In the diagram

$$\mathbf{TI}_i(g/i) = \mathbf{TI}_i(s/i),$$
$$\mathbf{TI}_i(s/i) \geqslant \varepsilon \quad \text{and}$$
$$t_g = t_s$$

But if, in the same process of gradual reduction of the original uncertainty or original level of entropy, it is enough for us to achieve what we have described as sufficient transmitted information, which is established by convention as the value ε, under the assumptions that $Syn_i g, s$ and $\mathbf{TI}_i(s/i) \geqslant \varepsilon$, then it is obvious that with gradual interpretation in graphic communication the requisite value for the reduction of uncertainty (indicated on the y axis of the diagram by the value $\mathbf{U}(i) - \mathbf{U}(i/g)$) can be achieved much earlier, i.e. in time t'_g, which is less than the time needed to achieve the same reduction of uncertainty under the gradual interpretation of a set of digital data, i.e. time t'_s. This means then that preference must be given to the graphic form of communication, even though it holds that

$$Syn_i g, s$$
$$\mathbf{TI}_i(g/i) = \mathbf{TI}_i(s/i) \geqslant \varepsilon,$$
$$t_g = t_s \text{ under the assumption that}$$
$$t'_g < t'_s.$$

It must be emphasised that the above discussion is not at all intended to imply any general conclusion that the graphic form of communication is to be given absolute priority over digital forms. Such priority is warranted only in certain areas of human activity, notably in areas which work with maps, graphic documentation in mechanical and construction engineering, in some areas of the natural and technical sciences, and in all areas where the data under consideration can be localised in time and space. The other justification for its priority is given by the fact that the fairly strict criteria linked to the concepts of 'informational synonymity' and 'sufficient transmitted information' can be satisfied.

NOTES

TO CHAPTER I

[1] by Brutjan amongst others [173].

TO CHAPTER II

[1] In Frege 'Sinn' and 'Bedeutung'. This corresponds to Russell's differentiation between 'meaning' and 'denotation', Black's 'sense' and 'reference' and others. In the account that follows we shall use the term 'meaning' as the more general term with respect to 'sense' and 'denotation' in instances where the two modes are not distinguished.
[2] In the original version of Tarski's conception [147] the semantic concept of truth was related only to formalised languages.
[3] This is the line taken also by Tarski [147], but see also [106] and others.
[4] This may be doubted. We shall have more to say later about the reservations that may be held against Tarski's empirical premiss and that are to do with the criterion of synonymy.
[5] The proof will be given in the section on synonymy.
[6] Within Czech writings on logic, an analysis of this antinomy is to be found in [170].
[7] Taken historically, metalanguage was first formulated in syntactic terms. This corresponds in particular to Hilbert's conception of metamathematics as the 'theory of proofs' in mathematics.
[8] Carnap's scepticism, expressed especially in [22], grew less later. He expressed his new standpoint in a paper on certain concepts of pragmatics [29].
[9] Admittedly it can be demonstrated that with greater or lesser changes this kind of schema can be found in earlier authors. Moreover, this schema best corresponds to the naive view of the role of words and the naive approach to the problems of semantics generally.
[10] For example, in natural languages we distinguish phonemes, morphemes, words, sentences, etc., and from elements of one category we can create elements of a higher category. Although the significance of formation rules in logic is rather different, it is nevertheless clear that there is a certain categorisation here and that expressions of one category are formed into expressions of another.
[11] The original schema of linguistic communication is taken over from C. Cherry's work [33].
[12] This is the set of elements of the psyche that are relevant with respect to the given conversation, usually expressed in mentalist terminology as 'ideas', 'impressions' and the like.
[13] This is by no means to suggest that the domain of objects may not also include as an element certain psychological or somatic states of either or both of the persons between whom the conversation takes place. That kind of situation is obviously hard to express in diagram form, and therefore we make no attempt to distinguish whether the domain of objects does or does not contain phenomena of that kind.

[14] The concept of 'average level of risk' is taken from the theory of statistical decision functions. At the same time it is assumed that any conversation that is supposed to ensure an effective exchange lays certain demands of quality and reliability on the process of reaching an understanding and therefore tolerates only those losses that do not exceed the level set by the aim of the conversation.
[15] The nature of the relationship between the two aspects will be elaborated at a later stage.
[16] This fact, which is natural and has its own objective foundations, cannot be expounded in subjectivist terms. Any such subjectivist account is usually put forward in the form of the thesis of the 'basis of one's own psyche' (die eigenpsychische Basis) for the 'logical structure of the world'. This kind of account figured, amongst others, in the period of the evolution of Carnap's thinking that is represented by his book *Der logische Aufbau der Welt* [18].

TO CHAPTER III

[1] In his later work [74] Kotarbiński confined himself to the three terms express, signify and denote.
[2] This discovery accords with the Sapir–Whorf hypothesis of linguistic relativism.
[3] From the account so far it only follows that if 'signify' is a two-place predicate and an expression of language is one of its arguments, then what we have been calling 'sense' is its other argument.
[4] A question also asked sometimes is 'What does "what does ... signify" signify?'. However, this is a rather different question which we shall provisionally leave to one side. This means that for present purposes we shall not be seeking a conception of sense partly also for the simple reason that the sheer variety of situations that give rise to the quest for the sense of an expression of language and seek criteria of sense have led, and will continue to lead, to an equal variety of conceptions about sense. We are concerned here instead with the procedure by which we elucidate sense for ourselves – the procedure employed in a situation where we do not know what this or that signifies.
[5] Russell sees this category in very broad terms, including within it such words as 'near', 'far', 'present', 'past', 'future', 'this', and in addition he assigns to this category a special importance in connection with the conceptual structure of our image of the objective world. In this context his views on language and its semantic problems have clearly been coloured by his philosophical standpoint.
[6] In *Meaning and Necessity* [23, p. 97] Carnap notes that the term 'denotes' is also sometimes used in a sense that is not that of a semantic relation: in the sentence '*X* denotes *y*', *X* may also be the name of a property and *y* an entity that has that property. For this kind of obviously non-semantic relation there is clearly every justification for using such expressions as 'pertains to' or 'relates to'. Misgivings arising out of this duality of 'denotes' have sometimes led to terms like 'naming relation' or 'name relation' being used to characterise the semantic relation associated with 'denotes'. Therefore we are here using 'denotes' for what Carnap [23] calls the 'name relation' and for what Frege uses the term 'bezeichnet'. Russell and Church both use 'denotes'.
[7] This is the range of problems that has been paid most attention in logical semantics as well, which explains why the conception of logical semantics that is based on the 'denote' relation has usually been described as 'denotational semantics' (see e.g. [89]). On the other hand, the semantics that was concerned not only with denotation, but also with sense, i.e.

intensional semantics, has not been so fully developed, most attempts and any development having gone back to Frege's conception.

[8] The conditions under which language expressions may represent each other, where the conditions are connected with the nature of the context, will be elucidated in greater detail in the chapter on sense and denotation. In the exact determination of the semantic characteristic that is preserved we shall distinguish between the concept of 'substitution', for where an expression's extensional characteristic is preserved, and 'L-substitution' for preservation of the intensional characteristic.

[9] Here will belong the various conventions governing the use of abbreviations or the special code phrases used in international trade and the like.

[10] The stricter of the two requirements is also of more significance for natural languages. This is particularly so in the area of the semantic foundations of natural language translation theory, which also has to be based on other considerations to do with the problems of context as well as certain pragmatic aspects.

[11] The term 'modes of meaning' is borrowed from C.I. Lewis [83], [84]. Taking 'meaning' as a more general term, we shall be able to specify it more closely by distinguishing between 'sense' and 'denotation', and possibly certain other modes of meaning as well.

[12] For an account of the logical problems of formalised language see, for example, the introduction to Church's work [39] and Scholz and Hasenjaeger's [141] among others.

[13] Church uses the term 'logistic system' in the same sense [39].

[14] The concept of 'primitive sign' assumes, as Church points out [39, p. 49], the indivisibility of such signs in two respects: (a) at no place in the building up of $\mathscr{L}$ are parts of them used; (b) any finite linear sequence of primitive signs can only be treated as a finite linear sequence of primitive signs in one way.

[15] Generally speaking, these are language expressions where there is justification for relatively independent semantic analysis. Δ therefore corresponds to Carnap's concept of 'designator' [23].

[16] 'Concatenation' corresponds to S.C. Kleene's term 'juxtaposition' [72, p. 71]. R.M. Martin uses the term used here [89, p. 72ff.]. The term 'concatenation' should be kept distinct from 'contextual concatenation' ($\overset{c}{\frown}$), which will be introduced later in the semantic analysis of context.

[17] *OMNIA* and *EXIST* are used in the metalanguage to record the names of the quantifiers, hence they cannot simply be translated into natural language as 'all' and 'some' – they must always be connected to the appropriate variables. For this reason we do not use the notation *OMNIA* $\dot{\frown}\xi\dot{\frown}$ ... and so forth.

[18] The differentiation of signs is not itself the direct concern of semantics. Logical semantics generally distinguishes two kinds of signs: signs in the concrete sense and signs in the abstract sense. (The terminological distinction of the two types of signs differs from author to author considerably, although the essentials of the distinction largely coincide. C.S. Peirce [109] used the terms 'token' and 'type', Carnap [22] 'sign-event' and 'sign-design', Reichenbach [124] 'token' and 'symbol', Dürr [40] 'Zeichen in konkreter Bedeutung' and 'Zeichen in abstrakter Bedeutung', etc.) When we have in mind a certain actual occurrence in time and space, i.e. a certain sign event, we refer to sign in the concrete sense, otherwise we refer to sign in the abstract sense. In logic and logical semantics the usual viewpoint is that which takes signs in the abstract sense as the starting point for the semantic analysis of formalised language. Conceptions can, however, be found, and they are usually associated with an extreme nominalist viewpoint, that take signs in the concrete sense as the starting point. (This possibility has been expanded in particular by R.M. Martin [89].)

[19] The present account will note in particular the conception of model developed in the works of J.G. Kemeny [67], [69]. The concept of 'model' had already been used in the earlier works of Tarski, especially [148].
[20] We shall point out some of the basic methods of introducing semantic concepts at a later stage.
[21] To be more precise, we shall, for the needs of the present analysis, declare certain wff to be meaning postulates.
[22] In this respect the reader is referred to the above-mentioned literature and other publications and findings of modern logic.
[23] This should obviously not be identified with the fact that $\mathcal{ML}$ has its own syntax. Moreover a metalanguage may be a formalised language, but it may also be a part of some, indeed any, natural language.
[24] This problem will be dealt with further as L-semantics.
[25] The heart of this problem is the semantic analysis of non-logical constants.
[26] We treat this as a pragmatic concept. This means that it includes a relation to the aims and tasks of a particular scientific activity.
[27] These views of Tarski's come from his work [146] of 1935 when the ideas of the Vienna Circle enjoyed considerable authority. In explanation Tarski pointed out that semantic concepts are neither physicalist – in the sense of the original physicalism of the Vienna Circle – nor descriptive – in the sense of Carnap's later standpoint of the period of his *Introduction to Semantics* [22].
[28] In this respect particular importance attaches to the works of R.M. Martin, especially his [89].
[29] The second way is, if possible at all, apparently closely connected with the logical theory of types, according to Tarski.
[30] This proceeding is followed by Carnap [22, p. 27], Stegmüller [144, p. 44], Martin [89, p. 119] and others.
[31] This is actually acknowledged by Carnap [22, p. 48], who admits that Tarski's approach is preferable to the one he himself upholds.
[32] This relates to 'semantic systems' S_1 to S_5 in [22].
[33] For example, if in $\mathcal{L}$ we also consider the individual constants p and t with the interpretation given previously, the sentences P_1p, P_2p, t are instances of true sentences in $\mathcal{L}$.
[34] But in any case this was obvious from the earlier definition of satisfaction.
[35] Generally speaking, language expressions 'denote' (it would be better to say 'relate to' or 'designate') both what in predicates corresponds to class and what corresponds to property. Carnap uses the term 'designation' in precisely this context. An expression that 'denotes' in sense (3), i.e. a 'designator' in Carnap, relates to both extension and intension. This is moreover the point of departure for Carnap's method of extension and intension in semantic analysis, to which we shall return later.
[36] This is a consequence of Martin's nominalist conception. It means that what is denoted is always an individual.
[37] This operator differs from Church's λ-operator in that it relates to a nominalist system of language. In contrast, the abstractum "$(\lambda x)(\ldots x \ldots)$" is 'semantically neutral' and admits of the further interpretation of an abstractum as a class or as a property of all the elements of that class. The nominalist conception of abstraction, which does not allow the existence of classes, has only one way of interpreting an abstractum: as that which may denote an individual.

TO CHAPTER IV

[1] This fact was pointed out by Carnap in his fundamental work on L-semantics [22, p. 59].
[2] It will be obvious that this notation is simplified, since it does not use *SEQ* but, in the interests of simplicity and perspicuity, the logical signs of $\mathscr{L}$. We shall keep to this simplified form of notation in the rest of the account as long as the interpretation of the logical signs used is quite obvious.
[3] The fact that we are working with a conjunctive class of sentences is given by the intuitive requirement that the semantic characterisation of a class of sentences should depend on the semantic characterisation of each individual sentence component.
[4] Carnap himself attempted to solve some of the problems attaching to this viewpoint in [23].
[5] In Carnap's conception these are invariably sentences or classes of sentences. In what follows we shall show that Carnap's conception is heavily constrained by his conception of 'state description', which is not a semantic method in the true sense of the word, since the relevant universe is actually an image of the syntactic structure of the object language.
[6] Wittgenstein's concept of 'logical space' has to be understood in connection with his conception of the meaning of a sentence. A sentence's meaning is identified with the knowledge of its truth conditions. (4, 024: Einen Satz verstehen heisst wissen, was der Fall ist, wenn er wahr ist. Man kann ihn also verstehen, ohne zu wissen, ob er wahr ist. Man versteht ihn, wenn man seine Bestandteile versteht.) If we know the truth conditions of a sentence without having to know whether it is true or not, we are warranted in saying that we understand the sentence. (4.2 : Der Sinn des Satzes ist seine Übereinstimmung und Nichtübereinstimmung mit den Möglichkeiten des Bestehens und Nichtbestehens der Sachverhalte.) The set of conditions that the sentence allows constitute its 'logical space'. (4, 463: Die Wahrheitsbedingungen bestimmen den Spielraum, der den Tatsachen durch den Satz gelassen wird.) Each individual condition we can call a 'logical state'. Among the possible groups of truth conditions two extreme cases can be distinguished: in the first, a sentence is true for all its truth conditions, this type being called by Wittgenstein a tautology; in the second, a sentence is untrue for all its possibilities in terms of truth, in which cases it is a contradiction.

This conception comes from Wittgenstein's *Tractatus* [165]. It had a strong influence on Carnap's conception of 'logical range' and become the starting point for a refinement of the concept by means of the concept of 'state description'. It must be added that the same conception was also the starting point for the origin and evolution of the so-called empiricist or verifiability criterion of sense, which of course proved untenable. (This problem will be dealt with in later chapters.) That is to say that where Wittgenstein's conception was applied in connection with empiricist trends it proved unproductive. A further shortcoming of Wittgenstein's conception of 'logical space' is its total subservience to extensional formalised languages, in which the truth values of molecular sentences are determined exclusively by the values of the atomic sentences. This is clearly a severe constraint and so the conception of 'logical space' can have no claim to universality.

Wittgenstein's original conception of the sentence and of language generally effectively completely ignores the communicational use of language. By contrast, his later ideas, published posthumously [166], in which Wittgenstein rejected some of his earlier thoughts and made considerable changes to others, to take note of this function of language amongst other things. His main emphasis here is on the practice of the use of language and the work includes the hypothesis of 'language play', by which it is possible to explain many aspects of language acquisition and some of the properties connected with the semantics of the practical use of language.

7 This means, for instance, that the method cannot be used in relation to formalised languages where the values of individual variables are real numbers.

8 It must be stressed that this decomposition takes place within the sphere of the language itself; this should be apparent from this entire account.

9 Since LR and other concepts connected with it may be regarded as classes, the possibility of operations with classes is assumed.

10 For brevity's sake we shall omit the expression 'in $\mathscr{L}$' in further definitions.

11 We believe it more suitable to formulate this relation in this way and not as some reference to the 'empirical world'. The latter formulation is, for one thing, narrow, and, for another, may be associated with only a one-sided conception of empiricism. The actual world is of course one of the possible worlds with respect to $\mathscr{L}$. If this should not be so, if there is discord between the actual world and a possible world (with respect to $\mathscr{L}$), it becomes necessary to extend $\mathscr{L}$ (e.g. by constituting new terms) in such a way as to eliminate such discord, or to construct a new language system.

12 Carnap's viewpoint, to express it in the terms used in the previous note, resides in the fact that the concept of 'actual world' is identified with that of 'possible world' (with respect to $\mathscr{L}$). But this is only one of the situations possible. This is why we think it more proper to speak of 'discord' and the 'elimination of discord' between the actual and possible worlds.

13 These reservations also relate therefore to Carnap's well known classification of sentences in a semantic system (i.e. in a language). Carnap uses the following diagram to illustrate this [22, p. 142]:

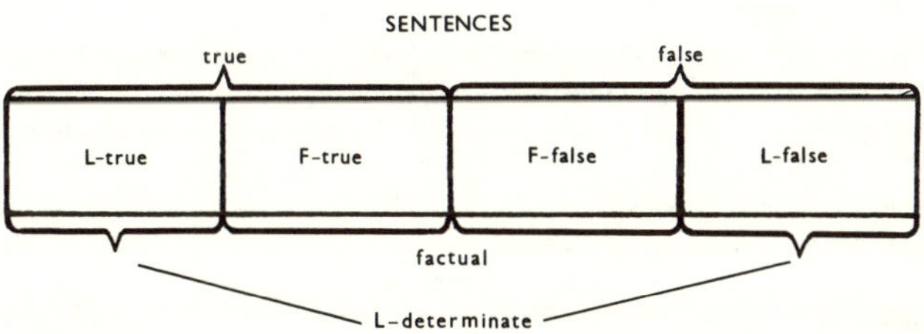

14 Carnap uses the term explication in the sense that it is by means of explication that we replace a more or less unclear concept (i.e. the explicandum) by an exact concept (the explicans). At the same time it is assumed that the explicans has exactly defined properties and that its use is governed by explicit rules. In his book on the logical foundations of probability and inductive logic [32, p. 15] he summarises the basic roles of concept explication in the following requirements:

(1) The explicans must be sufficiently *similar* to the explicandum for it to be possible to use the explicans in the majority of cases where the explicandum has been used.

(2) The rules governing the use of the explicans must be stated in an exact way to enable the explicans to be incorporated into a well constructed system of scientific concepts.

(3) The *explicans* must be *productive* in the sense that its formulation facilitates the greatest number of general statements.

(4) The explicans must be *as simple as possible* provided the previous requirements permit this.

[15] Carnap distinguishes two concepts of probability, which he denotes probability$_1$ and probability$_2$. The former is described as inductive probability or the degree of confirmation. Probability$_2$ is statistical probability interpreted in terms of frequency. This differentiation also corresponds to what has sometimes been described as subjective probability and objective probability.

[16] The diagrams are borrowed from [32].

[17] Carnap later constituted the properties of these **c**-functions axiomatically. Another method of axiomatically constituting **c**-functions is given by Stegmüller in an appendix to [32]. These systems of axioms essentially correspond to the contemporary axiomatic construction of probability theory.

[18] We shall show in what follows that these constraints may also be formulated by means of 'meaning postulates'.

[19] In order to give more complete information on Carnap's system of inductive logic we shall give some of the most important theorems, but using the kind of abbreviated notation already employed. We shall also use the sign $\Vdash$, which means that the formula following the sign is L-true. The most important theorems for **m**-functions are:

(1) if $\Vdash \mathbf{i} \equiv \mathbf{j}$ (i.e. if **i** is L-equivalent to **j**), then $\mathbf{m(i)} = \mathbf{m(j)}$
(2) If **i** is L-true, then $\mathbf{m(i)} = 0$
(3) If $\Vdash \mathbf{i}$, then $\mathbf{m(i)} = 1$
(4) $0 \leqslant \mathbf{m(i)} \leqslant 1$
(5) If $\Vdash \mathbf{i} \longrightarrow \mathbf{j}$, (i.e. if $\mathbf{i} \longrightarrow \mathbf{j}$), then $\mathbf{m(i)} \leqslant \mathbf{m(j)}$
(6) $\mathbf{m(i \vee j)} = \mathbf{m(i)} + \mathbf{m(j)} - \mathbf{m(i \cdot j)}$
(7) If $\mathbf{m(i \cdot j)} = 0$, then $\mathbf{m(i \vee j)} = \mathbf{m(i)} + \mathbf{m(j)}$
(8) $\mathbf{m}(\sim \mathbf{i}) = 1 - \mathbf{m(i)}$

In order to be able to formulate some of the basic theorems for **c**-functions, we have to accept certain constraints: If we consider the language $\mathscr{L}^{\pi}_{n}$, in which n is a finite number of individuals constants in $\mathscr{L}$, we have to eliminate the case where **e** in **c(h, e)** is an L-untrue sentence. The situation is more complicated when n approaches ∞. Carnap takes into consideration even these languages, i.e. $\mathscr{L}^{\pi}_{\infty}$, in which the sentences in **c(h, e)** must be such that each expression has a certain value.

Under these conditions the following theorems apply:

[1] $0 \leqslant \mathbf{c(h, e)} \leqslant 1$
[2] If $\Vdash \mathbf{e} \longrightarrow \mathbf{h}$ (i.e. if $\mathbf{e} \longrightarrow \mathbf{h}$), then $\mathbf{c(h, e)} = 1$
[3] If $\Vdash \mathbf{e} | \mathbf{h}$ (i.e. if **e** is L-incompatible with **h**), then $\mathbf{c(h, e)} = 0$
[4] If $\Vdash \mathbf{e}_1 \equiv \mathbf{e}_2$ (i.e. if $\mathbf{e}_1$ is L-equivalent to $\mathbf{e}_2$), then $\mathbf{c}(\mathbf{h}, \mathbf{e}_1) = \mathbf{c}(\mathbf{h}, \mathbf{e}_2)$
[5] $\mathbf{c(h \vee i, j)} = \mathbf{c(h, e)} + \mathbf{c(i, e)} - \mathbf{c(h \cdot i, j)}$
[6] If $\mathbf{c(h \cdot i, e)} = 0$ (i.e. if $\mathbf{e \cdot h \cdot i}$ is L-untrue), then $\mathbf{c(h \vee i, j)} = \mathbf{c(h, e)} + \mathbf{c(i, e)}$
[7] $\mathbf{c(h \cdot i, e)} = \mathbf{c(h, e)} \times \mathbf{c(i, e \cdot h)} = \mathbf{c(i, e)} \times \mathbf{c(h, e \cdot i)}$
[8] $\mathbf{c}(\sim \mathbf{h}, \mathbf{e}) = 1 - \mathbf{c(h, e)}$
[9] $\mathbf{c(i, e)} = \mathbf{c(h \cdot i, e)} + \mathbf{c}(\sim \mathbf{h \cdot i}, \mathbf{e})$
[10] If $\mathbf{h}_1 \underset{L}{\longrightarrow} \mathbf{h}_2$, then $\mathbf{c}(\mathbf{h}_1, \mathbf{e}) \leqslant \mathbf{c}(\mathbf{h}_2, \mathbf{e})$
[11] If $\mathbf{e \cdot h} \underset{L}{\longrightarrow} \mathbf{j}$, then $\mathbf{c(h \cdot j, e)} = \mathbf{c(h, e)}$

This sketch form of the system of basic theorems of inductive logic (i.e. the system of **c**-functions and **m**-functions whose arguments are sentences in $\mathscr{L}$) was originally formulated in such a way as to be based on the method of state description, the starting point of which was the concept of ***LR***. But since, as has been said, the 'contentual or semantic power' of a sentence is governed not by what the sentence (in the extensional respect) includes, but by what it excludes from the sphere of the 'possible worlds' with respect to a given $\mathscr{L}$, it is also possible to conceive the same system on the basis of ***LC***.

[20] **m(i)** + **m**(~ **i**) must be equal to the total for all individual ***SD***. This total is equal to 1.

[21] The following may be considered the most important theorems for **cont**:

(1′) If ||–**i** ≡ **j**, then **cont**(**i**) = **cont**(**j**)
(2′) If **i** is L-untrue, then **cont**(**i**) = 1
(3′) If ||–**i**, then **cont**(**i**) = 0
(4′) 0 ⩽ **cont**(**i**) ⩽ 1
(5′) If ||–**i** ⟵ **j**, then **cont**(**i**) ⩾ **cont**(**j**)
(6′) **cont**(**i**·**j**) = **cont**(**i**) + **cont**(**j**) – **cont**(**i** ∨ **j**)
(7′) If ||–**i** ∨ **j**, then **cont**(**i**·**j**) = **cont**(**i**) + **cont**(**j**)
(8′) **cont**(~ **i**) = 1 – **cont**(**i**)

[22] A more detailed analysis of procedures of reduction and constitution will be given in due course.

[23] In the interests of keeping the notation simple we shall again be writing **m(i)** instead of **m**(Φ_i), **c(h, e)** instead of **c**(Φ_h, Φ_e), **cont(i)** instead of **cont**(Φ_i) etc.

[24] Cf. Leibniz: 'Il y a deux sortes de vérités, celles de *raisonnement* et celles de *fait*. Les vérités de raisonnement sont nécessaires et leur opposé est impossible, et celles de fait sont contingentes et leur opposé est possible.' (*Princ. philos.* § 33).

[25] The concept of A-truth, defined on the basis of the concept of 'interpretation', cannot, of course, be identified with the concept of L-truth, which has been defined on the basis of the concept of 'state description'.

[26] Of course this viewpoint is nowhere formulated explicitly by Kemeny. Yet it is logical to assume that this viewpoint is compatible with the original intention of Kemeny's 'new approach' to semantics. Moreover it makes it possible to apply that approach both in situations where the empirical context is decisive and in those where an abstraction is made from that context.

[27] Understandably, this definition is not a general definition of synonymity, concerning as it does only the synonymity of constants. As a general definition it would clearly be too narrow.

[28] The procedures of reduction and constitution from the point of view of semantics will be dealt with in due course. From the point of view of the present account of logical semantics it is important to appreciate that the so-called meaning postulates are a sort of expression of reduction and constitution procedures.

[29] What is meant here by 'formal characterisation' is the full implementation of all the steps that fall within the syntactic stage of the build-up of $\mathscr{L}$. It goes without saying that both the syntactic and the semantic stage assume a $\mathscr{ML}$.

[30] This is where Kemeny's approach to the introduction of semantic concepts differs from that other conception evolved by Quine. Quine includes the first kind of semantic concepts within the theory of meaning, the second within the theory of reference. He considers the two theories to be different in principle. He treats semantics in the proper sense of the word as a theory of meaning, while he includes what Tarski described as semantics in the theory of reference (see especially [119]). The most important concepts of the theory of reference in Quine's sense are, in addition to the concepts of truth and untruth, those of naming, denotation and extension.

[31] This definition too has, as we shall show in the more detailed analysis of synonymity to be presented in due course, only a limited range.

TO CHAPTER V

[1] Frege did not introduce the distinction of sense and denotation until his study *Ueber Sinn und Bedeutung* [45] of 1892. By contrast his earlier works, such as [43], [44] and others, do not have this assumption.

[2] In due course we shall show, in connection with some of the ontological problems of language, that these problems too are closely connected with one's conception of the ontological starting point of language and may be resolved in different ways in nominalist and Platonist systems.

[3] In these formulations, and unlike Tarski's practice, we use the sign $\underset{\mathrm{id}}{=}$ for the relation of identity, while for the identity of classes and therefore also for numerical equality we have used the sign $=$. The last of the formulations given here may also be expressed in a formalised notation. If either of the two things has properties $A_1, A_2, \ldots A_n$ and if $F = [A_1, A_2, \ldots A_n]$, then

$$(\forall x)(\forall y)[(x \underset{\mathrm{id}}{=} y) \rightarrow (Fx \equiv Fy)],$$

which is no more than an adaptation of the notation for Whitehead and Russell's transcription of Leibniz's law [164, p. 23].

[4] Whitehead and Russell likewise formulate the identity law more in a way to capture the second approach: "If x and y are identical, either can replace the other in any proposition without altering the truth-value of the proposition." [164, p. 23.]

[5] On this point Frege remarks that even today it is not always easy to ascertain whether a newly discovered asteroid or comet is identical with some object that has already been discovered.

[6] In Frege 'Bedeutung', in the English terminology 'nominatum'.

[7] In Frege 'Sinn', connotation, in Russell 'meaning'.

[8] Some authors, such as C.I. Lewis, speak of 'modes' of meaning.

[9] In this connection Frege appeals to grammar and speaks of expressions that are formed properly according to grammar. However two things should be noted: (1) In formalised languages the rules that come into the syntactic stage of the build-up of a formalised language are of a formal character; (2) By contrast the grammatical rules of a natural language do not and cannot have this formal character. For instance, if we have rules of declension differentiated by gender, we are in fact respecting a certain decomposition of the universe of the given language; we are respecting certain considerations of a semantic nature. Hence the term 'a grammatically well formed expression' cannot be defined without consideration of the language's semantics. In this respect Frege's viewpoint is imprecise and cannot be fully applied in natural languages.

[10] It goes without saying that this does not apply only to natural languages. In formalised languages it is also possible to operate with names which have no denotation. Two situations are possible: (1) Denotation is only presupposed abstractly as being possible without having to be explicitly stated. This applies to all the non-logical signs in a formalised language where no set interpretation is explicitly stated. (2) We have to allow for the possibility of names whose denotation cannot be determined. The expression 'a prime number greater than two and divisible by two' has a sense but no denotation. It may be added that in the use of language in the general sense it is apparently never possible to do without names that have sense but not denotation.

[11] The problem of the non-existence of denotation in names of this kind is not statable unambiguously, as we shall see at a later stage, primarily in connection with the theory of descriptions. One can imagine actors in children's theatre and the like.

[12] Scholastic logic distinguished several forms of these suppositiones. In addition there was a differentiation of 'suppositio', 'significatio', 'copulatio' and others.

[13] We shall use the term 'proposition' when referring to sense or (in terms of the method of extension and intension) the intension of a sentence. Thus the term 'proposition' must be distinguished from the term 'statement', which we shall use whenever it is not clear whether a sentence or its sense is meant. Frege himself did not use the term 'proposition', but in the same sense he used 'Gedanke'. Since the literal translation of Frege's term might give an impression of psychologism, which is totally at variance with Frege's entire conception of logic and semantics, we deem it appropriate to adhere to modern logical terminology. Since we are going to use the term 'proposition' as the sense of a sentence, we need to choose another term for Russell's 'proposition'. Therefore for Russell's term 'proposition' we shall use 'sentence' for an expression of language, or 'statement' if it is not stated explicitly whether it is an expression of language or its sense.

[14] Frege argues along rather different lines, speaking, for example, not of name components but of sentence components, but the tenor of his argument is in agreement with the procedure suggested, which is based on the more exact formulations for the sense and denotation of compound names introduced by Church but fully in the spirit of Frege's conception.

[15] In logical thinking Frege was not the first to use the term 'truth value'. It is customary in this connection to point to the works of Peirce (see for example his *Collected Papers*, vol. 3, Cambridge, Mass., 1932, pp. 210–238) and other logicians. Frege himself did also use the term in his earlier works (e.g. [44]), but not directly relating to the problems of sense and denotation.

[16] We shall show in due course that this kind of context can be described, from the semantic aspect, as non-extensional.

[17] This method is evidently less suitable for imperatives and interrogatives, but it may on the other hand be of use in the semantic analysis of so-called normative sentences. Normative sentences of the type 'is supposed to be A', 'must not be B', 'is forbidden to C', 'must be D' etc. (they are either different kinds of normative sentences or combinations of sentences with certain special types of context) need not differ greatly from declarative sentences in their grammatical aspect, but they show considerable differences in the logical and semantic aspect. It is possible to draw up formalised systems of normative sentences and then it may be worthwhile postulating certain abstract objects, such as 'legal validity', which can be considered the denotations of these sentences.

[18] The possibility of a metatheoretical approach to this kind of system and so too the possibility of its syntactic and semantic analysis has been pointed out by V.A. Uspenskij [179].

[19] We shall return to some of these questions in the discussion of what will be called extensional and intensional context.

[20] The term 'meaning' is used here in the broader sense. Hence a 'description' in Russell's conception may have both sense and denotation.

[21] In this connection Russell criticises the idealist conception whereby any grammatically well formed phrase represents some object. By this theory, therefore, phrases like 'round square' or 'the present king of France' are treated as representing objects which have to be treated as such despite the fact that they do not occur. Clearly this is a theory of an objective

idealism and is based on hypostatising the denotation of any expressions whatsoever if they are formally admissible in terms of the syntax of the given language (and possibly also in terms of its grammar).

[22] The sequel to Russell's polemic with Meinong in the *Introduction to Mathematical Philosophy* [133] is interesting in this context. Russell is emphatic here that in logic we are no more justified in admitting the existence of 'unicorns' than we would be in zoology. Logic is just as related to the real world, although in a rather more abstract way. To claim that 'unicorns' exist in heraldry, literature or fantasy is, as Russell puts it, a paltry evasion: "There is only one world, the 'real' world." [133, p. 169].

[23] In Russell 'propositional function'. It is to be remembered that the term 'proposition' has a different sense as used by Russell from that in which it is used in the present work, which follows on from Frege. This is why Russell's term 'propositional function' is replaced here by 'statement function'.

[24] In a natural language that has no articles definite and indefinite descriptions are not generally distinguished explicitly, the difference being deducible from the overall context. To take an example from Czech: the two sentences (1) 'Karel Čapek byl český spisovatel' and (2) 'Karel Čapek byl autorem dramatu RUR', contain respectively an indefinite description in the phrase 'česky spisovatel' (a Czech writer) and a definite description in the phrase 'autorem dramatu RUR' (the author of the play RUR).

[25] At this point in the discussion we have to operate with the concept of 'statement function', which explains why we shall use the symbolism of the *Principia Mathematica* system in our analysis of the theory of descriptions. In that symbolism statement functions are denoted by small Greek letters, e.g. φx, ψx. The sign $\vdash$ is the 'assertion sign', which can be read as 'we assert that...' and which is to be strictly distinguished from the sign $\Vdash$. The symbolism of the *Principia Mathematica* system uses dots instead of the customary brackets. It is further necessary to distinguish $\varphi\hat{x}$ as the name of a property which constitutes a statement function, and $\hat{x}(\varphi x)$ as the class made up of $\varphi\hat{x}$. A class is therefore determined by the fact that all the individuals which have the property $\varphi\hat{x}$ make up the class $\hat{x}(\varphi x)$.

[26] In the *Principia Mathematica* notation, where existence is expressed by the sign E!, (D_1 – V 2 C) appears as:

$$\mathrm{E}\,!\,(\imath x)(\varphi x) = :(\mathrm{E}c):\varphi x \equiv_x \cdot x = c$$

[27] This is undoubtedly the positive side of the formula, which is recorded in *Principia Mathematica* as:

$$\vdash :\psi(\imath x)(\varphi x)\cdot \supset \cdot \mathrm{E}!\,(\imath x)(\varphi x)$$

[28] Elsewhere Russell also uses the term 'ultimate constituents'.

[29] This attitude need in no way be altered by the fact that the original conception of the name relation as suggested by Frege is associated with certain platonising tendencies, whereas Russell explicitly subscribed to the view that the starting point in logic is the real world, meaning that concepts for 'non-existent' objects were to be excluded.

[30] In the system of *Principia Mathematica* Whitehead and Russell resolved this problem by distinguishing a class and that which determines a class. If $\varphi\hat{x}$ is a statement function which is satisfied by any x which is mortal, then $\varphi\hat{x}$ is the name of the property which determines the class of all individuals that are mortal, and $\hat{x}(\varphi x)$ is the name of that class. At the same time it is accepted that the class might be empty. Thus a class consists of all the terms which satisfy a particular statement function, so each statement function determines a

class. If $\hat{\alpha}$ is the class determined by the statement function φz, then $x \in \hat{\alpha}$ if and only if $x \in \hat{z}(\varphi z)$. Then the definition of class is

$$\text{Class} = \text{df}\,\hat{\alpha}[(\exists\varphi)\cdot\alpha = \hat{z}(\varphi z)]$$

[31] This formulation of the principle of comprehension is analogous to the previously given formulation in the system of *Principia Mathematica*, which is to be found in the preceding footnote.

[32] The procedure described by Church as λ-conversion clarifies one of the important ways of forming concepts. The problems of forming concepts by abstraction by means of λ-conversion have been analysed from the gnoseological point of view by D.P. Gorskij [176].

[33] If we can treat the expressions 'man' and 'intelligent being capable of manufacturing tools and using language for communication' as classes, then both classes have the same elements.

[34] This viewpoint is also projected in various conceptions of the formation and differentiation of concepts, for example in the differentiation of what have been called substantial concepts, functional concepts, relational concepts and others. This viewpoint is also partially projected in C.G. Hempel's *Fundamentals of Concept Formation*, Univ. of Chicago Press, 1952.

[35] Meaning the principle of extensionality for classes.

[36] In this connection Carnap uses the term 'predicator'. Predicator is an expression for predicates in the broader sense of the word, including classes. A more general concept is 'designator', which is any language expression to which semantic analysis can be applied.

[37] The term 'signify' may appear unusual in this context. But its use is necessary and in accordance with the previously quoted interpretation of the term 'signify' in natural language. Carnap himself avoids these difficulties by using for both cases the term 'express' or the neutral 'predicators for classes' and 'predicators for properties'. Since these are different semantic relations, it would be useful to distinguish the two terminologically as well, although that does not quite correspond to linguistic awareness.

[38] This also means that the so-called verifiability criterion of sense cannot suffice. At a later stage we shall also try to show that the verifiability criterion itself leads to results that essentially refute it.

[39] In this connection Carnap recommends the term 'individual concept'.

[40] Instead they speak of extensional functions.

[41] R.M. Martin's attempt [90] cannot be considered a consistent analysis of this kind. Although attempting to produce a formalised pragmatics, he did so at the cost of reducing the system of pragmatic concepts to a handful of elementary concepts. It is also interesting that Carnap, who originally (e.g. in [22]) expressed considerable scepticism about the possibilities of theoretical pragmatics, was later, in [26], to voice a pressing demand for a system of theoretical pragmatics to be elaborated.

[42] This situation is analogous to Russell's classical example in which the King of England asks whether Walter Scott is the author of Waverley. Here, however, Russell was not investigating the context of the two sentence components, but the substitution of an expression with the same extension.

[43] We shall return to some of the problems of context when we analyse synonymity. For the problem of synonymity encounters difficulties at precisely the point when we have to respect not just the connection between two constants (in the sense of (D_9 – IV 2 B)), but also a particular concatenation of expressions to each of which a relatively independent meaning can be assigned.

[44] In linguistic terms this is obviously the substantivisation of adjectives.

[45] Lewis gives no specially clear-cut definition of 'term'. For him a 'term' is a language expression which names something or relates to a thing or things of the most diverse kinds, whether actual things or thoughts about things. In this respect Lewis's viewpoint may be said to resemble that of Meinong (which had already been criticised by Russell), according to which even expressions like 'unicorn', 'centaur' and the like relate to things, albeit fictitious things. According to this viewpoint language does not of itself have available means for distinguishing between real 'things' and 'things' which are merely fictitious or imaginary.

[46] Lewis abides by Russell's interpretation of the term 'proposition'.

[47] In this connection Lewis actually speaks of linguistic meaning.

[48] Here Lewis speaks of analytical relations and of definitions. These terms, which may give rise to some uncertainties, can, as we have shown, be replaced by the concept of L-equivalence.

[49] Hence the analysis of linguistic intension could be split into two parts: the context-free and the context-bound analysis of intension.

[50] For example, Quine uses in the same sense the term 'designatum', and uses 'designans' for name [115].

[51] This second remark could be controversial. It hints at the question of how far semantic analysis can be 'ontologically neutral' or how closely is it tied to a particular conception of ontology. This controversy and the related controversy between the nominalist conception and others will be dealt with in greater detail in due course.

[52] Carnap describes this an ambiguity in the method of the name relation.

[53] This is also done, as we have pointed out, in the system of *Principia Mathematica* by the distinction between $\varphi\hat{x}$ as the name of a property and $\hat{x}(\varphi x)$ as the name of a class.

[54] It is interesting to recall that Carnap, who supported the second view, later abandoned it and expressed himself (e.g. in [23]) unequivocally in favour of the third view. He also stressed that the semantic rules in respect of a particular expression should be primarily understood as the determination of the expression's intension.

[55] These ideas were applied by A.D. Getmanova [175] in an analysis of the relations between a logical system and its models. In analysing the relations between a language system and systems of extralinguistic entities we are in the same situation.

[56] This label cannot be deemed very appropriate, for synonymity, as we shall show, presupposes more than just identity of extension, i.e. the fact that two names relate to the same nominatum. Synonymity must be linked primarily to the intension, not the extension, of expressions.

[57] The most striking attempt along these lines is Carnap's extensionality thesis whereby it is assumed that any non-extensional system of language is translatable into an extensional system. (This thesis was reasoned out especially in [20].) For example, the modal sentence 'it is necessary that Σ'' is, according to this conception, L-equivalent to the sentence 'Σ is L-true'. Later, however, Carnap reached some partial conclusions which restricted translatability into an extensional language system. This is notably the case of belief sentences

[58] We have to resort to a dispositional term here because we need not be dealing with translation of one expression into another with the meaning preserved under all circumstances. Dispositional factors are usually tied here to conditions that are pragmatic in nature, such as the facilities or knowledge of the language users. If a specialist in linguistics declares that some archaic expression which has passed out of use and which therefore few people

understand is a synonym of an expression in the modern language, it is assumed that translation is only feasible under certain circumstances, i.e. given the relevant knowledge.

59 The concept of 'translation', and with it that of 'the same meaning', is of course also ambiguous. Therefore the relativisation suggested for 'the same meaning' also applies to 'translation'.

60 This characterisation of the concept of 'interpretation' is not entirely at one with the term's use in common parlance, where it is often given a much broader sense. We speak, for instance, of the 'interpretation of historical events', the 'interpretation of observed facts' and so forth. We cannot fail to agree with B. Mates [91] that the use of the term 'interpretation' in such a broad sense may lead to confusion. First and foremost 'description' must be kept separate from 'interpretation'. In interpretation both the interpretandum and the interpretans are a language expression or a sequence of expressions – a 'text'. In the case of description, that which is the object of a description is a set of objects or events which are empirically accessible to the observer who is equipped with the right means of observation, measuring and experimentation. Then we have to separate interpretation from explanation (in [152], where the most important types of explanation were indicated, we used the less suitable term 'explication', which in the present work is used in another sense). It is particularly the confusion of explanation and interpretation that can lead, and in history has led, to misunderstandings. As opposed to description, what we normally require of a scientific explanation is that it fix some general dependence between objects or events, whether that dependence be of the determinist or the statistical type. It is also on this that the predictive value of scientific explanation is based (see [152] for more detailed discussion of this). But the translation of one set of expressions into another set is not an explanation, although it may appear to answer the question 'why' which is often associated with explanation. A famous case of this kind of apparent explanation, which in fact is no more than a translation or unsuitable interpretation, is the question and answer quoted from Molière: "Cur opium facit dormire? Quia habet virtutem dormitivam."

61 Quine treats synonymity as a concept of the theory of meaning [119] along with analyticity and others.

62 Relatively the most exact formulation of them in the logical literature comes from Quine [118].

63 This viewpoint was expressed by Quine [121], who replaced the *salva veritate* criterion by that of *salva analyticitate*. According to Quine two expressions are synonymous if their equivalence is L-true (in Quine's terminology: if their biconditional is analytical; hence the criterion of salva analyticitate).

64 Lewis does not deal with the quantification of intension in general terms, but only in relation to sentences. An analytical sentence has zero intension, while a contradiction has universal intension. This means that his conception of it roughly corresponds to the concept of 'logical content'. It also means that the extreme values on a possible scale of intension are reserved for what Carnap calls L-determined sentences, i.e. L-true sentences (which in Carnap coincide with analytical sentences) and contradictions. Carnap himself attempted to extend this distinction of determinedness and L-determinedness to the other expressions to which a relatively independent meaning can be assigned. From this point of view the abstractum $(\lambda x)(P_1 x \vee \sim P_1 x)$ is a L-determined abstractum. According to Lewis's conception this kind of abstractum has zero intension.

65 At the same time we assume that these are complex expressions. This assumption corresponds to Lewis's condition of analytical comparability.

[66] This weakness of the concept of intensional isomorphism has been noted in particular by Mates [91], Church [38] and Scheffler [139]. Carnap replied to some of the objections raised in [30].
[67] This presupposes that one kind of criteria of sense will be taken into account with empirical predicates and another kind with abstract expressions.
[68] By contrast, the conception of 'intensional isomorphism' disregards pragmatic aspects and confines itself to the L-equivalence of substitutable components.
[69] Carnap agreed in part with the main burden of Quine's criticism: "If I understand him correctly, he believes that, without this pragmatical substructure, the semantical intension concepts, even if formally correct, are arbitrary and without purpose" [28, pp. 234–235]. But he concedes that this is a possible viewpoint that is even fruitful, although, as he says, there is not always the need to require the priority of the pragmatic correlates of the usual semantic concepts.
[70] This formulation has been somewhat simplified, since it takes no account of relativisation to a particular object language or to any capacities, memories or delays available to the language user.
[71] In this conception of synonymity it is then necessary to presuppose some criterion of sense or, to be more exact, a criterion on the basis of which a conclusion can be reached that two expressions do have the same sense. This does not mean that such a criterion must necessarily be an empirical one.
[72] The term 'marginal references' was introduced into the literature of logical semantics by A. Naess [97] [98]. The book [98], as opposed to [97], provides an all-round analysis of numerous examples of synonymity.
[73] This is also the case in the quoted examples of expressions taken in marginal conditions.
[74] A more detailed analysis of 'ambiguity' will be given at a later stage in connection with the problem of vagueness.

TO CHAPTER VI

[1] In the $\mathscr{L}$ quoted these were individual names, predicates, abstracta and sentences.
[2] In the case of the language to which we have already referred it can easily be shown that, for instance, $P_1 p \cdot P'_1 p$ is a well formed sentence. By contrast, the expression '$P_1 p \cdot \longrightarrow$' (Prague is in Bohemia and if) is not a well formed sentence. The examples traditionally quoted in this connection ever since the days of Carnap's 'Logical Syntax of Language' are the expressions 'Caesar is a prime number', which is formally a well formed sentence, and 'Caesar is when', which is not. While the former is well formed, it has no sense (sinnleer), and the second is nonsensical (sinnwidrig). It is obviously not the nonsensical sentences that are at the focus of attention, since their elimination is mostly a much easier matter, but the differentiation of those expressions (or rather sentences) that have a sense from expressions that do not. In purely linguistic terms we can visualise the reverse situation: We have so far assumed that sentences are made up of elements that themselves have a sense, the net result being either nonsense (if the relevant rules are broken), or a well-formed sentence that may then either have a sense or not. But it is perfectly possible to imagine the situation where while respecting all the relevant rules, the constituent elements do not themselves have a sense. Examples of expressions composed of elements that have no sense but apparently observing correctness of syntax, grammar and style are usually used to demon-

strate the relative independence of language form and lexical content. Examples of this kind (one such, adopted from L.V. Shcherba, is analysed in [129]) lie beyond the competence of semantic analysis and any criterion of sense.

[3] This possibility was pointed out by P. Marhenke [87], who himself indicated no more concrete realisation of the possibility.

[4] A significant feature of unitary theories of sense is the view that these means are the same for all sentences that have a sense, or at least all those that can be described as synthetic.

[5] This fact needs to be emphasised in the context of the justified criticisms of subjectivistically or positivistically conceived empiricism. Such criticisms do not of course mean that the system of human knowledge can be built up other than on an empirical basis; what they do mean is that scientific philosophy abandons the objective conception of the empirical basis of knowledge.

[6] At the same time, what we obviously have in mind here are only those expressions to which semantic analysis can be applied or a relatively independent meaning assigned.

[7] This was far from fully appreciated in the original viewpoint of Carnap and other adherents of the verifiability criterion during the period of the Vienna Circle. They loved to demonstrate the lack of sense of sentences on chosen texts of the German idealists: they found the sentences of these authors to be empirically unverifiable and therefore lacking in sense (meaningless). To this 'sequitur' there is the ready objection that there have existed or do exist disciples or followers of those philosophers who are convinced that they have understood the sense of the texts and are capable of expounding them, i.e. translating them into other expressions. At the same translatability 'salva intensione' (i.e. with the associated possibility of preserving the sense or intension of the sentence) is clearly another necessary, though likewise inadequate, condition of the decidability of the sense of a sentence.

[8] The best known attempt to link the narrowly defined empirical criterion with these other means is Carnap's [21].

[9] P.W. Bridgman originally formulated his views in his *Logic of Modern Physics* [16], where he laid the foundations for the whole conception that has been described as operationalism.

[10] This schema was later formalised by means of what were called reduction sentences and the reduction characteristic of dispositional predicates [21]. The logical difficulties of this formalisation gave rise to a number of quite different attempts to find other formalisations of procedures that are analogous to Bridgman's schema of operational definition.

[11] We shall ignore here the various philosophical interpretations of these procedures. We do, however, take it as a matter of course that Einstein's denial of the absolute nature of 'simultaneity' cannot be interpreted as a denial of its objective nature, just as relativisation is not to be understood as subjectivisation.

[12] The homocentric version of operationalism is an analogue of the 'physicalism' of the Vienna Circle period, or of the reductionist viewpoint of Carnap in [18]. Bridgman himself was more inclined towards homocentric operationalism, which for him was an extension of Einstein's 'observer's viewpoint'.

[13] This also means that the situation expressed by Heisenberg's uncertainty relation is a special instance of the fact that on the basis of certain properties of the channel (as macroscopic measuring instrument) and of the source of messages (as a microprocess), the decision-making that is possible at the output of the channel has its limitations.

[14] In this context it is again important to remember the semantic analogy of Occam's razor : nomina non sunt multiplicanda.

[15] It is here that we can also appreciate the untenability of the subjectivist or homocentric version of operationalism: it is scarcely acceptable to treat statements about the objective chemical structure of a source of radiation and statements describing the process of spectral analysis as synonymous.

[16] These requirements correspond to the deductive nomological model of scientific procedures (see e.g. [54]).

[17] In fact there are two kinds of translatability: translatability from esoteric language into language that is more broadly comprehensible, e.g. to people accustomed to think in terms of the natural sciences, and translatability from the language of such texts into other natural languages.

[18] The concept of observation sentence assumes another important concept: that of observational or empirical predicate. The problems of defining an empirical predicate will be shown in due course.

[19] This kind of analogy is clearly quite justified. At the same time it is interesting that many adherents of the Vienna Circle frankly admitted the connection between their viewpoint and the early version of positivism, while limiting the connection to just the early part of their work. They usually claimed to have moved far away from positivism in their later ideas. For example, Schlick [140] conceded that in the early days of the Vienna Circle the desire to connect the concept of 'sense' with that of 'immediately given' was very strong. In particular he mentioned Carnap's *Der Logische Aufbau der Welt* of 1928. Carnap too, in his remarks on positivism and physicalism incorporated into [21], expressly renounced this positivist conception. He pointed out that the positivists believed (and he too in the work referred to) that any descriptive or empirical concept in science is definable through the medium of perceptions or observations of an actual observer. This view proved to be too narrow and had to be 'extended'. The expansion is rooted largely in the fact that the class of actual observers was joined by a class of potential observers past, present or also future.

[20] We have already pointed out that in a language in which account has to be taken of the semantically important relations of the language's non-logical constants, it is actually possible to construct two concepts of analyticity. The same applies to the concept of contradiction.

[21] This question was first uttered by R. Ingarden at the Eighth International Congress of Philosophy in Prague in 1934 [63].

[22] This critical reservation may also be formulated as follows (see [151, p. 39]): The sentence 'the sense of a sentence is given by the logical possibility of its verification' is obviously not analytical, i.e. its truth does not follow from the logical impossibility of taking as true the opposite of it, e.g. the sentence 'the sense of a sentence is not given by the logical possibility of its verification'. If it is a synthetic sentence, then even it has to be verified. And that is clearly not feasible in the semantic metalanguage in which the verifiability criterion of sense is itself formulated.

[23] In Schlick's terminology this is the class of 'Konstatierungen'.

[24] In the ample literature devoted to the problem of the empirical criterion of sense the most serious critical reservations against the requirement of complete verification were formulated by Hempel [55], Pap [106] and Stegmüller [144].

[25] We are omitting intuitionist systems for obvious reasons.

[26] Especially in the foreword to the later edition of his book *Language, Truth and Logic* of 1946 [4].

[27] Similar reservations were expressed against Ayer's version of the verifiability criterion by Church [36].
[28] The language that we have been using as an example in the foregoing account is, in formal terms, coincident with the schema of empiricist language. But unlike empiricist language we have not assumed that the predicates are of necessity empirical. Hence the competence of the meaning postulates was considerably broader and did not have to be restricted to their relations with empirical predicates only.
[29] The actual link between the two predicates has not been fully elucidated in the literature. The first could be assumed not to be an empirical predicate. The second is obviously dispositional. The systematic application of the verifiability criterion would lead to the conclusion that the two predicates are synonymous. Such a conclusion is unacceptable.
[30] This was demonstrated by A.C. Ewing [42] as early as 1937.
[31] Without being expressly stated, this view is obviously that of P. Marhenke [87].
[32] In particular this applies to theoretical concepts, as we shall show later.
[33] In this connection it should be recalled that this criticism of the verifiability criterion of sense cannot be interpreted as a total underestimation of efforts to find the connections between the concepts of 'having a sense' and 'empirical'. Despite all the reservations against the verifiability criterion, it has to be appreciated that it did raise a serious problem, although the solutions put forward, usually claiming universality, could not, and still cannot, be considered satisfactory.
[34] This applies, for example, to sentences that operate with the predicates 'be a square' and 'be round'. It is easy to conceive a meaning postulate that captures their incompatibility with respect to all individuals of a given universe. But if we were to use this method to fix incompatibility in all situations and tasks, such as in poetry, we would have to treat such expressions as "the emotions of his dead mistress" (K.H. Mácha) as inadmissible on principle. All that can be added here is that certain metaphysical systems that used to be taken as examples of sentences that have no sense (Carnap, for instance, analysed texts from Heidegger) are closer to poetry than to empirical science.
[35] It is usually the names of observable objects and their observable properties that are quoted as examples of empirical terms, and names like 'electron', 'spin', 'gene', 'energy' and so forth as examples of theoretical terms. There are doubtless many instances where there might be hesitation as between 'observable' or empirical terms and theoretical terms. This goes for the movement of elementary particles as observed through a Wilson chamber, to quote just one example.
[36] This conception is developed in [18].
[37] It should be stressed that Carnap himself originally preferred the phenomenalist interpretation. Only later did he say that semantic analysis is neutral in ontological controversies.
[38] This definition is not expressed in the normal shape. The same will apply to certain other definitions to come when of a like nature.
[39] Much has been written on the problems surrounding dispositional predicates. Of the more important works we might quote those of Pap [106], [107], Sellars [142] and Goodman [48].
[40] Carnap original also included theoretical concepts in the dispositional predicates, but he abandoned the idea later.
[41] The term 'open concept' was introduced by Pap [105], [108].
[42] This was rightly pointed out by Shvyrev [180].

[43] If the class '$(\hat{x})(P'x)$' is an empty class, this does not yet mean that (a) the class of objects under investigation does not have the property Q (for we know that many substances do have the property 'be soluble' without having to put them in a liquid), or (b) that decisions about Q based on other circumstances are precluded. Many of the properties that can be expressed by dispositional predicates can be assumed on the basis of theoretical analysis. Hence the statement that $(\hat{x})(P'x)$ precludes decisions about Q merely refers to empirically substantiated decisions and does not exclude other ways of deciding.

[44] The concept of causal implication was introduced by A. Burks (*The Logic of Causal Propositions*, Mind, V. 60, 1951). A detailed analysis of the properties of causal implication is given by V.S. Shvyrev (*K voprosu o kauzalnoj implikacii*, Logicheskije issledovanija, Moscow, 1959).

[45] This kind of probability approach to the problems of dispositional predicates and reduction sentences will be expounded in the section on reduction.

[46] The subjectivist interpretation of empiricism rests then in the fact that this decision-making is understood as decisions about states at the input of the 'observer channel'.

[47] This term was introduced by Carnap in [31].

[48] Other conceptions of theoretical concepts will be treated in greater detail in due course.

[49] The view that assumes this kind of reducibility will be described and criticised in what follows as empiricist reductionism.

[50] This kind of empiricist language $\mathscr{L}_0$ is linked by Carnap to a number of requirements, of which the most important are: (1) the requirement of observability for primitive non-logical terms, (2) the requirement of constitutability (by means of explicit definitions or reduction sentences for other terms), (3) the requirement of nominalism, i.e. the requirement that the values of variables be concrete observable entities, (4) the requirement of finitism, (5) the requirement of constructivism, i.e. the requirement that each value of any variable be denoted by some expression of $\mathscr{L}_0$, and (6) the requirement of extensionality. Clearly the requirements are not all equally specific to $\mathscr{L}_0$. Besides the first two, whose interpretation is obvious in the light of what has gone before, the third requirement, linking nominalism with empiricism, is characteristic.

[51] In defining the admissibility of a domain of individuals Carnap formulated three conventions about a domain of entities $\boldsymbol{D}$ admissible as values of variables in the theoretical language $\mathscr{L}_T$: (1) The domain or set of entities $\boldsymbol{D}$ contains a numerous subset $\boldsymbol{J}$. (2) Any ordered n-tuple of entities in $\boldsymbol{D}$ (for finite n) also belongs to $\boldsymbol{D}$. (3) Any class of entities in $\boldsymbol{D}$ also belongs to $\boldsymbol{D}$. [31, p. 43]. It is clear that these conventions are particularly appropriate for theoretical languages where the values of their variables are not only empirically ascertainable and discernible entities, but also constructive entities.

[52] It is interesting that for many thinkers inclined towards the natural sciences the only way out from the difficulties of instrumentalism has been apriorism in the interpretation of theoretical concepts, whether a platonising apriorism or apriorism of the Kantian type. The impossibility of 'empirically justifying' theoretical concepts, especially some of the concepts of theoretical physics, led many theoretical physicists from empiricist instrumentalism to apriorism. This was so with Poincaré, Eddington, and partly so with Heisenberg and others. The instrumentalism-apriorism dilemma is of course only the semblance of a dilemma, as is that of instrumentalism-naive realism, which arises out of a simplification of the possibilities and a narrowing of the interpretation of theoretical terms.

[53] In the days of the Vienna Circle this feature led to attempts to justify 'physicalist' psychology, 'physicalist' sociology, biology etc. Also the main principles of the unionist movement

and the attempts to 'unify science' and 'unify the language of science' were based on an assumption of some absolute 'reduction base'.
[54] Quine is right to see in this feature one of the most telling shortcomings of 'the dogma of reductionism' [121, p. 43]. He does not think that it is viable to consider the 'empirical content' of a sentence from some scientific discipline taken in isolation.
[55] A criticism of this aspect of empiricist reductionism is given in [150]. It is noted in particular that unilateral reduction, i.e. reduction in the direction of empirical predicates, is certainly not the only direction possible and in the majority of cases is not even the scientifically most advantageous direction for reduction. Some of the conclusions of this work, which came about independently of Carnap's stand in [31], coincide with this departure from empiricist reductionism.
[56] But let it be emphasised that a scientific portrayal of the world cannot be any kind of compromise between the two extreme positions, claiming that 'the truth is somewhere in between'. The aim of science has always been to search for some order in a world that has always appeared chaotic (in this respect we may speak of science as a negentropic activity of man), to find unity in diversity and to defend the principle of the material unity of the world. The principle of the material unity of the world need not be associated with the old mechanism and its belief that the world consists of a handful of basic building blocks and can be explained by a handful of basic laws. This principle, if associated with the principle of 'qualitative infinity' (in the sense given the term by Bohm [14]) or of the 'inexhaustibility of matter' (in the sense of Lenin [79]), while not denying the qualitative differences of different domains, does point to the essential significance of the two-way transitions between them and so too to the importance of reduction in science. This is by no means to suggest that just any version of reduction is scientifically useful and scientifically admissible, or that reduction is possible in an absolute sense. On the contrary, it raises the question of the *scientific admissibility of reduction*, which the old mechanism never asked.
[57] This brief outline of the theory of levels clearly cannot be treated as complete. The theory of levels is also connected with a number of methodological questions, such as those of determinism, dynamic and statistical laws, scientific explanation and prediction, hidden parameters and so forth. The analysis of these and other questions is beyond the scope of the present work, which mentions the theory of levels only inasmuch as it is connected with the problems of the language of science and semantic analysis.
[58] A more detailed exposition with indications of the broader connections is given in [110].

TO CHAPTER VII

[1] This idea of Russell's is a kind of analogy of Einstein's idea mentioned previously.
[2] By way of example we might quote the original formulation of the paradox of the *Liar*. If we say: 'All Cretans are liars', we need not have in mind that which is expressed by the general quantifier, i.e. we need not exclude the possible existence of at least one Cretan who does not lie or sometimes does not lie. One might put forward similar ideas in the case of other expressions that correspond to the logical terms of formalised languages.
[3] This does not preclude the possibility of constructing formalised languages that operate with some vague logical terms. Certain possibilities along these lines were shown by Kubiński [76].
[4] This concept can be characterised in a more abstract way than other concepts used as the

bases of certain logico-pragmatic conceptions, e.g. 'accept' (Martin [90]), 'belief' or 'claim' (Carnap [26]), 'recognise' or 'reject' (Ajdukiewicz [1]).

[5] The attribute 'discreteness', applied to a language, i.e. a system characterised by having a vocabulary and semantic and syntactic rules, cannot be applied to just any means for the transmission and storage of messages, for we cannot eliminate from the latter those that are continuous in character. On the other hand language – and we mean both natural and formalised languages – can be said to be discrete. If we were to try to transfer the term language to such continuous means as graphs or artistic representations, we would be guilty of an unwarranted generalisation of the concept of language.

[6] Vagueness, as we shall see in due course, cannot be identified with ambiguity or denotational or referential opacity. The name 'Prague' is ambiguous insofar as it denotes the capital of Czechoslovakia, a restaurant in Moscow and a suburb of Warsaw on the right bank of the Vistula. Yet the name is not vague in any of these three interpretations. The relations between vagueness and denotational opacity are analogous.

[7] In a similar sense Carnap [28] uses the term 'intensional vagueness', but against the background of 'pragmatic intension' rather than intension in the purely semantic sense.

[8] This is 'extensional vagueness' in Carnap's terminology.

[9] This distinction corresponds to the well known differentiation of the rules of sense given by Ajdukiewicz [1]. In the spirit of his entire conception of logical semantics and philosophy Ajdukiewicz did of course emphasise the conventional character of all rules of sense.

[10] See (D_3 – V 5 C).

[11] The definition here following does not in principle conflict with (D_3 – V 5 C).

[12] The term 'transparency' had already been used by Whitehead and Russell in *Principia Mathematica*. The term 'referential opacity' was used by Quine [122, p. 141].

TO CHAPTER VIII

[1] The problem of ontic commitments is discussed by Quine in [115], [117], [120] and [122].

[2] The more commonly occurring term in the literature is 'realism'. But because this term has several meanings, which are moreover quite heterogeneous, we shall use the term 'Platonism' and mean by it realism of a Platonist type.

[3] The more detailed analysis of the unity of gnoseological and ontological issues and the connections between them and semantics is a matter for philosophy and cannot be encompassed by the present work.

[4] The term 'substantialisation' is taken here in the sense of the common phenomenalist criticism of substantialisation, for example in the sense of Locke's familiar criticism of the requirement of a 'bearer' of a property.

[5] The phenomenalist criticism of nominalism cannot be identified with the criticism lodged by platonic realism. Platonic realism does not assert the epistemological priority of a property, especially an empirical property, but that of an 'idea', i.e. an hypostatised property.

[6] This also means that even in modern logic it is not possible to deny the utility of the view point which Hao Wang describes as 'predicativism' [160, p. 44] and which largely corresponds to the conception of the epistemological priority of properties. Similarly of course one cannot deny the usefulness of nominalist systems in logic, although they are more suitable for languages of the first degree. This is also one reason why under the nominalist approach to logical semantics it is the analysis of first degree languages which is the centre of attention, as Martin in particular emphasises.

[7] In Marxist philosophical literature this question has been discussed in depth by A.I. Ujomov [178]. His worthy and in a sense pioneering work merits great appreciation, especially since Ujomov was the first to mention many of these questions in Marxist writings. Unlike Ujomov, we use the term 'property' here in a rather wider sense, as anything that can be expressed by a predicate of any kind. Hence in the present conception the concept of 'property' also embraces relations.

[8] In this connection mention must be made of the difficulties connected with the nominalist interpretation of expressions of natural language. For the nominalist the term 'tree' is a one-place predicate whose arguments are individual actual trees. Then the expression 'x is a tree' is satisfied by any actual tree. But if we have the sentence 'trees are plants', the expression 'x is a plant' clearly allows us to recognise even predicates as values of individual variables. But this cannot be admitted by the strict nominalist. Therefore he does not interpret the sentence quoted as the inclusion of classes, i.e. as '$(\hat{x})(x \text{ is a tree}) \subset (\hat{x})(x \text{ is a plant})$', but as the logically conceived dependence of two predicates which relate to individuals of one and the same universe, i.e. as '$(\forall x)\,[(x \text{ is a tree}) \rightarrow (x \text{ is a plant})]$'. But it must be added that inclusion is defined by means of membership in a class, so even in the case of the first interpretation we arrive at individuals.

[9] It is understandable that this view of platonism by which the ontological priority of universals or abstract entities is promoted can also be transfered to the sphere of the natural sciences, but even to the domain of sociology or psychology. Sociological platonism or sociological realism treats the human individual as the 'product' of general conditions, whether these be social, historical or other conditions. Then the individual can be managed according to the abstract scenery of the role assigned to him, universalist categories which, whether with justification or merely verbally, are associated with him. The problem of sociological platonism understandably has its moral consequences.

[10] See the previous note.

[11] Goodman sees this principle as one of the mainstays of nominalism [51].

[12] Any practical identification process presupposes that we only have limited facilities at our disposal, i.e. limitations of 'capacities', 'delays' and 'memories'. Were we to set up an infinite system of distinguishing criteria, we should have to abandon altogether the possibility of making a practical identification.

[13] It also goes without saying that these 'yardsticks' are always bound to the different distinguishing criteria: it is no use trying to determine lengths by using weights. In the formalised notation we shall consider the 'yardsticks' to be individual constants, the tie between 'yardsticks' and the relevant criteria being indicated by matching indices.

[14] The situation is similar under the phenomenalist system, in which likewise it is not individuals that are the starting point, but that which corresponds to properties or qualities. Even in this system, however, an individual may be determined as the intersection of certain qualities.

[15] This was noted by Quine [122, p. 117], who quotes this example from Korzybski [75, p. 194]. Korzybski is trying to assert an extreme nominalist view in so-called general semantics. The expression '$1 = 1$' is not true, because the two signs on either side of the equals sign are different spatially. This means that, using our terminology, spatial difference has to be excluded from the system of distinguishing criteria. In fact, of course, what we have here is numerical equality, which is given by the identity of the number of elements in the class.

It is generally symptomatic that most supporters of the nominalist method of constructing logical systems have had reservations about the concept of 'identity'. For example, Wittgenstein in his *Tractatus* [165, theorem 5.5302] claims that Russell's definition of the sign '=' is insufficient, because in the sense of that definition it cannot be said that two objects have *all* properties in common. In another theorem he adds that "Von *zwei* Dingen zu sagen, sie seien identisch, ist ein Unsinn, und von *Einem* zu sagen, es sei identisch mit sich selbst, sagt gar nichts" (theorem 5.5303).

[16] It can also be objected that definition (D_2 – VIII 2 C) cannot suit our purposes if the predicate Q is an analytical predicate (i.e. $Q_i = Q_j \vee \sim Q_j$) which by definition belongs to all individuals.

[17] It needs to be stressed that a statement of similarity is not the only grounds for concept formation. There are other patterns of concept formation which are based on other reasons than similarity. The advantage of constitution procedures based on similarity is that they are capable of embracing both the traditional (Lockean) conception of abstraction and the conception of conversion, which need not be based on merely a statement of a 'common' empirical property.

[18] This kind of idea would clearly create no problems in a conceptualist or Platonist system.

[19] This formulation need not be identified with the principle of tolerance which would mean a justification of any conventionalism in the construction of a language, its logical basis and its interpretation. Carnap himself did not abandon the principle of tolerance completely, which he had formulated in the early thirties, especially in [18]. Evidence of this is the addition to that formulation, made in 1950: "Let us be cautious in making assertions and critical in examining them, but tolerant in permitting linguistic forms" [27, p. 221].

[20] Leibniz, as we have seen, distinguished truth "in all possible worlds" (vérités de raison) and truth "in a certain world" (vérités de fait). Kant gave no uniform characterisation of analytical propositions. According to Kant an analytical proposition is characterised by three things: (1) the fact that they are based on principles of logic, especially the principle of tertium non datur (analytical propositions are such that their contradiction is unthinkable), (2) the fact that they relate to concepts, not things, (3) the fact that analytical propositions do not extend, but merely repeat our knowledge. An important matter for Kant's position is the distinction between analytical propositions and a priori propositions: a proposition is analytical if its truth can be deduced from logical principles; and a proposition is an a priori proposition if its truth can be ascertained independently of experience. Hence the question of whether there exist a priori synthetic propositions, to which Kant replies in the affirmative and which is of cardinal importance for an understanding of his position. By contrast modern logic has taken it for granted that the answer should be in the negative, although in recent years the certainty about this negative answer has not been so unshakeable as it appeared in the thirties.

[21] Quine's opposition was aimed in particular against Carnap's view which identified analyticity with L-truth and linked analyticity – in the spirit of the Leibniz tradition – with the method of state description. Quine's view has been supported particularly by M. White [162] and [163], N. Goodman [50] and others. A standpoint close to that of Carnap is taken by J.G. Kemeny [69] and [70], H. Putnam [114] and G. Maxwell [93]. A view which likewise argues in favour of the differentiation of analytical and synthetic sentences and therefore come close to Carnap's view has been taken by K. Ajdukiewicz [2], M. Przelecki [112] and E.D. Smirnova [177]. However, the last mentioned works do not really connect directly

with the discussions of the American logicians, although they do take up some of their results.

[22] Quine's criticism applies chiefly to Carnap's works.

[23] A review of the basic definitions is given by Stegmüller [144, p. 293], Smirnova [177, pp. 349–350], Mates [92] and others. These reviews could have been supplemented by the historical view of things, which also shows the huge variety of conceptions.

[24] In Carnap's semantics this distinction is not made. On the other hand Kemeny [69], while in principle not agreeing with Quine's scepticism, makes the distinction quite scrupulously.

[25] Ajdukiewicz describes this kind of analytical sentences in the following terms: "Nous dirons que la proposition *P* est une proposition analytique au sens sémantique de la langue $\mathscr{L}$ si elle est un postulat ou la conséquence d'un ou plusieurs postulats de cette langue" [2, p. 260].

[26] The problems of analytical sentences can also be used to demonstrate the previously stated conception of the relations between ontic and semantic decision.

[27] From the philosophical point of view this looks like a 'back-to-front' version of Kant's question about the possibilities for a priori synthetic propositions. In the present instance we are talking about analytical sentences whose analytical truth is mediatedly connected with the ascertained relations of extralinguistic objects. However, we think that it is not expedient to speak of a posteriori analytical propositions, since decisions of principle on analyticity are bound to semantic rules, not facts, the question of the genesis of those rules being left to one side.

[28] This problem was noted in particular by Ajdukiewicz [2].

[29] In this case the situation can be described ex post as 'hidden synonymity'.

[30] This concerns more than just the case of the 'black swans' quoted in logical text-books, although this case does provide a good example of the second situation. Before black swans were discovered in Oceania, white plumage could be treated as a generic mark of all swans and the sentence 'swans are white' could be considered analytical. Kemeny in [70] quotes an example to illustrate the first situation: the sentence which expressed Newton's second principle, i.e. '$\mathbf{f} = \mathbf{ma}$' could originally be considered synthetic. More careful analysis allows us to conclude that it is an analytical sentence.

TO CHAPTER X

[1] One example of a minimal system of axioms is that mentioned by R.D. Luce and H. Raiffa [85]. The concept of preference as based on a combination of the measures of probability and desirability was introduced by R.C. Jeffrey [64]. A preference ordering based on the distance measure was constructed by J.G. Kemeny and J.L. Snell [71]. There are other examples which might also have been quoted.

[2] See e.g. H.S. Houthakker [61].

[3] This means that the operation interpolated between expressions from the different spheres is connective, and not adjunctive in Reichenbach's sense [124]. According to Reichenbach, a logical operation is connective if the corresponding matrix can be read in one particular order, and adjunctive if the order is immaterial in an interpretation.

[4] See e.g. S. Halldén [52].

[5] One of the major pioneers of the logic of preference, G.H. von Wright, speaks of 'proposition-like entities' [167] [168].

[6] A survey of the properties of a comparative predicate and a reconstruction of a quasi-serial order is set out in [153].
[7] Among logicians, Chisholm and Sosa [34] explicitly reject sentences (6), (7) and (8) as admissible, but not on the grounds that they are at variance with the sphere of evaluations, but because of their own conception, which they describe as a hedonistic theory of preferability.
[8] The differentiation of connective and adjunctive operations has been made in logic by Reichenbach [124].
[9] Within modern logic these problems have already been noted quite early by Whitehead and Russell in their *Principia Mathematica* and later by Carnap in *Meaning and Necessity*. In recent years new conceptions of a logical theory of propositional attitudes have been put forward, notably by J. Hintikka [60] and N. Rescher [126].

TO CHAPTER XI

[1] A more detailed account of these difficulties along with a discussion of the problems of finitism in identification tasks is to be found in [156].
[2] A survey of some of the basic alternative possibilities for calculating the measure of uncertainty is given in [155].
[3] These measures correspond to formulae 6.7 and 6.8 in [102].
[4] These normalised measures of transmitted information correspond to formulae 6.10 and 6.11 in [102]. In addition, (14) corresponds to Hempel's measure of systematising power [57].

REFERENCES

[1] Ajdukiewicz, K. : 1934, 'Sprache und Sinn', *Erkenntnis* **4**, 100–138.

[2] Ajdukiewicz, K. : 1958, 'Le problème du fondement des propositions analytiques', *Studia logica* **8**, 259–272.

[3] Ajdukiewicz, K. : 1960, *Język i poznanie*, Tom I, Warszawa.

[4] Ayer, A.J. : 1946, *Language, Truth and Logic*, Dover Publ., New York.

[5] Bar-Hillel, Y. : 1951, 'A Note on State Descriptions', *Philosophical Studies* **2**, 72–75.

[6] Bar-Hillel Y., and R. Carnap: 1952, *An Outline of a Theory of Semantic Information*, Technical Report No. 247, MIT.

[7] Bar-Hillel Y., and R. Carnap: 1953, 'Semantic Information', in W. Jackson (ed.), *Communication Theory*, Butterworths Scientific Publications, London, pp. 503–511.

[8] Bergmann, G. : 1961, 'Physics and Ontology', *Philosophy of Science* **28**, 1–14.

[9] Bertin, J. : 1967, *Sémiologie graphique*, Mouton – Gauthier-Villars, Paris.

[10] Black, M.: 1949, *Language and Philosophy* (Studies in Method), Cornell Univ. Press, Ithaca, New York.

[11] Bloomfield, L.: 1944, *Linguistic Aspects of Science*, International Encyclopedia of Unified Science, Chicago.

[12] Bloomfield, L.: 1957, *Language*, George Allen and Unwin, London.

[13] Boas, F.: 1949, *Race, Language and Culture*, Macmillan, New York.

[14] Bohm, D.: 1957, *Causality and Chance in Modern Physics*, Routledge and Kegan Paul, London.

[15] Bréal, M.: 1904, *Essai de sémantique, Science des significations*, Librairie Hachette et Cie., Paris.

[16] Bridgman, P.W. : 1928, *The Logic of Modern Physics*, Macmillan, New York.

[17] Brillouin, L. : 1962, *Science and Information Theory*, 2nd edn., Academic Press, New York.

[18] Carnap, R. : 1928, *Der logische Aufbau der Welt*, Berlin, 2nd edn., F. Meiner, Hamburg 1961.

[19] Carnap, R. : 1929, *Abriss der Logistik*, Springer, Wien.

[20] Carnap, R.: 1937, *Logical Syntax of Language*, Routledge, London.

[21] Carnap, R.: 1936–1937, 'Testability and Meaning', *Philosophy of Science* **3**, 1936 and **4**, 1937, reprinted in H. Feigl *et al.* (eds.), *Readings in the Philosophy of Science*, Appleton Cent., New York, 1953, pp. 47–92.

[22] Carnap, R.: 1942, 'Introduction to Semantics', *Studies in Semantics*, Vol. I, Harvard Univ. Press, Cambridge Mass.

[23] Carnap, R. : 1947, *Meaning and Necessity*, Univ. of Chicago Press, Chicago.

[24] Carnap, R.: 1950, *Logical Foundations of Probability*, Univ. of Chicago Press, Chicago.

[25] Carnap, R.: 1952, 'Meaning Postulates', *Phil. Studies* **3**, 65–73, also printed as an appendix to the second edition of [23].

[26] Carnap, R.: 1955, 'On Some Concepts of Pragmatics', *Phil. Studies* **6**, 89–91, reprinted in the second edition of [23].

[27] Carnap, R.: *Empiricism, Semantics and Ontology*, as reprinted in the second edition of [23].

[28] Carnap, R.: 1955, 'Meaning and Synonymy in Natural Languages', *Phil. Studies* **7**, 33–47, reprinted in the second edition of [23].

[29] Carnap, R.: 1955, 'On Some Concepts of Pragmatics', *Phil. Studies* **6**, 87–89, reprinted in the second edition of [23].

[30] Carnap, R.: *On Belief-Sentences, Reply to Church*, as reprinted in the second edition of [23].

[31] Carnap, R.: 1956, 'The Methodological Character of Theoretical Concepts', *Minnesota Studies in the Philosophy of Science* **1**, 38–76.

[32] Carnap, R.: 1959, *Induktive Logik und Wahrscheinlichkeit*, Springer, Wien.

[33] Cherry, C.: 1957, *On Human Communication*, MIT.

[34] Chisholm, R.M., and E. Sosa: 1966, 'Intrinsic Preferability and the Problem of Supererogation', *Synthese* **16**, 321–331.

[35] Church, A.: 1941, 'The Calculi of Lambda-Conversion', *Annales of Math. Studies* **6**,

[36] Church, A.: 1949, 'Review of Ayer's Language, Truth and Logic', *Journal of Symbolic Logic* **14**, 52–53.

[37] Church, A.: 1951, 'A Formulation of the Logic of Sense and Denotation', in P. Hanke *et al.* (eds.), *Structure, Method and Meaning*, The Liberal Arts Press New York,

[38] Church, A.: 1954, 'Intensional Isomorphism and Identity of Belief', *Phil. Studies* **5**, 65–73.

[39] Church, A.: 1956, *Introduction to Mathematical Logic*, Princeton Univ. Press, Princeton N.J.; Russian translation: Введение в математическую логику, Изд. инностранной литературы, Москва, 1960.

[40] Dürr, K.: 1954, *Lehrbuch der Logistik*, Basel, Stuttgart, 1954.

[41] Einstein, A.: 1921, *Geometrie und Erfahrung*, Springer, Berlin.

[42] Ewing, A.C.: 1937, 'Meaninglessness', *Mind*, 1937.

[43] Frege, G.: 1879, *Begriffschrift*, Halle.

[44] Frege, G.: 1891, *Function und Begriff*, Jena.

[45] Frege, G.: 1892, 'Über Sinn und Bedeutung', *Zeitschrift für Philosophie und philosophische Kritik* **100**, 25–50; English translation 'On Sense and Nominatum', in H. Feigl *et al.* (eds.), *Readings in Philosophical Analysis*, Appleton-Century-Crofts Inc., New York, 1949, pp. 85–115.

[46] Frege, G.: 1893, *Grundgesetze der Arithmetik*, Bd. I, II, Jena.

[47] Gödel, K. 1931, 'Über formal unentscheidbare Sätze der Principa Mathematica und verwandter Systeme', *Monatshefte Math. Phys.* **38**. 173–198.

[48] Goodman, N.: 1947, 'The Problem of Counterfactual Conditionals', *Journal of Philosophy* **44**, 113–128; reprinted in L. Linsky (ed.), *Semantics and the Philosophy of Language*, Univ. of Illinois Press, Urbana, 1952, pp. 231–247.

[49] Goodman, N.: 1951, *The Structure of Appearance*, Harvard Univ. Press, Cambridge Mass.

[50] Goodman, N.: 1952, 'On Likeness of Meaning', in L. Linsky (ed.), *Semantics and the Philosophy of Language*, Univ. of Illinois Press, Urbana, 1952, pp. 67–76.

[51] Goodman, N.: 1956, 'A World of Individuals', in J.M. Bocheński *et al.* (eds.), *The Problem of Universals: A Symposium*, Univ. of Notre Dame Press.

[52] Halldén, S.: 1957, *On the Logic of 'Better'*, Library of Theoria, Uppsala.

[53] Hayakawa, S.I.: 1949, *Language in Thought and Action*, Harcourt Brace, New York.

[54] Hempel, C.G., and P. Oppenheim: 1948, 'Studies in the Logic of Explanation', *Philosophy of Science* **15**, 135–175; reprinted in H. Feigl et al. (eds.), *Readings in the Philosophy of Science*, Appleton-Century-Crofts Inc., New York 1953, pp. 319–352.

[55] Hempel, C.G.:1952, 'Problems and Changes in the Empiricist Criterion of Meaning', *Revue Internationale de Philosophie* **11**, 1950; reprinted in L. Linsky (ed.), *Semantics and the Philosophy of Language*, Univ. of Illinois Press, Urbana, 1952, pp. 161–185.

[56] Hempel, C.G.: 1958, 'The Theoretician's Dilemma,' *Minnesota Studies in the Philosophy of Science*, Vol. II, Univ. of Minnesota Press, pp. 37–98.

[57] Hempel, C.G.: 1965, *Aspects of Scientific Explanation*, The Free Press, New York.

[58] Hermes, H., and H. Scholz: 1952, *Mathematische Logik*, Enz. Math. Wiss. Bd. I, l. Teil, Leipzig.

[59] Hilpinen, R.: 1970, 'On the Information provided by Observations', in J. Hintikka *et al.* (eds.), *Information and Inference*, D. Reidel Publishing Company, Dordrecht and Boston, 1970, pp. 97–122.

[60] Hintikka, J.: 1962, *Knowledge and Belief*, Ithaca, New York.

[61] Houthakker, H.S.: 1965, 'On the Logic of Preference and Choice,' in A.T. Tymieniecka (ed.), *Contributions to Logic and Methodology in Honor of J.M. Bocheński*, Amsterdam, 1965, pp. 193–207.

[62] Husserl, E.: 1913, 1921, *Logische Untersuchungen*, Bd. 1, 2, Halle.

[63] Ingarden, R.: 1936, 'Der logistische Versuch einer Neugestaltung der Philosophie', *Actes du 8ème Congrès International de Philosophie*, Prague 1936, pp. 203–208.

[64] Jeffrey, R.C.: 1965, *The Logic of Decision*, McGraw-Hill, New York, London and Sydney.

[65] Kaneff, S. (ed.): 1970, *Picture Language Machines*, Academic Press, London and New York.

[66] Kazemier, B.H., and D. Vuysje (eds.): 1962, *Logic and Language*, Studies dedicated to Professor Rudolf Carnap, D. Reidel Publishing Company, Dordrecht.

[67] Kemeny, J.G.: 1948, 'Models of Logical Systems', *Journal of Symbolic Logic* **13**, pp. 16–30.

[68] Kemeny, J.G.: 1951, 'Review of Logical Foundations of Probability', *Journal of Symbolic Logic* **16**, 205–207.

[69] Kemeny, J.G.: 1956, 'A New Approach to Semantics', *Journal of Symbolic Logic* **21**, 1–27.

[70] Kemeny, J.G.: 1963, 'Analycity versus Fuzziness', *Synthese* **15**, 57–80.

[71] Kemeny, J.G., and J.L. Snell: 1962, *Mathematical Models in the Social Sciences*, Blaisdell Publishing Company, New York, Toronto and London.

[72] Kleene, S.C.: 1952, *Introduction to Metamathematics*, Van Nostrand Comp., New York.

[73] Kotarbiński, T.: 1961, *Elementy teorii poznania, logiky formalnej i metodologii nauk*, Zakl. im. Ossolińskich, Wrocław, Warszawa and Kraków (first ed'n. 1929).

[74] Kotarbiński, T.: 1955, *Kurs logiky*, Państwowe wydawnictwo naukowe, Warszawa.

[75] Korzybski, A.: 1933, *Science and Sanity*, Science Press, Lancaster.

[76] Kubiński T.: 1958, 'Nazwy neostre', *Studia Logica* **7**, 115–149.

[77] Ledley, R.S., and L.B. Lusted: 1959, 'Reasoning Functions of Medical Diagnosis', *Science* **130**, No. 3366, pp. 9–21.

[78] Lee Whorf, B.: 1957, *Language, Mind and Reality*, Selected Writings, MIT Press.

[79] Lenin, V.I.: 1957, 'Materialismus a empiriokriticismus', *Spisy* sv. 14, Státní nakladatelství politické literatury, Praha (Czech translation).

[80] Lévy-Bruhl, L.: 1936, *Morceaux choisis*, Gallimard, Paris.
[81] Lewis, C.I.: 1918, *Survey of Symbolic Logic*, Berkeley.
[82] Lewis, C.I.: 1934, 'Experience and Meaning', *Philosophical Review* **43**; reprinted H. Feigl *et al.* (eds.), *Readings in Philosophical Analysis*, Appleton-Century-Crofts Inc., New York, pp. 128–145.
[83] Lewis, C.I.: 1944, 'The Modes of Meaning', *Philosophy and Phenomenological Research* **4**; reprinted in L. Linsky (ed.), *Semantics and the Philosophy of Language*, Univ. of Illinois Press, Urbana, 1952, pp. 49–63.
[84] Lewis, C.I.: 1946, *An Analysis of Knowledge and Valuation*, Open Court Publishing Company, La Salle, Ill.
[85] Luce, R.D., and H. Raiffa: 1957, *Games and Decisions. Introduction and Critical Survey*, McGraw-Hill, New York, London and Sydney.
[86] Malinowski, B.: 1936, 'The Problem of Meaning in Primitive Languages', published as an appendix to C.K. Ogden and I.A. Richards, *The Meaning of Meaning*, Routledge and Kegan Paul, London, pp. 296–336.
[87] Marhenke, P.: 1950, 'The Criterion of Significance', *Proceedings and Addresses of The American Philosophical Association* **23**; reprinted in L. Linsky (ed.), *Semantics and the Philosophy of Language*, Univ. of Illinois Press, Urbana, pp. 139–162.
[88] Marković, M.: 1961, *Dijalektička teorija značenija*, Nolit, Beograd.
[89] Martin, M.R.: 1958, *Truth and Denotation. A Study in Semantical Theory*, Univ. of Chicago Press and Kegan Paul, London.
[90] Martin, R.M.: 1959, *Toward a Systematic Pragmatics*, Amsterdam.
[91] Mates, B. 1951, 'Synonymity', *Univ. of California Publications in Philosophy* **25**, reprinted in L. Linsky (ed.), *Semantics and the Philosophy of Language*, Univ. of Illinois Press, Urbana, 1952.
[92] Mates, B.: 1951, 'Analytic Sentences', *Philosophical Review* **60**.
[93] Maxwell, G.: 1962, 'The Necessary and the Contingent', *Minnesota Studies in the Philosophy of Science*, vol. III, Univ. of Minnesota Press, Minneapolis.
[94] Maxwell, G.: 1962, 'The Ontological Status of Theoretical Entities', *Minnesota Studies in the Philosophy of Science*, Vol. III, Univ. of Minnesota Press, pp. 3–27.
[95] Morris, C.W.: 1938, *Foundations of the Theory of Signs*, International Encyclopedia of Unified Science, Chicago.
[96] Morris, C.W.: 1946, *Signs, Language and Behaviour*, Prentice-Hall, New York.
[97] Naess, A.: 1949, 'Towards a Theory of Interpretation and Preciseness', *Theoria* **15**, 1949; reprinted in L. Linsky (ed.), *Semantics and the Philosophy of Language*, Univ. of Illinois Press, Urbana, pp. 248–271.
[98] Naess, A.: 1953, *Interpretation and Preciseness. A Contribution to the Theory of Communication*, Skrifter det Norske Videnskaps-Akademi, Oslo.
[99] Nagel, E.: 1961, *The Structure of Science*, Harcourt Brace, New York.
[100] Nake, F., and A. Rosenfeld: 1972, *Graphic Languages*, North Holland Publ. Company, Amsterdam.
[101] Narasimhan, R.: 1970, 'Picture Languages', in [65], pp. 1–30.
[102] Niiniluoto, I., and R. Tuomela: 1973, *Theoretical Concepts and Hypothetical-Inductive Inference*, D. Reidel Publishing Company, Dordrecht and Boston.
[103] Ogden, C.K., and I.A. Richards: 1936, *The Meaning of Meaning*, 4th edn., Routledge and Kegan Paul, London.
[104] Osgood, C.E.,Suci, G.J., and P.H. Tannenbaum: 1957, *The Measurement of Meaning*, Univ. of Illinois Press, Urbana.

[105] Pap, A.: 1953, 'Reduction Sentences and Open Concepts', *Methodos* **5**.

[106] Pap, A.: 1955, *Analytische Erkenntnistheorie*, Springer, Wien.

[107] Pap, A.: 1958, 'Disposition Concepts and Extensional Logic', *Minnesota Studies in the Philosophy of Science*, Vol. II, Univ. of Minnesota Press, Minneapolis, pp. 196–224.

[108] Pap, A.: 1958, *Semantics and Necessary Truth*, Yale Univ., New Haven, Conn.

[109] Peirce, C.S.: 1931–1935, *Collected Papers*, Harvard Univ. Press, Cambridge, Mass.

[110] Perez, A., and L. Tondl: 1965, 'Modely některých vědeckých procedur z hlediska logiky a teorie informace', in J. Kožešník (ed.), *Problémy kybernetiky*, Nakl. ČSAV, Praha, pp. 9–22.

[111] Pietarinen, J.: 1970, 'Quantitative Tools for Evaluating Scientific Systemizations', in J. Hintikka *et al.* (eds.), *Information and Inference*, D. Reidel Publishing Company, Dordrecht and Boston, pp. 123–147.

[112] Przełęcki, M.: 1963, 'O pojęciu zdania analyticzego', *Studia logica* XIV.

[113] Putnam, H.: 1960, 'What theories are not', in E. Nagel *et al.* (eds.), *Logic, Methodology and the Philosophy of Science*, (Proceedings of the 1960 International Congress), Stanford Univ. Press, Stanford, 1962.

[114] Putnam, H.: 1962, 'The Analytic and Synthetic', *Minnesota Studies in the Philosophy of Science*, Vol. III, Univ. of Minnesota Press, Minneapolis, pp. 358–397.

[115] Quine, W. van O.: 1934, 'Notes on Existence and Necessity', *Journal of Philosophy* **40**, 113–123; reprinted in L. Linsky (ed.), *Semantics and the Philosophy of Language*, Univ. of Illinois Press, Urbana, 1952, pp. 77–91.

[116] Quine, W. van O.: 1953, *From a Logical Point of View, Harvard Univ. Press, Cambridge*, Mass.

[117] Quine, W. van O.: 1953, 'Logic and the Reification of Universals', in [116], pp. 102–129.

[118] Quine, W. van O.: 'The Problem of Meaning in Linguistics', in [116], pp. 47–64.

[119] Quine, W. van O.: 1953, 'Notes on the Theory of Reference', in [116], pp. 130–138.

[120] Quine, W. van O.: 1953, 'On What There Is', in [116], pp. 1–19.

[121] Quine, W. van O.: 1953, 'Two Dogmas of Empiricism', in [116], pp. 20–46.

[122] Quine, W. van O.: 1960, *Word and Object*, Harvard Univ. Press, Cambridge, Mass.

[123] Ramsey, F.P.: 1931, *The Foundations of Mathematics*, New York.

[124] Reichenbach, H.: 1947, *Elements of Symbolic Logic*, Macmillan, New York.

[125] Rescher, N.: 1966, 'Notes on Preference, Utility and Cost', *Synthese* **16**, 332–343.

[126] Rescher, N.: 1968, *Topics in Philosophical Logic* (Synthese Library), D. Reidel Publishing Company, Dordrecht and Boston.

[127] Rogers, R.: 1963, 'A Survey of Formal Semantics', *Synthese* **15**, 17–56.

[128] Rosenfeld, A.: 1969, *Picture Processing by Computer*, Academic Press, New York.

[129] Rovenskij, Z., Ujomov, A., and J. Ujomova: 1962, *Stroj a myšlení, filosofický nástin kybernetiky* (Czech translation of a Russian original), Orbis, Praha.

[130] Russell, B.: 1940, *Inquiry into Meaning and Truth*, George Allen and Unwin, London, 1951 (first edn. 1940).

[131] Russell, B.: 1905, 'On Denoting', *Mind* **14**, pp. 479–493.

[132] Russell, B.: 1912, *The Problems of Philosophy*, Czech translation *Problémy filosofie*, Nakl. Čin, Praha, 1927.

[133] Russell, B.: 1920, *Introduction to Mathematical Philosophy*, 2nd edn., London.

[134] Russell, B.: 1921, *Analysis of Mind*, London.

[135] Russell, B.: 1923, 'Vagueness', *Australian Journal of Philosophy* **1**; quoted by Black [10].

[136] Russell, B.: 1948, *Human Knowledge, its Scope and Limits*, George Allen and Unwin, London.

[137] Sapir, E.: 1921, *Language*, Harcourt Brace and Co., New York.

[138] Schaff, A.: 1965, *Úvod do sémantiky* (Czech translation), Nakladatelství politické literatury, Praha.

[139] Scheffler, I.: 1955, 'On synonymy and Indirect Discourse', *Philiosophy of Science* 22, 39–44.

[140] Schlick, M.: 1946, 'Meaning and Verification', *The Philosophical Review* **45**; reprinted in H. Feigl *et al.* (eds.), *Readings in Philosophical Analysis*, Appleton-Century-Crofts Inc., New York, 1949, pp. 146–170.

[141] Scholz, H., and G. Hasenjaeger: 1961, *Grundzüge der mathematischen Logik*, Springer, Berlin.

[142] Sellars, W.: 1958, 'Counterfactuals, Dispositions and Causal Modalities', *Minnesota Studies in the Philosophy of Science*, Vol. II, Univ of Minnesota Press, Minneapolis, pp. 225–307.

[143] Stebbing, L.S.: 1945, *A Modern Introduction to Logic*, Methuen, London.

[144] Stegmüller, W.: 1957, *Das Wahrheitsproblem und die Idee der Semantik*, Springer, Wien.

[145] Tarski, A.: 1937, *Einführung in die mathematische Logik*, Springer, Wien; English translation: *Introduction to Logic and to the Methodology of Deductive Sciences*, Oxford Univ. Press, New York, 9th print, 1961.

[146] Tarski, A.: 1936, 'O ugruntowaniu naukowej semantiki', *Przegląd Filosoficzny* **39**, 30–57; English translation: 'The Establishment of Scientific Semantics', in A. Tarski, *Logic, Semantics, Metamathematics*, Oxford Univ. Press, 1956.

[147] Tarski, A.: 1936, 'Der Wahrheitsbegriff in den formalisierten Sprachen', *Studia Philosophica* **1**, 261–405; English translation: 'The Concept of Truth in Formalised Languages', in A. Tarski, *Logic, Semantics, Metamathematics*, Oxford Univ. Press, 1956.

[148] Tarski, A.: 1936, 'Über den Begriff der logischer Folgerung', *Actes du Congrès International de Philosophie Scientifique*, Vol. 7, 1936; English translation: 'On the Concept of Logical Consequence', in A. Tarski, *Logic, Semantics, Metamathematics*, Oxford Univ. Press, 1956; Polish original: 'O pojęciu wynikania logicznego', *Przegląd filosoficzny* 39, 1936, pp. 58–68.

[149] Tarski, A.: 1944, 'The Semantic Conception of Truth,' *Philosophy and Phenomenological Research* **4**; reprinted in H. Feigl *et al.* (eds.), *Readings in Philosophical Analysis*, Appleton-Century-Crofts Inc., New York, 1949, pp. 52–84.

[150] Tondl, L.: 1957, 'Gnoseologická úloha abstrakcé, in L. Tondl (ed.), *Otázky teorie poznání*, Státní nakl. politické literatury, Praha, pp. 77–131.

[151] Tondl, L.: 1958, 'Novopositivismus', in L. Tondl, *Současná západní filosofie*, Orbis, Praha, pp. 17–54.

[152] Tondl, L.: 1959, 'Kauzální analýza a kauzální explikace', in O. Zich *et al.*, *K metodologii experimentálních věd*, Nakl. ČSAV, Praha, pp. 139–278.

[153] Tondl, L.: 1972, 'Prerequisites for Quantification in the Empirical Sciences', *Theory and Decision* **2**, 238–261.

[154] Tondl, L.: 1972, 'Evaluation of Components in Ledley and Lusted's Diagnostic Model', *Kybernetika* **8**.

[155] Tondl, L.: 1972, 'K probleme semantičeskoj informacii' (in Russian), *Kybernetika* **8**, pp. 189–212.

[156] Tondl, L.: 1973, *Scientific Procedures*, D. Reidel Publishing Company, Dordrecht and Boston.

[157] Ullmann, S.: 1957, *The Principles of Semantics*, Blackwell, Oxford.

[158] Vigier, J.P.: 1962, 'Determinism and Indeterminism in a new "level" Theory of Matter', in E. Nagel *et al.* (eds.), *Logic, Methodology and Philosophy of Science*, Stanford Univ. Press, pp. 262–264.

[159] Walpole, H.: 1941, *The Nature of Words and Their Meanings*, W.W. Norton and Comp. Inc., New York.

[160] Wang Hao: 1953, 'What is an Individual', *Philosophical Review* **62**, 413–420.

[161] Wang Hao: 1962, *A Survey of Mathematical Logic*, Science Press, Peking.

[162] White, M.G.: 1956, *Toward Revision in Philosophy*, Harvard Univ. Press Cambridge, Mass.

[163] White, M.G.: 1952, 'The Analytic and the Synthetic: an Untenable Dualism', in L. Linsky (ed.), *Semantics and the Philosophy of Language*, Univ. of Illinois Press, Urbana, pp. 272–286.

[164] Whitehead, A.N., and B. Russell: *Principia Mathematica*, reprinted Cambridge University Press, 1962.

[165] Wittgenstein L.: 1933, *Tractatus Logico-Philosophicus*, Kegan Paul, London.

[166] Wittgenstein, L.: 1953, *Philosophical Investigations*, Oxford.

[167] von Wright, G.H.: 1963, *The Logic of Preference*, Edinburgh.

[168] von Wright, G.H.: 1972, 'The Logic of Preference Reconsidered', *Theory and Decision* **3**, 140–169.

[169] Zich, O.: 1947, *Úvod do filosofie matematiky*, Jednota českých matematiků a fyziků, Praha.

[170] Zich, O. *et al.*: 1958, *Moderní logika*, Orbis, Praha.

[171] Баженов, Л. Б., Усмов, А. И., and В. Г. Фарбер (eds.): 1961, *Логико-грамматические очерки*, Высшая школа, Москва.

[172] Бирюков, В. Б.: 1960, 'Теория смысла Готлоба Фреге', in *Применение логики в науке и технике*, Москва.

[173] Брутян, Г.: 1959, *Теория познания общей семантики*, Изд. АН Армянской ССР, Ереван.

[174] Вижье, Ж. П.: 1962, 'Теория уровней и диалектика природы', *Вопросы философии* **10**.

[175] Гетманова, А. Д.: 1959, 'О соотношениях математики и логики в системах типа Principia Mathematica', in *Логические исследования*, Изд. АН СССР, Москва, pp. 189–217.

[176] Горский, Д. П.: 1961, *Вопросы абстракции и образование понятий*, Изд. Мысль, Москва.

[177] Смирнова, Е. Д.: 1962, 'К проблеме аналитического и синтетического', in *Философские вопросы современной формальной логики*, Изд. АН СССР, Москва, pp. 323–363.

[178] Уемов, А. И.: 1963, *Вещь, свойство, отношение*, Изд. АН СССР, Москва.

[179] Успенский, В. А.: 1959, 'К проблеме построения машинного языка', in А. А. Ляпунов (ed.), *Проблемы кибернетики* Т. 2, Москва.

[180] Швырев, В. С.: 1962, 'Неопозитивистская концепция эмпирического значения и логический анализ научного знания', in *Философские вопросы современной формальной логики*, Изд. АН СССР, Москва.

INDEX OF NAMES

Ajdukiewicz, K. 3, 268, 272, 374, 376, 377
Ayer, A. J. 4

Bar-Hillel, Y. 3, 68, 74, 76, 81, 84, 290, 331
Bekhterev, V. M. 4
Bergmann, G. 214
Bertin, J. 340–341
Birjukov, B. V. 112
Black, M. 5, 234, 237, 354
Bloomfield, L. 7
Boas, F. 6
Bohm, D. 222, 223, 373
Boltzmann, L. 68
Bréal, M. 1
Bridgman, P. W. 171–172, 175, 369
Brillouin, L. 277–280
Brouwer, L. 258
Brutjan, G. 354
Burks, A. 372

Carnap, R. 3, 4, 16, 43–44, 49, 53–55, 62–65, 67–68, 74, 76–79, 81–82, 87, 90–92, 94, 108, 110, 123, 125, 128–129, 136–139, 141, 144–145, 147, 157–161, 185–186, 189, 192–193, 195–197, 200–201, 206, 208, 214, 217, 225, 241, 247, 252–254, 258, 264, 265, 269–270, 272, 280, 290, 294–295, 313–314, 328, 331, 336, 354, 355, 356, 357, 358, 359, 360, 365, 366, 367, 368, 369, 370, 371, 372, 373, 374, 376, 377, 378
Chase, S. 5
Cherry, C. 21, 25, 30, 68, 354
Chisholm, R. M. 378
Chomsky, N. 114, 345
Church, A. 10, 106–107, 109–110, 112, 124, 258, 355, 356, 357, 363, 365, 368, 371
Chwistek, L. 11

Descartes, R. 119
Ducasse, C. J. 131
Dürr, K. 356

Eddington, A. 372
Einstein, A. 6, 171–172, 232, 369, 373
Ewing, A. C. 371

Frege, G. 10, 35, 50, 63, 66, 102–116, 122–123, 128–130, 132–134, 138–139, 141, 143, 150, 258, 328, 346, 354, 355, 356, 362, 363, 364
Freud, S. 5

Getmanova, A. D. 366
Gödel, K. 122
Goodman, N. 145, 257, 371, 375, 376
Gorskij, D. P. 365

Halldén, S. 311, 319, 377
Hao Wang 262, 374
Hasenjaeger, G. 10, 356
Hayakawa, S. I. 5
Heidegger, M. 371
Heisenberg, V. 222, 369, 372
Hempel, C. G. 212, 285, 291, 365, 370, 378
Heyting, A. 258
Hilbert, D. 354
Hilpinen, R. 297
Hintikka, J. 378
Houthakker, H. S. 377

Ingarden, R. 370

Jeffrey, R. C. 322, 377

Kant, I. 265, 376, 377
Kemeny, J. G. 46, 62, 84, 92–96, 99–102, 357, 361, 376, 377
Kleene, S. C. 356
Korzybski, A. 5, 375
Kotarbiński, T. 3, 31–32, 38, 40, 42, 355
Kubiński, T. 373

Ledley, R. S. 285
Lee-Whorf, B. 6, 355
Leibniz, G. W. 10, 12, 24, 46, 61, 66, 95–97, 103–104, 106, 112, 134, 146, 265, 325–327, 330, 361, 362, 376
Lenin, V. I. 254, 373
Leśniewski, S. 11
Lévy-Bruhl, L. 6
Lewis, C. I. 123, 138–142, 156–158, 169, 180, 356, 362, 366, 367
Locke, J. 374, 376
Luce, R. D. 377
Lukasiewicz, J. 12

Mach, E. 214
Mácha, K. H. 371
Malinowski, B. 6
Marhenke, P. 188, 369, 371
Markov, A. A. 258
Martin, R. M. 17, 48, 57–58, 60, 92, 356, 357, 365, 374
Marx, K. 149, 303
Mates, B. 153, 308, 367, 368, 377
Maxwell, G. 210, 213, 376
Meinong, A. 140, 364, 366
Mill, J. S. 3
Morris, C. W. 4, 7, 16

Naess, A. 368
Nagel, E. 214
Newton, I. 171, 183, 268, 377

Ogden, C. K. 5, 17
Osgood, C. E. 7

Pap, A. 370, 371
Pavlov, I. P. 4
Peano, G. 50, 253
Peirce, C. S. 4, 16, 356, 363
Perez, A. 225
Pietarinen, J. 288
Poincaré, A. 372
Przelecki, M. 271, 376
Putnam, H. 197, 376

Quine, W. van O. 92, 101–102, 110, 145–146, 149, 155–161, 217, 248–252, 257, 265–266, 268–270, 328, 330, 335–336, 361, 366, 367, 368, 373, 374, 375, 376, 377

Raiffa, H. 377
Reichenbach, H. 356, 377, 378
Rescher, H. 310, 313–315, 378
Richards, I. A. 5, 11, 14, 17–19
Russell, B. 10–11, 32–35, 38–40, 50, 105–106, 115–122, 131–132, 136, 138–139, 147, 150, 190–191, 196, 211 215, 233–234, 240, 248–249, 258, 270,–271, 326, 328, 354, 355, 362, 363, 364, 365, 366, 373, 374, 376, 378
Ryle, G. 4

Sapir, E. 6, 355
Schaff, A. 2
Scheffler, I. 368
Schlick, M. 3, 179–183, 191, 193, 212–214, 370
Scholz, H. 10, 356
Sellars, W. 371
Shannon, K. 3, 20, 68, 81, 92
Shcherba, L. V. 369
Shvyrev, V. S. 371, 372
Smirnova, E. D. 268, 376, 377
Snell, J. L. 377
Sosa, E. 378
Stebbing, L. S. 122
Stegmüller, W. 268, 357, 360, 370, 377

Tarski, A. 4–5, 11–14, 16, 50–57, 59, 92, 94, 104, 154, 247, 354, 357, 361, 362

Ujomov, A. I. 375
Ullmann, S. 152
Uspenskij, V. A. 194, 246, 325, 329, 363

Vigier, J. P. 222–223

Walpole, H. 5
Weyl, H. 258
White, M. 376
Whitehead, A. N. 136, 147, 248, 258, 326, 362, 364, 374, 378
Wittgenstein, L. 3, 59, 67, 91, 148–149, 177, 191, 217, 358, 376
von Wright, G. H. 310, 319, 323, 377

INDEX OF SUBJECTS

A-truth (*see also* analytical truth) 97, 100, 361
 –, definition of 97
absolute intelligence 279
absolute time 171
abstract terms 139, 141
abstraction 40, 47–49, 58, 139, 209, 211, 215, 259, 357, 365, 376
 –, geometrical 49
abstractum 47, 58, 124, 127–130, 139, 255, 367, 368
 –, empty 58
 –, many-placed 58
 –, non-empty 58
 –, one-placed 58
acquaintance 119–122, 249
actual world 46, 61, 66, 95-98, 112, 138, 359
adjective 123–124
 –, substantivisation of 366
alphabet 44
alternative 304–306, 321–322
ambiguity 2, 144, 164, 242-246, 368, 374
 –, definition of 164, 243
analysis of context 134, 151
analysis of language 4, 15–18, 19–25
analytical comparability, *see* comparability
analytical relations, *see* relations
analytical truth, *see* truth
analyticity 61, 74, 86–87, 92, 95-97, 100, 161, 168, 266–270, 367
 –, degrees of 267–269
anthropology 5, 16
 –, cultural 6, 239
antinomy 11–12, 14
 –, Berry's 14
 –, Grelling-Nelson 14
 –, semantic 10–12
 – of the Liar 12–14
 – of the name relation 149–150
 – of synonymous names 10, 149, 328
aposteriorism 265
apriorism 265
arithmetic 50
 –, axiomatisation of 50
arithmetical concepts 50
article, definite 117, 364
 –, indefinite 117, 364
assertion (= Konstatierung) 370
assignment 7–8, 18, 275, 290
 –, subjectivist 203
 – of values 93
association 39
associationalist approach 39
atom 172, 213–214
atomic expression (*see also under* sentences; statement) 310
axiom 58–60, 94, 99, 304, 311-312, 323, 360, 377
axiom schema 58
axiomatic approach 50

behaviourism 49
behaviourist approach 7, 160
behaviourist trends 4
belief 34–35, 131–132, 137, 191, 329
belief sentences 132, 136–138, 156-157, 366
black box 195, 205-206

C-function 78–79, 81, 90, 360
 –, regular 79
calculability 212
Cambridge school 4
cartographic shapes 341
causality 221
channel 174ff, 257
 –, input of 176, 204–207
 –, operation of 176, 204–207
 –, output of 176, 206ff

- of an actual observer 178
- of communication 15, 20
- of conversation 21
- of interpretation 154, 367
- of the language user 55, 105, 178
- of observation 226-227
- of the observer 21, 25, 27, 105, 121, 126, 174, 176, 197-198, 203-205, 208, 213, 222, 257

class(-es) 57, 75, 84, 123-128, 139, 144, 147-148, 234, 240, 249-250, 255-257, 262, 275-276, 307, 325, 331, 359, 364, 365, 366, 375
-, definition of 365
-, existence of 250
-, identity of 126-127, 260
- of individuals 94, 115, 124
- of objects 275
- of sentences 45, 56, 62-64, 67, 75, 77
- - conjunctive 63, 67
- - empty 63, 117, 124
- - universal 63
- of state descriptions 72

code 43ff
-, cable 43
-, highway 277
-, semaphore 43

code phrases 356
code system 43-44
coding 22, 44
communicability (*see also* transmissibility) 20, 29, 164
communicant 21-30
communication 8, 15-23, 31, 36-37, 44, 251
-, emotional 2
-, graphic 339-342, 344-347, 350-353
-, intellectual 2, 340
-, language 15-23, 26, 134, 151, 192
-, process of 1, 11, 17, 21-22, 60, 71, 144, 167, 188, 245

communication theory 44, 178, 198, 226
comparability, analytical 156
-, principle of 306-308
- - in the narrow sense 309

complement 75, 168, 302
complementary class 234
completeness 48, 254
concatenation 45, 322, 341-3, 356
-, contextual 134, 329, 356

concept 123, 128, 130, 172
-, comparative 318
-, individual 123, 134, 365
-, open 201-202, 209, 371
- of truth in semantics 11, 51, 354
-s of L-semantics, definition of 92

conception of identity (Leibniz's) 12
conception of language, nominalist 57, 357
- -, "semantically neutral" 57

conception of sense 355
conception (psychological) of the sense of sentences 190
conceptualism 258-259
concrete terms 141
conditional connections 199-202
conditioned reflex 39
conjunction 45
connotation 3, 141
consistency 254, 324
consistency graph 237
constant(s) 44, 100, 278, 357
-, definition of synonymity of 98
-, individual 88, 198, 271, 357
-, logical 46, 79, 93, 96-101, 157
-, non-logical 46, 53, 84, 86, 93, 96-101, 188, 268-270, 357, 370
- -, primitive 208
-, primitive 93
-, semantic 57

constitution 99, 201, 208, 218, 223, 227
constructivism 60, 258-260, 372
content (comprehension) of a concept or term 3, 139-140
contentual power 74-80, 360
context 112, 134-138, 145-147, 151, 245, 328, 356
-, basic 113
-, empirical 102
-, extensional 114, 121, 134-135, 137, 146, 363

–, intensional 114, 136–137, 363
–, non-extensional 136, 329, 363
–, reduction of 151
–, sentential 328
– of propositional attitudes 157
contingency 265
contingent 95
contradiction 12, 74, 77, 97, 100, 168, 229, 232, 323, 358, 367, 376
convention(s) 54–55, 167, 276
–, language 271–272
– for C-function 79
– for same extension 127–128
– for same intension 127–128
– for the admissibility of entities as the values of variables in theoretical language 209
conversation 21–30
–, aim of 23–30
conversion 376
correspondence rules 207–210, 284
covering law 275
criterion for distinguishing 261–264
criterion of identity 112, 134, 146, 325–326
– – of expressions 24
– of material adequacy 51, 54
– of ontic commitments 250
– of preservation of sense 225
– of 'salva analycitate', *see* 'salva analycitate'
– of sense 143, 166–167, 177, 180–181, 187, 208, 214, 217, 228, 355, 368
– –, empirical 178, 370
– –, objective 170
– –, operationist 143, 166, 170–177, 197, 224
– –, pragmatic 143
– –, translatability 187–192, 194
– –, verifiability 143, 166, 177–183, 188–194, 206
– for the specification of individuals 257
– of synonymity 101, 354
– –, pragmatic 159
cybernetics 5, 195, 338
data, a posteriori 275–277, 282–283, 285–286, 291–293, 296–298, 303
data, a priori 275–277, 282–283, 285–286, 291, 295–298, 303
data object 329–331, 334
deciding power 287–288, 292, 294–296, 300
decision, 2-valued 70, 82, 84
–, *n*-valued 70
–, ontic 247–250, 253, 256, 270–271, 377
–, semantic 236–239, 241, 246–250, 253, 256, 265, 270–271, 377
decision base 286–287, 290–292, 295–296, 298–300
– –, adequate 292–293
decision functions space 226
decision space 226
decoding 22–23, 226
definition 11, 14, 56, 268, 271–272, 284, 366
–, explicit 51, 211, 218, 284, 372
–, operational 171, 201, 219
–, ostensive 38–40, 133, 211
degree of confirmation 360
delicacy of discrimination 126
delusions of the senses 204
denotate 18, 39, 114, 133, 307, 321–322
denotation 3, 10–11, 24, 42–43, 57, 102–114, 143–145, 150, 235–236, 243, 280, 328, 330, 346, 354, 355, 356, 361
–, compound (*Den*) 57–59
–, direct 109
–, identity of 102
–, indirect 108, 114, 141
–, singular 140
–, zero 140–141, 271
– of a shape 345
– of a term 139ff
– of theoretical concepts 213
denote 31, 41–43
description 115–118, 367
–, definite 118
–, individual 118–120, 122, 133, 240, 248–249
– –, interpretation of 248
–, state (*SD*) 64, 67–72, 77–90, 280,

313-314, 324, 361
– –, definition of 88
–, structural 69, 82-83
descriptive level 344
descriptum 133
designans 366
designator 356, 357
designatum 366
determinism 221, 373
–, Laplace's 202, 221
determinist relations 36
diagnosis 276-277, 280-281, 283, 285-286, 289, 303
diagrams 341
dialectology 239
dialogue 21-22, 27, 30
differentiation of classes and properties 125
– of sense and denotation 125
– of types of meaning 2
disjunction 45, 63, 76, 79, 313, 333
dogma of reductionism 217, 373
dogmas of empiricism 217, 266

economics 4
economy, linguistic or terminological 259, 264
eigenpsychische Basis 355
electron 172, 214
elements of a universe 10, 14, 280
elementary level 217, 221
emotionality 29
empirical 61, 159, 194, 275, 279, 282, 288, 294-296
– component of a theory 276
– concepts 210, 216
– evidence 13
– generalisations 284
– pragmatics 16
– procedure 176-177, 201, 208
– terms 195
– theoretical relevance 294-296
– theory of sense 3
– world 98, 359
empiricism 49, 122, 170, 173, 176, 187, 194, 197, 203-204, 206, 226, 369, 372
–, logical 49, 218
–, naive 211-212
–, materialist interpretation of 203
–, subjectivist interpretation of 203
energy 371
entelechy 208
entities, abstract 41, 250-258
– –, existence of 250, 252
– –, hypostatisation of 256-257
–, concrete 250
–, constructive 322
–, extralinguistic 7, 146, 271, 325
–, hypothetical 278-9
–, linguistic 325
–, theoretical 281
entropy 215, 281, 303, 346, 351-352
epistemic gain or loss 298, 300-302
epistemological operator 321
equipollence 156
equivalence 45, 63, 97, 126, 134, 141, 155, 305-306, 328
essentia 141
essential parameters 219, 223
esthetics 1, 5, 8
ethnography, comparative 6
–, historical 6
exactness 244ff.
existence 118-121, 140, 214-215, 248
–, accessible 119
–, definition of 118
– in the objective sense 121
expectation 179
experience 35, 38, 132, 171, 181, 202-203, 275, 376
–, elementary 195
–, immediate (= acquaintance) 249
–, religious 171
explanans 286
explanation 212, 221, 224, 276, 279-281, 285-287, 291, 367
explicandum 152, 359
explicans 76, 152, 359
explication 359, 367
express 31-35, 365
expression(s) of language 11-12, 16, 31-42, 59, 66, 74, 103-104, 107-109, 110, 123, 127-129, 134, 144, 151-154, 157, 228-229, 270

– –, contradictory 368
– – without a sense 368
extension 127–151, 337, 361
–, pragmatic 158ff.
–, primary 145
–, same 130, 134, 146
– –, definition of 130
–, secondary 145
– of individual expressions 132
extensionality of an expression with respect to context 135
external questions of ontology 252–254

F-concepts (*see also* factual concepts) 74–76
F-empty class 141
factual concepts 62, 72
fairy tales 192
finitisation 60, 326
finitism 260, 303, 326–327, 378
fluidum 208
formalised language, *see* language, formalised
formalised pragmatics 365
formula 46
–, atomic 52
–, molecular 52
–, Shannon's 68, 81
–, well-formed (wff) 46–54
– –, holding of 94
fringe 233–235, 237–238, 244
function, estimate (**est**) 81
–, evaluation 311ff
–, extensional 136, 326
–, intensional 326
–, propositional 364
–, statement 117
–, weight or loss 227
– of communication 32ff
– of expression 32ff
functional concepts 365

game theory 304
gene 371
genetics 5, 213
goal complex 281–283, 285, 289, 291, 295–296, 298–300
grammar 362–364
–, generative 345
graphs 339, 348
gravitation 209–210
grids 339, 341–342

heteronymity 242–243
–, definition of 164
hierarchisation of expressions 10, 129
– of languages 10
– of viewpoints in the analysis of expressions 129
homoionym (*see also* pseudosynonym) 152
homonymy 2, 42
hygiene, mental 5
–, social 5
hypothesis 172ff, 207–208, 210, 241, 272, 283, 297, 331
–, atomist 214
– of language play 358
– of linguistic relativity 6, 239
hypotheses in science 275–277, 279–283, 288, 291, 296, 302

idealisation 209, 215, 232
idealism, objective Platonic 256
–, subjective 220
identification of the indistinguishable 261–262, 327
identity 10, 24, 41, 103–104, 326, 376
–, conception of 103
–, criterion of 104, 325–326
– of individuals 116
– of names 103ff
– of objects in a universe 103
imperatives 110–111
implication 45, 97
–, causal 135, 202, 206ff, 218
–, material 199ff, 201–203, 218, 285
–, strict 135
inclusion 64
incompatibility 45, 63
indirect speech 108, 114, 121, 134, 136
indisputability 48, 50
individual 44, 53–58, 67–71, 82–84, 93–94, 115–116, 123–130, 133–134, 144, 147, 208–209, 234, 249, 256–264, 371

–s, aggregation of 260
–s, difference between 116
–s, distinguishability of 71, 260
–s, distribution of 82
–s, domain of 52, 58
–s, similarity of 260, 263
individual constant, *see* constant, individual
individual description, *see* description, individual
individualisation 71
induction 202
inference, *see* rules of inference
infinity, denumerable 67
–, qualitative 373
information (semantic) 76, 81–83, 126, 274–279, 304–305, 351
–, absolute 279
–, exchange of 26, 30
–, processing of 20, 304
–, storage of 20, 198, 304
–, transmission of 20, 153–154, 198, 281, 341, 351
–, transmitted (transinformation) 279, 285–286, 297, 325, 330–331, 334, 336, 350, 378
– –, sufficient 330, 334–335, 348–349, 353
information explosion 274
- science 304, 325
- system 283, 344, 346–347
- theory 1, 3, 15, 20, 44, 81, 83, 225
– –, mathematical 3, 20, 225
– –, semantic 3, 67–68, 74, 81, 274, 285, 289, 325
informational evaluation 277, 279–280, 287–288, 293, 298–303, 339, 346, 350–351
- irrelevance 333
- relevance 335ff
- synonymity, *see* synonymity, informational
injunction 114, 286
input space 226, 257
instrumentalism 211–216
intellectual exchange 22
intension 127–134, 138–151, 159–161, 357
–, linguistic 142–143, 161, 169
– –, analysis of 143
– – –, context-bound 366
– – –, context-free 366
–, pragmatic 158–161
–, same 130, 134, 142, 161
– –, definition of 130
–, sense 143, 161, 169
- of a term 140
- of individual expressions 132–134
intensional isomorphism 157–159, 329
intensionality of an expression 136
intercommunication (*see also* mutual understanding) 60, 245
interjections 32
internal questions of ontology 252–253
interpretandum 153, 163
interpretans 153, 163
interpretation 19, 46, 62, 77, 93, 96, 124, 163, 242, 267, 367
–, concept of 96–98, 100
–, impossibility of 243
–, 'official' 96, 98
–, phenomenalist 371
- of a sign 46
- of a message 20
interpretational norm 231, 234, 238, 246
interpreter 16
intuitionism 258

λ-conversion 124, 139, 365
λ-operator 357
L-concepts, introduction of 64–66
L-determinacy 73–74, 367
L-disjunction 61
L-disjunctive 62, 64
L-empty class 141
L-equivalence 61, 73, 97, 136, 141, 155–157, 368
L-equivalent 75, 79
L-implication 61, 64–65, 73, 78, 87, 97
–, definition of 87
L-semantics 67–68, 73–75, 357
L-substitutability 134–135, 142

–, defined with respect to contextual concatenation 134
L-true 61–66, 76, 87
L-truth 87, 93, 97–98, 267–270, 376
L-untrue 61–66, 76
language, axiomatic structure of 63
–, discreteness of 239, 374
–, elementary forms of 32
–, empiricist 185, 189, 192, 194, 207–208, 252, 372
–, extensional 135, 142
– –, definition of 135
–, formalised 2, 16–20, 43–50, 54–60, 74, 93–94, 98, 102, 124, 133, 135–136, 145, 147, 167, 188, 193, 220, 234, 244, 252, 338, 344, 354, 359, 373, 374
– –, extensional 358
– –, non-interpreted 44
–, functions of 31–33, 240
–, graphic 339–340, 345
–, intensional 171
– –, definition of 171
–, internal structure of 59
–, natural 5, 7–8, 11–13, 17ff, 50, 74, 100–102, 103, 107–108, 115, 117, 124, 129–134, 138–139, 142, 145–146, 151, 167, 238, 243–245, 269, 338, 344, 354, 356, 362, 364, 375
–, picture 339
–, properties of 32
–, role of 31
–, semantically closed 12–15, 46, 154
–, theoretical 372
–, use(s) of 5, 7, 71, 91, 99, 113, 152, 168, 192–194, 245, 362
– –. emotive 5
– –, intellectual 5
– –, referential 5
– and thought 35
– of finite order 74
– of infinite order 74
– of primitives 6
– of science 4, 49, 61, 172–173, 193, 206, 216–217, 222, 230, 271–273
– –, non-pragmatic 194, 246
language user 2, 16–19, 31–41, 60, 71, 106, 132, 142–143, 152, 154, 158–161, 166–167, 169–170, 188, 190, 192–195, 229, 234–237, 245, 258, 269–270
–, state of 32–38, 161
law of identity (Leibniz's) 10
– of participation 6
– of the theory of identity 104
laws (in science) 172, 183, 207, 221, 241, 272, 275–282, 288, 291, 296, 302, 373
learning process 239
level, *see under* descriptive, elementary, pragmatic, semantic, syntactic
level of risk 23, 281, 355
linguistic analysis 4
– framework 251–254
– relativity, *see* relativity
linguistics 1, 5, 8, 16, 103, 151
–, comparative 239
–, historical and comparative 6
logic 1, 43, 47, 103, 114, 233–234, 265, 323, 328, 348, 356, 357, 364, 378
–, Aristotelian 2
–, assertion 322
–, cartesian 305, 314
–, classical 13, 136, 233
–, deductive 78–80
–, extensional 217
–, inductive 72, 76–80, 90, 293, 331, 359, 360
–, mathematical 3, 5, 117
–, medieval 2, 109
–, modern 9, 10, 112–113, 218, 255, 266, 328, 374, 378
–, preference 310–311, 377
–, probability 293, 331
–, scholastic 141
–, traditional 112, 131, 265
– of epistemic modality 322
logical 66
– analysis 4
– concepts 61, 63–66, 72–74, 93
– consequence 47, 65, 87
– constants, *see* constants
– content 74–77, 361, 367

- –, definition of 75
- equivalence 336
- impossibility 181
- necessity 265
- positivism, *see* positivism
- possibility 179–181
- pragmatics 17
- probability 36, 67–68, 72, 359
- –, theory of 3
- range (*LR*) 64, 67, 72, 74–80, 83, 87, 92, 324, 336, 358
- –, empty 72
- –, universal 72
- semantics, *see* semantics
- signs 45–46, 55, 265, 358
- state (*LS*) 72, 74, 76–78, 314
- structure of the world 195, 217, 355
- syntax 4, 17, 323–324
- terms (*see also* logical concepts) 87–88, 208, 216–218, 234–235, 266–269, 373
- truth, *see* truth
- world 61
logicism 131, 258
logico-semantic analysis 4
logistic system 356
Lwow-Warsaw school 3, 51

m-function, regular 80–83
m'-function, regular (for L') 90
m-function, symmetrical 83
- –, definition of 83
many-valued logics 112
macrolevel 222
map(s) 339, 341–3, 347, 349, 353
marginal (fringe) conditions 162–165, 228, 297
- references 162, 368
material adequacy 51–52, 54
- unity of the world 373
mathematical concepts, application of 233
mathematical models of decision 304
mathematics 103, 114, 255, 328
meaning 13, 115, 151–152, 158, 354, 363
–, assignment of 96
–, changes of 2, 7
–, measurement of 7
–, preservation of 2
–, cognitive 2, 152
–, emotional 2, 22
–, emotive 152
–, linguistic 7
–, referential 2, 22
–, relatively independent 129–130, 134, 144, 149, 153–154, 157, 365, 369
–, same 151–154, 367
- –, relativisation of 152–153
–, social cultural 7
–, stimulus 158–160
- –, negative 158–159
- –, affirmative 158
–, syntactic 216
meaning of meaning 1
meaning postulates 64, 83–91, 94, 99, 181, 185, 188, 192, 210, 242, 265, 267–272, 357, 360, 361, 371
means of communication 15, 49
measure of content (**cont**) 80–81, 290, 298, 300, 331–333, 349
- –, conditional [**cont(j/i)**] 81
- – for sentences in L' [**cont'(i)**] 91
- of credibility 299–300
- of deciding power 288
- – –, definition of 289
- of information [**inf(i)**] 81–83, 331, 333–335
- –, definition of 83
- of probability 290, 298, 300, 305, 333, 349, 377
- of relevance 294
- of transmitted information $\mathbf{TI}_i$ 330–334, 336, 348–349, 351
- – –, definition of 331–332
mechanism 7, 122, 221–222
mentalist approach 6–7
- terms 7, 160
metachannel 21
metalanguage 11–15, 26, 44–60, 66, 78, 93–95, 124–125, 133, 154, 182, 187, 346, 354, 370
–, components of 14, 47

–, neutral 125
–, semantic 182, 266, 344
metalanguage signs 45, 62, 78
– transcriptions (of pictorial devices) 344
metamathematics 354
method, axiomatic 56
–, behaviourist 160
–, direct 142
– of extension and intension 122–123, 134, 143–144, 150–151, 169, 363
– of identifying the indistinguishable 260–262
– of state description 71–72, 76–77, 79, 83–96, 265, 324, 376
– of structural analysis 161
methodological analysis 49
– problems of science 48, 91
methodology, scientific 49
microlevel 222
modality 136, 270
model 46, 62, 86, 93–98, 340, 357
–, definitions of 94
–, communication 20, 126, 174–176, 197, 205, 224
–, intended 93, 95
– of decision 304
– of reduction 224–228
–, deductive nomological 225, 285, 291, 295, 302, 370
– of science 218, 222, 257
modes of meaning 10, 43, 139, 142, 151–152, 158, 356, 362
modus ponens 47
molecular expressions (*see also* sentences, molecular) 310, 312
monologue 30
morpheme 354
morphology of language 51, 56, 59
multidimensional space 49
multiplication of entities 257, 264
Münster school 10
mutual understanding (*see also* intercommunication) 23, 25, 37
– –, effective 24–25
mythology 271
–, Greek 120
name(s) 13, 39, 41–42, 103–108, 115, 123, 129, 144–147, 258
–, denotation of 105–106
–, identity of 103–105, 122
–, sense of 105–107, 108–110, 258
–, use of 110, 146–148
– –, autonymous 110–111
– –, purely denotational 146, 149
– –, in the sense of name relation 110
–, compound 109, 112
– –, denotation of 363
–, metalanguage 52
–, proper 108, 115, 133, 145
names, concatenation of 108
– –, contextual 134
–, generic 40
– of classes 40, 148
– of classes of classes 148
– of classes of properties 148
– of individuals 103, 115
– of properties 68, 148, 364
– of properties of properties 148
– of qualities 40
– of relations 40, 68
– of types of behaviour 40
naming 361
natural numbers, system of 252–253
negation 45, 63, 77, 184
noise 21, 154, 174, 176, 227, 231
nomenclature, *see* scientific nomenclature
nominalism 41, 250–252, 255–262, 372, 375
nominalist conception of language 357
nominatum 144–146, 174, 366
nomological statement 276, 282–286, 290–291, 293, 302
non-empty class 141
non-logical 66
– basis 257
– constants, *see* constants
– signs 46, 55, 84, 265, 362
– terms 74, 87–92, 94, 208, 211, 216, 233ff, 266–269, 272
– –, interpretation of 233ff
– –, primitive 372
– –, dependences between 89–91

- world 61
norm, *see* interpretational norm
norms 275-277, 286

object, abstract 122
-, accessible 120
-, actual 140
-, constructive 284
-, non-actual impossible 176
-, non-actual possible 176
-, non-existent 120, 364
object channel 21-22
object language 11, 13, 21, 26, 31, 43-44, 54, 57, 93-94, 135, 137, 154, 162, 168, 182, 187, 242, 262, 270, 344, 358
objective probability 360
- world 6, 149, 204, 220, 247, 253-254, 355
observability 197, 372
- in principle 166, 198
observable 194-198, 205, 213
observation 189, 197-198, 211, 278-281, 285, 288, 297
-, actual 178
-, compass of 205
observation sentences (Beobachtungssätze) 178, 183-184, 187, 195, 198, 203, 211-212
observer, absolute 279
-, actual 178-179
-, potential 370
observer channel, *see* channel
observer's viewpoint 369
Occam's razor 125, 147, 220, 256-258, 264, 369
ontic commitments 248-252, 257, 270
- - of an entire language system 252
ontological controversies 371
- priority of universals 375
ontology 249, 252
-, problems of 41, 251, 362
-, traditional 254
operationism 170-176, 206, 226, 369
-, homocentric 173, 226
operator of abstraction 45, 47, 58, 124, 139
- of individual description 45, 47
output space 226, 257
Oxford group 4

paradox, the Burali-Forti 11
-, Cantor's 11
-, Richards's 11
-, Russell's 11
- of the 'bald man' 230
- of the 'heap' 230
- of the 'liar' 11-12, 373
paradoxes (*see also* antinomies) 3, 11
- of material implication 218
parameters, *see* essential parameters
pattern, *see* picture shape
phenomenalism 256-257
- and substantialisation 374
philosophy 1, 4, 247
-, analytical 4
-, linguistic 4
-, 'school' 177
-, semantic 5, 247
phlogiston 208, 271
phoneme 354
phrase 100-101
-, definitions of 100-101
-s, definition of synonymity of 101
-, denoting 106, 115-117
- -, denotation of 116
- -, sense of 116
physicalism 7, 193, 218, 357, 369, 370, 372
physics 4, 172, 214, 222, 277
-, classical 69
-, modern 173
picture (pictorial) shape (pattern) 338-341, 343, 346, 350-351
Platonism 41, 232, 250-252, 258-260, 374
-, sociological 375
Platonist-nominalist controversies 250-252, 255-256, 258
poetry 371
popularisation (of a scientific text) 153
positivism 3, 179, 203, 214, 370
-, logical 148
possible state 313-315, 317, 320
- world(s) 46, 61, 66-67, 74-75, 79, 93, 95-96, 113, 313, 359, 361

postulate of admissible risk 24
- of identity of referent 24
pragmatic concepts 158, 235–236
– –, types of 235–236
– level 4, 7, 15–28, 98, 233, 235
– relations 18
pragmatics, empirical 16
–, formalised 365
–, logical 17
–, theoretical 365
pragmatism 4, 171
predicate 44, 56–57, 71, 124–133, 138–141, 145, 147, 168, 198–199, 205, 220, 234, 249–250, 255–257, 281, 325, 368, 371
–, definition of pragmatic intension for 160
–, comparative 308–310, 378
– –, two-place 309
–, dispositional 40, 198–202, 206–207, 369
– –, proper 206
– –, theoretical 206
– – –, definition of 206
–, empirical (*see also* observational p.) 40, 182, 194–213, 216–223, 226–231, 240, 264, 368, 370, 371
–, metalanguage 162–164, 168, 181–182, 228–229, 269
–, molecular 82–83
–, observational 185, 281, 370
–, primitive 82–83
–, theoretical 281
– with 1 argument, *i.e.* 1-place predicate 40, 58, 67–68, 76, 82, 87, 93, 125, 129, 132, 168, 192, 270, 325–326, 375
– with 2 arguments, *i.e.* 2-place predicate 31–33, 37, 45, 68
predicate calculus 281
predicativism 374
predicator 365
– for classes 365
– for properties 365
prediction 277, 279–281, 283, 286, 291, 299
preferability 323, 378
preference 304–305, 307–308, 310, 315, 319, 321–324, 329, 353, 377
– ceteris paribus 315–318, 320, 323–324
–, differential 316–318, 320–321, 323
–, inclusive 315, 317–318, 320, 323–324
–, relativised 316, 318, 324
– under explicitly defined circumstances 317–318, 320, 323–324
primitive concepts 321
– mentality 6
– signs 44–46, 56
– terms 57–66
principium rationis sufficientis 207
principle of analysis of propositions containing descriptions 122
– of comprehension 365
– of extensionality for classes 365
– of identity (Leibniz's) 24, 116, 325
– of indifference 207
– of interchangeability (substitutivity) 144–147, 149–151
– –, reduced 146
– of $n + 1$ non datur 70
– of statistical independence 207
– of subject matter 144–147, 245
– of tertium non datur 13, 116, 234, 376
– of tolerance 376
– of univocality 144–147, 164, 245
prinzipielle Koordination 179
probability 298, 304, 314, 335, 360
–, conditional 285
–, inductive 78–80
–, logical 36, 67–68, 72, 359
–, objective 360
–, statistical 360
–, subjective 335, 360
probability$_1$ 360
probability$_2$ 360
probability theory 67, 290, 305, 360
probabilities, a priori distribution of 227
–, two-dimensional distribution of 227

prognostics 304
property 57–58, 84, 104, 115, 118–120, 123–128, 144, 147, 240, 250, 257, 262–263, 276, 307–308, 364, 366, 375
- of "have a sense" 180–182, 186, 188, 192–193, 208
- of "not have a sense" 180–182
- in the semantic sense 125–126, 130
properties, identity of 127–128, 260, 325–326
proposition 112, 123, 130–132, 142, 158, 265, 322, 363, 366
proposition-like entities 310, 377
propositional attitudes 136, 157–158, 268, 270, 321–322, 324, 378
Protokollsätze 183
pseudosynonym (*see also* homoionym) 152
psyche 29, 32–35, 119
–, "atoms" of 195
–, level of 29, 34
–, state of 32–35
psychiatry 1, 5
psychoanalysis 5, 34–35
psychologism 131, 138
psychology 2, 5, 7, 16, 29, 160, 354
–, behavioural 159, 195
–, gestalt 221
–, perception 341
–, physicalist 372
–, social 5, 29

Q-predicate 82–83
quality of decision-making 204ff, 227
quantifier 88, 356
–, existential 45ff, 184
–, general 45ff, 184–185
questions (interrogatives) 2, 110–111

radical concepts 62–64, 66, 73, 92
range of entities, "basic" 99
real numbers 307, 319, 359
- state 72, 76
- world 232, 364
realisable 195ff
realism 213
–, naive 173, 211–214, 256
–, Platonic 250–258
- sociological 375
reduction 99, 171, 208, 217–228, 292, 361
- of data 338–339
- base (for theoretical concepts) 217–220
– –, absolute 217–219
- pair 200–201, 206
- theorems 284
reductionism, empiricist 214, 217–220, 372
–, radical 217
reference 17–19, 213
referent 18, 24
regress, infinite 187
relation 123
–, analytical 366
–, asymmetrical 89, 260
–, Heisenberg's 92, 222, 369
–, irreflexive 89, 260
–, name 10, 13, 108, 114, 122, 139, 143–151, 154, 245, 355, 364
–, reflexive 73, 164, 308
–, symmetrical 42, 64, 73, 164, 308
–, transitive 42, 64–65, 73, 89, 164, 260, 308
- of indifference 307, 313
relational concepts 365
relativism, linguistic 6, 355
relevance, empirical theoretical 293–295
–, test 293, 297–298
– –, definition of 298
- quotient 294
requirement of deducibility 224
- of invariance of sense 224
resemblance 262–264
–, definitions of 262–263
–, abstract 264
–, structural 264
risk, average 225, 227, 329
–, semantic 176
rule(s), *see* semantic, syntactic r.
–, formation 345
–, generative 345

–, transformational 345
- of inference 47
- of sense 242, 374
– –, axiomatically formulated 242
– –, empirical 242
- of substitution 47

"Salva analyticitate" criterion 328, 330, 335–336
"salva relatione" criterion 329ff
"salva veritate" criterion 12, 327–330
"salva veritate" test 104, 106, 112, 116, 155, 328
satisfaction 11, 51–58, 62
–, definition of 53–54
schema of the communication process 21–22
- of conversation 21–22
- of preference 304
- of reference (Richards's) 5, 17–18
scientific nomenclature 172, 174
scope of a concept 3
second signalling system 39
semantic analysis 8, 83–84, 114, 122–123, 128–130, 138, 142, 265, 345–347, 369
– - of picture shapes 345
– – of sentences 112–114, 151, 191–192
– - of a text 347
- approach 19
- concepts 11, 14, 31, 47, 50–51, 55–60, 73, 76, 84, 92–96, 160, 354, 357
– –, introduction of 14, 50–60
– –, axiomatic introduction of 57–60, 92
– –, definitional introduction of 50–56, 92
- constants, *see* constants
- decision, *see* decision
- determinateness 100
– –, definition of 100
- differential 7
- identity 325–328
- level 4, 15–30, 46, 98, 107, 233, 344
- localisation 347
- product 59
- rules 269–273, 340
- sum 59
- system 43, 54, 357, 359
semantics:
–, basic categories of 8
–, denotational 355
–, general 3, 5, 247, 375
–, intensional 256
–, linguistic 2, 7
–, logical 7, 10, 17, 68, 99, 252, 255, 266, 323–324, 328, 355, 356
–, text 347
–, theoretical 3
- of art 8
- of logical concepts (*see also* L-semantics) 61–66, 91ff, 97, 99
- of pictures 340, 346
- of the Lwow-Warsaw school 3
semantisation 16, 27–28, 33
semasiology 1
semi-model 93–94
semiotics, empirical 16
–, Morris's 4–5, 7, 16
sense 3, 10, 24, 33–43, 68, 101–109, 115–117, 122–123, 129, 138, 143–144, 150, 152–153, 194, 243, 328, 330, 346, 354, 370
–, cognitive 168–169, 177
–, direct 109
–, empirical 169, 177
–, indirect 108, 114
–, same 228
– –, preservation of 43, 228
–, transfer of 173
sentence 2, 43–45, 110, 115, 128–132, 152, 328, 342, 347, 368
–, atomic 66–69, 76, 84
–, bilateral reduction 200–201, 206–207
, compound 111
–, declarative 111–113, 129
–, imperative 111, 113, 363
–, indicative 129
–, interrogative 113, 363
–, L-true 156, 366, 367
–, metaphysical 183–184, 186, 190
–, molecular 184, 358

–, normative 2, 363
–, pictorial 122
–, reduction 89, 185, 200, 206, 218, 369, 372
–, true 54, 132, 358
–, untrue 131
–, denotation of 66, 110–114
–, extension and intension of 130–131, 148–149
–, meaning of 77, 358
–, sense of 110–112, 130–132, 149, 169, 180–182, 184–185, 189–194, 369
–, truth of 131, 236, 269, 358
–, truth value of 66, 111, 155, 217
sentences, analytical 74, 95, 100, 150, 168, 182, 186, 265–273, 367, 370
– – in the broad sense 268–269
– – in the nattow sense 268–269
–, factual (F-determinate) 74, 77
–, L-determinate 73–74, 77, 367
–, modal 149
–, synthetic 74, 169, 182, 185–186, 265–273, 369
– of the elementary level 218–219
– with sense 167–168, 184–188, 193, 217, 369
– without sense 168, 186–187, 193
–, classification of 359
–, linguistic classification of 111
sign(s) 17–18, 20, 45–46, 125, 275, 328, 338, 340–341, 344–345, 347–348
– in the abstract sense 356
– in the concrete sense 356
signal 39
signification of a term 140–141
signify 31, 35–41, 43, 127, 365
simultaneity 171–172
sociology 1, 5, 16, 221, 247
–, physicalist 372
sources of semantic problems 1ff
space, *see* input space, decision s., logical s., multidimensional s., output s.
state, *see* real state, logical s., possible s.
state of affairs 265, 309–310, 313, 321, 323–324, 345, 347, 350
state of a person 138
state description, *see* description
state that 31–35
statement, atomic 313
–, general 282
–, law-like 282
–, metaphysical 166
–, nomological, *see* nomological statement
–, singular 282–283, 286
–, meaningful synthetic 288
–, theoretical 282
– of fact 67
– of probability 78
statistical probability, *see* probability
stimulus 158–160
stochastic relations 36
subconscious 34
subjective image 107
– probability, *see* probability
– world 22, 28–29, 38, 107
subordinate clauses 114
substantial concepts 365
substitutability (substitution, interchangeability) 10, 24, 134, 149–150, 152–155, 268, 328, 330, 334, 348–349
subsumption 275–276
suppositio (*and* supposition) 3, 109, 363
suppositio materialis 110
symbol 17
synonyms 154–155
–, pure 152
synonymity 2, 41, 98, 151–165, 174–175, 188, 228, 242, 265, 268, 272, 328–330, 346, 353–354, 366
–, cognitive 156
–, hidden 272, 377
–, informational 325, 329–330, 334–337, 339, 346–350
– –, definitions of 334–335, 349
– –, strict 335–337
–, interlingual 156
–, linguistic 156, 330
–, stimulus 158–159
–, structural 159
–, criteria for 154

–, definitions of 98, 160–161
– – on the basis of interpretation 163
syntactic analysis 49
- approach 19
- level 4, 15–25, 46, 51, 234, 334–335
- rules 19, 273, 340, 345
syntax (of a language) 93, 94, 339, 364
–, logical, *see* logical syntax
–, picture 339–340
synthetic truth, *see* truth
syntheticity 100, 265–266
system of aims 28
- of assignments 16
- of communication 15, 17
- of data 281
- of ideas 17
- of preferences 312, 319
- of things 17
- of values 322–323
- of words 17
–, language 43, 74
systematisation 212, 274–275
–, deductive 212–213
systematising type, tasks of the 275–277, 279–281, 286, 289, 295–296, 298, 302
systematising power 288, 378

tautology 77, 80, 287–290, 297, 319–320, 324, 358
testing 185
theorems of incompleteness 95, 97, 99
- of a preference system 311, 324
theoretical concepts (*see also* theoretical terms) 190, 201, 206–216, 226, 227, 232, 284, 371
– –, admissibility of 208–210
– –, a priori nature of 215–216, 372
– –, absolute a priori nature of 216
– –, relative a priori nature of 216
– –, categories of 215
– –, cognitive sense of 215
– –, interpretation of 210–215
– –, hypothetical realist interpretation of 214
– –, instrumentalist interpretation of 211–213
– –, naive realist or empirical interpretation of 211
– –, realist or materialist interpretation of 214–215
– –, neutralist conception of 214
– –, ontological problems of 214
– –, systemic character of 216
– –, vagueness of 241
theoretical terms (*see also* theoretical concepts) 194, 371
theory 209–210, 212, 222, 241, 251, 283–285
–, complete 283–284
- of cognition 247
- of descriptions 10, 114–122, 248
- of extensional and intensional context 114
- of identification 257
- of individual descriptions 118, 248
- of levels 222–223
- of proofs 354
- of reference 101, 160, 361
- of reflexes 5
- of relativity 6, 92, 210
- of semantic categories 11
- of similarity 257
- of statistical decision 302, 304
- of translation 356
- of types 11, 43, 255, 357
– –, simple 11
thermal matter 208
thermodynamics 92
thought (thinking), prelogical 6
- of primitives 6
translatability 153, 169–170, 177, 189, 194, 370
- salva intensione 369
translatable 189–190
translation 2, 100, 153, 166, 169, 173, 190, 258, 366–367
–, free 101–102
– –, definition of 102
–, literal 101
–, machine 114
- into metalanguage 124
transmissibility (*see also* communicability) 16, 174
transmitted information, *see* information

transparency 244–245, 374
triangle of reference 5, 17
true 54–66
  –, definition of 54
truth 43, 100, 112–113, 131, 190–191, 266, 272
  –, analytical (*see also* A-truth) 95, 97, 217, 266–270, 272
  –, logical (*see also* L-truth) 217, 268–269
  –, synthetic 217
truth conditions 43–44
  - value 24, 57, 63, 66, 85, 104, 112–114, 121–122, 130, 138, 146, 328

uncertainty 280–281, 285–291, 295, 303, 329–331, 346, 351–353, 378
  –, conditional 287–290, 295, 331
understand 107, 132, 148–149
understanding, mutual, *see* mutual understanding
unified science 193, 221
unitary theory of sense 166–170, 186–187
univers dc discours 2
universals 250, 255, 259
universe 2–3, 10, 19, 22, 27–28, 39, 41, 46–47, 58–61, 66–67, 71–72, 74, 78, 82–83, 103, 137, 170, 198, 207, 218, 227, 232, 241, 247, 259, 264, 265, 268, 280, 284, 315, 342, 345, 347, 350, 358, 371, 375
  –, non-empty 309
  –, decomposition of 82, 362
  –, discreteness of 239–240, 245

untruth 100, 112, 131, 191

vagueness, denotational or referential 241, 243–245
  –, extensional 374
  –, measure of 237
  –, sources of 235
  - of logical and non-logical terms 233–234
  - of sense 241
  - of theoretical concepts 241
value (in political economy) 210, 215
variable 44, 51, 54–59, 88, 117, 139, 157, 209, 257, 277–278, 325, 372
  –, free 47, 58, 95, 100, 124
  –, individual 93, 257, 260, 359, 375
verifiability 3, 169, 177, 180, 186
verifiable 179, 186
verification 3, 166, 169, 177, 179, 183, 185
  –, complete 183–185
  –, conceivability of 179–180, 182
  –, possibility of 179–180
  –, logical possibility of 179–180, 182
Vienna Circle 4, 51, 149, 177–178, 195, 203, 217, 253, 369, 370, 372

Wilson chamber 126, 214, 371
word 1–2, 8, 17–18, 38–39, 328
  –, egocentric 40
  –, magic effects of 5
  –, meaning of 2
  –, use of 109–110
work of art 8

yardstick 228, 261, 375

BOSTON STUDIES IN THE PHILOSOPHY OF SCIENCE

1. Marx W. Wartofsky (ed.), *Proceedings of the Boston Colloquium for the Philosophy of Science 1961-1962.* 1963.
2. Robert S. Cohen and Marx W. Wartofsky (eds.), *In Honor of Philipp Frank.* 1965.
3. Robert S. Cohen and Marx W. Wartofsky (eds.), *Proceedings of the Boston Colloquium for the Philosophy of Science 1964-1966. In Memory of Norwood Russell Hanson.* 1967.
4. Robert S. Cohen and Marx W. Wartofsky (eds.), *Proceedings of the Boston Colloquium for the Philosophy of Science 1966-1968.* 1969.
5. Robert S. Cohen and Marx W. Wartofsky (eds.), *Proceedings of the Boston Colloquium for the Philosophy of Science 1966-1968.* 1969.
6. Robert S. Cohen and Raymond J. Seeger (eds.), *Ernst Mach: Physicist and Philosopher.* 1970.
7. Milic Capek, *Bergson and Modern Physics.* 1971.
8. Roger C. Buck and Robert S. Cohen (eds.), *PSA 1970. In Memory of Rudolf Carnap.* 1971.
9. A. A. Zinov'ev, *Foundations of the Logical Theory of Scientific Knowledge (Complex Logic).* (Revised and enlarged English edition with an appendix by G. A. Smirnov, E. A. Sidorenka, A. M. Fedina, and L. A. Bobrova.) 1973.
10. Ladislav Tondl, *Scientific Procedures.* 1973.
11. R. J. Seeger and Robert S. Cohen (eds.), *Philosophical Foundations of Science.* 1974.
12. Adolf Grünbaum, *Philosophical Problems of Space and Time.* (Second, enlarged edition.) 1973.
13. Robert S. Cohen and Marx W. Wartofsky (eds.), *Logical and Epistemological Studies in Contemporary Physics.* 1973.
14. Robert S. Cohen and Marx W. Wartofsky (eds.), *Methodological and Historical Essays in the Natural and Social Sciences. Proceedings of the Boston Colloquium for the Philosophy of Science 1969-1972.* 1974.
15. Robert S. Cohen, J. J. Stachel and Marx W. Wartofsky (eds.), *For Dirk Struik. Scientific, Historical and Political Essays in Honor of Dirk Struik.* 1974.
16. Norman Geschwind, *Selected Papers on Language and the Brain.* 1974.
18. Peter Mittelstaedt, *Philosophical Problems of Modern Physics.* 1976.
19. Henry Mehlberg, *Time, Causality, and the Quantum Theory* (2 vols.). 1980.
20. Kenneth F. Schaffner and Robert S. Cohen (eds.), *Proceedings of the 1972 Biennial Meeting, Philosophy of Science Association.* 1974.
21. R. S. Cohen and J. J. Stachel (eds.), *Selected Papers of Léon Rosenfeld.* 1978.
22. Milic Capek (ed.), *The Concepts of Space and Time. Their Structure and Their Development.* 1976.
23. Marjorie Grene, *The Understanding of Nature. Essays in the Philosophy of Biology.* 1974.

24. Don Ihde, *Technics and Praxis. A Philosophy of Technology.* 1978.
25. Jaakko Hintikka and Unto Remes, *The Method of Analysis. Its Geometrical Origin and Its General Significance.* 1974.
26. John Emery Murdoch and Edith Dudley Sylla, *The Cultural Context of Medieval Learning.* 1975.
27. Marjorie Grene and Everett Mendelsohn (eds.), *Topics in the Philosophy of Biology.* 1976.
28. Joseph Agassi, *Science in Flux.* 1975.
29. Jerzy J. Wiatr (ed.), *Polish Essays in the Methodology of the Social Sciences.* 1979.
32. R. S. Cohen, C. A. Hooker, A. C. Michalos, and J. W. van Evra (eds.), *PSA 1974: Proceedings of the 1974 Biennial Meeting of the Philosophy of Science Association.* 1976.
33. Gerald Holton and William Blanpied (eds.), *Science and Its Public: The Changing Relationship.* 1976.
34. Mirko D. Grmek (ed.), *On Scientific Discovery.* 1980.
35. Stefan Amsterdamski, *Between Experience and Metaphysics. Philosophical Problems of the Evolution of Science.* 1975.
36. Mihailo Marković and Gajo Petrović (eds.), *Praxis. Yugoslav Essays in the Philosophy and Methodology of the Social Sciences.* 1979.
37. Hermann von Helmholtz: *Epistemological Writings. The Paul Hertz/Moritz Schlick Centenary Edition of 1921 with Notes and Commentary by the Editors.* (Newly translated by Malcolm F. Lowe. Edited, with an Introduction and Bibliography, by Robert S. Cohen and Yehuda Elkana.) 1977.
38. R. M. Martin, *Pragmatics, Truth, and Language.* 1979.
39. R. S. Cohen, P. K. Feyerabend, and M. W. Wartofsky (eds.), *Essays in Memory of Imre Lakatos.* 1976.
42. Humberto R. Maturana and Francisco J. Varela, *Autopoiesis and Cognition. The Realization of the Living.* 1980.
43. A. Kasher (ed.), *Language in Focus: Foundations, Methods and Systems. Essays Dedicated to Yehoshua Bar-Hillel.* 1976.
46. Peter L. Kapitza, *Experiment, Theory, Practice.* 1980.
47. Maria L. Dalla Chiara (ed.), *Italian Studies in the Philosophy of Science.* 1980.
48. Marx W. Wartofsky, *Models: Representation and the Scientific Understanding.* 1979.
50. Yehuda Fried and Joseph Agassi, *Paranoia: A Study in Diagnosis.* 1976.
51. Kurt H. Wolff, *Surrender and Catch: Experience and Inquiry Today.* 1976.
52. Karel Kosík, *Dialectics of the Concrete.* 1976.
53. Nelson Goodman, *The Structure of Appearance.* (Third edition.) 1977.
54. Herbert A. Simon, *Models of Discovery and Other Topics in the Methods of Science.* 1977.
55. Morris Lazerowitz, *The Language of Philosophy. Freud and Wittgenstein.* 1977.
56. Thomas Nickles (ed.), *Scientific Discovery, Logic, and Rationality.* 1980.
57. Joseph Margolis, *Persons and Minds. The Prospects of Nonreductive Materialism.* 1977.
58. Gerard Radnitzky and Gunnar Andersson (eds.), *Progress and Rationality in Science.* 1978.

59. Gerard Radnitzky and Gunnar Andersson (eds.), *The Structure and Development of Science*. 1979.
60. Thomas Nickles (ed.), *Scientific Discovery: Case Studies*. 1980.
61. Maurice A. Finocchiaro, *Galileo and the Art of Reasoning*. 1980.
62. William A. Wallace, *Prelude to Galileo*. 1981.
63. Friedrich Rapp, *Analytical Philosophy of Technology*. 1981.
64. Robert S. Cohen and Marx W. Wartofsky (eds.), *Hegel and the Sciences*. (Forthcoming).
65. Joseph Agassi, *Science and Society*. (Forthcoming).
66. Ladislav Tondl, *Problems of Semantics*. 1981.
67. Joseph Agassi and Robert S. Cohen (eds.), *Scientific Philosophy Today*. (Forthcoming).
68. Władysław Krajewski (ed.), *Polish Essays in the Philosophy of the Natural Sciences*. (Forthcoming).
69. James H. Fetzer, *Scientific Knowledge*. (Forthcoming).

59. Gerard Radnitzky and Gunnar Andersson (eds.), *The Structure and Development of Science.* 1979.
60. Thomas Nickles (ed.), *Scientific Discovery: Case Studies.* 1980.
61. Maurice A. Finocchiaro, *Galileo and the Art of Reasoning.* 1980.
62. William A. Wallace, *Prelude to Galileo.* 1981.
63. Friedrich Rapp, *Analytical Philosophy of Technology.* 1981.
64. Robert S. Cohen and Marx W. Wartofsky (eds.), *Hegel and the Sciences.* (Forthcoming).
65. Joseph Agassi, *Science and Society.* (Forthcoming).
66. Ladislav Tondl, *Problems of Semantics.* 1981.
67. Joseph Agassi and Robert S. Cohen (eds.), *Scientific Philosophy Today.* (Forthcoming).
68. Władysław Krajewski (ed.), *Polish Essays in the Philosophy of the Natural Sciences.* (Forthcoming).
69. James H. Fetzer, *Scientific Knowledge.* (Forthcoming).